Springer Series in Computational Neuroscience

Volume 13

Series Editors
Alain Destexhe
Unité de Neuroscience, Information et Complexité (UNIC)
CNRS
Gif-sur-Yvette
France

Romain Brette
Equipe Audition (ENS/CNRS)
Département d'Études Cognitives
École Normale Supérieure
Paris
France

More information about this series at http://www.springer.com/series/8164

Ana Rita Londral • Pedro Encarnação
José Luis Pons Rovira

Editors

Neurotechnology, Electronics, and Informatics

Revised Selected Papers from
Neurotechnix 2013

 Springer

Editors
Ana Rita Londral
Faculty of Medicine
Universidade de Lisboa
Lisboa, Portugal

Pedro Encarnação
Católica Lisbon School of Business
and Economics
Universidade Católica Portuguesa
Lisboa, Portugal

José Luis Pons Rovira
Bioengineering Group
Spanish Research Council
Arganda del Rey, Madrid, Spain

ISSN 2197-1900   ISSN 2197-1919   (electronic)
Springer Series in Computational Neuroscience
ISBN 978-3-319-15996-6   ISBN 978-3-319-15997-3   (eBook)
DOI 10.1007/978-3-319-15997-3

Library of Congress Control Number: 2015940533

Springer Cham Heidelberg New York Dordrecht London

Printed on acid-free paper

Springer International Publishing AG Switzerland is part of Springer Science+Business Media (www.springer.com)

Preface

The present book includes extended and revised versions of a set of selected papers from the *First International Congress on Neurotechnology, Electronics and Informatics* (NEUROTECHNIX 2013), held in Vilamoura, Algarve, Portugal, from 18 to 20 September, 2013.

The purpose of the International Congress on Neurotechnology, Electronics and Informatics is to bring together researchers and practitioners in order to exchange ideas and develop synergies towards new advancements of neurotechnology either in general or regarding a particular case, application or pathology.

NEUROTECHNIX 2013 was co-sponsored by INSTICC (Institute for Systems and Technologies of Information, Control and Communication) and MedinRes – Medical Information and Research, held in cooperation with Neurotech Network, Nansen Neuroscience Network, Associação Portuguesa de EEG e Neurofisiologia Clínica, Sociedade Portuguesa de Neurologia and Fp7 Project Enlightenment, and held in collaboration with Antibodies Online.

The congress received submissions from 21 countries, in all continents. To evaluate each submission, a double-blind paper review was performed by the Program Committee, whose members are highly qualified researchers in the NEUROTECHNIX topic areas.

NEUROTECHNIX's program included panels and four invited talks delivered by internationally distinguished speakers, namely Kevin Warwick (University of Reading, United Kingdom), François Hug (University of Queensland, Australia), Aldo Faisal (Imperial College London, United Kingdom) and Alexandre Castro-Caldas (Universidade Católica Portuguesa, Portugal).

We would like to thank the authors, whose research and development efforts are recorded here for future generations.

Lisboa, Portugal
Lisboa, Portugal
Madrid, Spain

Ana Rita Londral
Pedro Encarnação
José Luis Pons Rovira

Contents

Contributors

Axel Blau Department of Neuroscience and Brain Technologies (NBT), Neurotechnologies (NTech), Fondazione Istituto Italiano di Tecnologia (IIT), Genoa, Italy

Fernando Brunetti Catholic University of Asunción, Asunción, Paraguay

Bioengineering Group, Spanish Research Council (CSIC), Madrid, Spain

Antonio J. del-Ama Biomechanics and Technical Aids Unit, National Hospital for Spinal Cord Injury, Toledo, Spain

David Feess Robotics Innovation Center, German Research Center for Artificial Intelligence (DFKI GmbH), Robert-Hooke-Str. 1, 28359 Bremen, Germany

Chair of Global Business, Faculty of Business Administration and Economics, University of Augsburg, Universitaetsstr. 16, 86159 Augsburg, Germany

Diego Galeano Catholic University of Asunción, Asunción, Paraguay

Ángel Gil-Agudo Biomechanics and Technical Aids Unit, National Hospital for Spinal Cord Injury, Toledo, Spain

Asiyeh Golabchi Department of Neuroscience and Brain Technologies (NBT), Neurotechnologies (NTech), Fondazione Istituto Italiano di Tecnologia (IIT), Genoa, Italy

Rouhollah Habibey Department of Neuroscience and Brain Technologies (NBT), Neurotechnologies (NTech), Fondazione Istituto Italiano di Tecnologia (IIT), Genoa, Italy

Hans-Jochen Heinze Department of Neurology, University Medical Center A.ö.R., Magdeburg, Germany

Leibniz Institute for Neurobiology, Magdeburg, Germany

German Center for Neurodegenerative Diseases (DZNE), Magdeburg, Germany

Georg Hettich Neurocenter, Neurological University Clinic, Freiburg, Germany

Hermann Hinrichs Department of Neurology, University Medical Center A.ö.R., Magdeburg, Germany

Leibniz Institute for Neurobiology, Magdeburg, Germany

German Center for Neurodegenerative Diseases (DZNE), Magdeburg, Germany

Matthias Kennel Fraunhofer Institute for Factory Operation and Automation IFF, Magdeburg, Germany

Rudolf Kruse Department of Knowledge and Language Processing, Otto-von-Guericke University, Magdeburg, Germany

Vittorio Lippi Neurocenter, Neurological University Clinic, Freiburg, Germany

Thomas Mergner Neurocenter, Neurological University Clinic, Freiburg, Germany

Juan C. Moreno Bioengineering Group, Spanish Research Council (CSIC), Madrid, Spain

Stefano Piazza Bioengineering Group, Spanish Research Council (CSIC), Madrid, Spain

José Luis Pons Rovira Neural Rehabilitation Group, Cajal Institute, Spanish Research Council (CSIC), Madrid, Spain

Christoph Reichert Department of Neurology, University Medical Center A.ö.R., Magdeburg, Germany

Department of Knowledge and Language Processing, Otto-von-Guericke University, Magdeburg, Germany

Jochem W. Rieger Department of Applied Neurocognitive Psychology, Carl-von-Ossietzky University, Oldenburg, Germany

Ulrich Schmucker Fraunhofer Institute for Factory Operation and Automation IFF, Magdeburg, Germany

Anett Seeland Robotics Innovation Center, German Research Center for Artificial Intelligence (DFKI GmbH), Robert-Hooke-Str. 1, 28359 Bremen, Germany

Sirko Straube Robotics Group, Faculty of Mathematics and Computer Science, University of Bremen, Robert-Hooke-Str. 1, 28359 Bremen, Germany

Robotics Innovation Center, German Research Center for Artificial Intelligence (DFKI GmbH), Robert-Hooke-Str. 1, 28359 Bremen, Germany

Diego Torricelli Bioengineering Group, Spanish Research Council (CSIC), Madrid, Spain

Kevin Warwick School of Systems Engineering, University of Reading, Reading, UK

Posturography Platform and Balance Control Training and Research System Based on FES and Muscle Synergies

Diego Galeano, Fernando Brunetti, Diego Torricelli, Stefano Piazza, and José Luis Pons Rovira

Abstract Balance control plays a key role in neuromotor rehabilitation after stroke or spinal cord injuries. Computerized Dynamic Posturography (CDP) is a classic tool to assess the status of balance control and to identify potential disorders. In this paper, we present the development of a low cost system and tool for the assessment and training of balance based on static posturography and functional electrical stimulation (FES). The assessment features are built upon a classic a CDP basis, while for training routines, the system uses bioinspired FES patterns and algorithms based on muscle synergies. This system includes low cost technology like the Wii Fit Balance Board and the Kinect. The work described is this paper includes the implementation of the system and first results as a CDP tool.

Keywords Posturography • FES • Muscle synergies • Kinect • Wii balance board

1 Introduction

There are several diseases that can affect human balance and posture control. Such diversity requires the participation of different specialists in the diagnosis and treatment process like neurologists, otolaryngologists and ophthalmologists among others. Posturography is defined as an objective assessment technique of postural control based. In this way, the monitoring of the center of pressure of the

D. Galeano (✉)
Catholic University of Asunción, Asunción, Paraguay
e-mail: diego.galeano@uca.edu.py

F. Brunetti
Catholic University of Asunción, Asunción, Paraguay

Bioengineering Group, Spanish Research Council (CSIC), Madrid, Spain

D. Torricelli • S. Piazza
Bioengineering Group, Spanish Research Council (CSIC), Madrid, Spain

J.L.P. Rovira
Neural Rehabilitation Group, Cajal Institute, Spanish Research Council (CSIC), Madrid, Spain

© Springer International Publishing Switzerland 2015
A.R. Londral et al. (eds.), *Neurotechnology, Electronics, and Informatics*,
Springer Series in Computational Neuroscience 13, DOI 10.1007/978-3-319-15997-3_1

1

person has proven to be an effective tool complementary to clinical diagnosis in order to quantify this neuromotor disorder. This technique also can be used as a complementary tool to help clinicians with the diagnosis of vertigo.

Posturography evaluates each of the sensory systems (visual, somatosensory and vestibular) involved in the complex balance system. Its purpose is to isolate the contribution of each of these systems to evaluate the status of each one separately. It also assesses movement strategies for maintaining balance, and examines the stability limits of the person and the ability to control voluntary movement.

Balance control is an important functional component of human gait. After spinal cord injury (SCI) or stroke, balance control is one of the first rehabilitation objectives towards the restoration of functional gait. In this scenario, posturography plays a key role to evaluate the progress of the affected subject. Classic therapies of posture control rehabilitation include exercises to improve stability limits or guided movements to reinforce control efforts of patient.

Over the last years muscle synergies have been described for several composed movements like those exerted during normal postural control. Muscle synergies can be understood as functional muscle co-activation patterns [1]. This theory proposes the existence of simplified mechanisms and signals that can control several muscles at the same time. The most interesting aspect of this theory is the consistency of these synergies among subjects, and its stability intra subject. The use of this knowledge for rehabilitation is still a research goal, as well as the assessment of muscle synergies in functional tasks after stroke or SCI [2].

The use of Functional Electrical Stimulation (FES) to interact with muscle synergies during the rehabilitation of balance is a novel approach proposed by Piazza et al. in 2009 [3]. This paper presents a low cost system that enables the implementation of this novel rehabilitation paradigm. It is a posturography tool to help with the assessment of postural control and its rehabilitation. The main contribution of the work lies in its simplicity and its potential use in rehabilitation. It is an exploratory device to study new rehabilitation approaches of balance control while monitoring the status of human balance and postural control system. The presented tool enables the evaluation of the effectiveness of current treatments and the design of new ones. The paper presents technical details of the system and preliminary results.

Further stages of this work include the validation of the designed posturography system comparing to similar ones like the NedSVE/IBV® [4] of the Institute of Biomechanics of Valencia, or the SMART Balance Master of Neurocom®. After this validation, the design of new therapies based on FES and muscle synergies will be possible and its evaluation in clinical environment.

2 Assessment Methodology and Postural Control Rehabilitation

In this section, balance assessment methods used in posturography are reviewed, as well as the tests designed for this purpose. Finally existing proposals for rehabilitation based on synergies are described.

2.1 The Computerized Dynamic Posturography (CDP)

Computerized Dynamic Posturography (CDP) was designed and developed by Nashner. It was clinically studied in collaboration with Black [17] and Wall and marketed in 1986 as Equitest by Neurocom Inc. [5].

The CDP is a technique that analyzes subject's postural control in static standing and his/her response to destabilizing conditions. It is based on the idea that the center of gravity (COG) oscillations reflect postural instability. Generally CDPs are based on dynamometric platforms. These systems analyze the postural oscillations by recording the vertical projection of gravity force, known as Center of Pressure (COP). More frequent tests made with similar platforms are:

- **Sensory System or Romberg's Test.** It is aimed at determining the ability of the patient to integrate the three systems responsible for assessing standing balance and body sway while different sensory conditions are applied. The results of this test are compared with results of normal subjects. It is performed with eyes open and eyes closed, with and without foam on which the subject stands. It can also be performed with the patient's head retroflexed, causing distortion in neck proprioceptors. These tests can also be used to evaluate proprioceptive information by making patients to rely in vestibular information to maintain the balance [6].
- **Stability Limit Test.** It assess the capacity of the subject to bring his COP to the border of his/her stability limit. Basically, this test is used to assess the maximum distance the patient can move his/her COP without changing the base of support, i.e. without moving his/her feet. During the test, the subject can see his/her COP representation on a computer screen in front of him, and he/she should move it toward the stability limits without moving its base of support. The test includes up to eight sequential different targets located around theoretical stability limits (according to previous measurements with healthy subjects).
- **Rhythmic and Directional Tests.** These tests try to assess subject's ability to perform rhythmic movements around of its center of gravity (COG). The subject is a asked to follow with his/her COP moving targets whose speed and range are configurable. The target is moved to a percentage of the stability limit previously calculated for the subject. This test is usually performed in the anteroposterior and mediolateral directions.

2.2 *Hybrid Approaches in Assisted Neuromotor Rehabilitation*

Hybrid exoskeletons have emerged as a way to improve motor assistance using the benefits of FES and robotic exoskeletons. They overcome individual limitations of the methods used separately. The FES uses natural muscles as actuators to generate a movement, which provides not only functional benefits but also physiological. Robotic exoskeletons artificial actuators are used to move the lims that can not be fully or partially controlled voluntarily.

Generally, people affected by stroke and SCI have healthy muscles. The hybrid approach proposes the use of their own muscles to complement the action of the robotic exoskeleton. Muscles are activated coordinately with the exoskeleton controller by means of an electrical stimulation system [7]. This approach results in a reduction of energy demand and allows the exoskeleton to use lighter and less powerful actuators. Moreover, this solution is considered more natural and help to preserve existing biological structures. Main problem when using FES is that it can produce muscle fatigue after long periods of stimulation. This problem limits the time of use. This is not a problem when using exoskeletons, which can be used for a longer time.

Balance control is an important not only when rehabilitating after stroke or SCI but also when using exoskeletons. Hyper is a Spanish research project aimed at developing new neurorobotics and neuroprosthetics therapies for people affected by stroke or SCI. First clinical interventions include the rehabilitation of balance and postural control. The use of hybrid approaches is well considered by clinicians but they way they are used in this rehabilitation process and it effectiveness is not clear yet for the scientific community.

The control of the assistive device is also not clear in terms of compensation actions and movement routines. The most common approach over the last years was the so called "assist-as-needed" (AAN) paradigm [8]. Following this paradigm, the interaction between the assistive device and the natural involved mechanisms in the considered task is given in terms of the final results, and not considering the underlying status of biological control mechanisms. In this way, Hyper encourages the study and development of new therapies to support classic ones. These novel treatments are mainly driven by bioinspired mechanisms for better and deeper interaction between the assistive device and remaining neuromotor control structures, in order to reinforce and rehabilitate them in a more natural way.

2.3 *Muscle Synergies*

The study of human control system is a open research field where there are still many questions to answer. One of them is how is coded the information to control the large number of degrees of freedom of human movements. More specifically, this problem states that to generate a specific motor task, there are multiple combinations of muscle activations that can generate similar results. Muscle synergies theory is a proposed answer to this question. The central nervous system can solve the complex task by choosing a specific set of muscle activations through a combination of a small set of neural patterns, called *synergy* [1].

Each muscle receives as input a modulating signal from higher neural centers, and outputs a weighted activation signal to activate a set of muscles. The activation of each muscle can be seen is a weighted sum of all synergies commands connected to it [9]. Then, muscle synergies can translate small sets of variables coming from the central nervous system into higher dimensional signals. They are strictly correlated to the functional performance and their modulation are related with user workspace. The most interesting characteristic of muscle synergies is that they are consistent inter healthy subjects [1, 3].

Mathematically, as indicated in [9], muscle synergies are explained by the following equation that describes the activation of a single muscle m,

$$m(t) = \sum_{i}^{K} c_i(t) w_i, \tag{1}$$

where m is the activation of the muscle function of time, c is the neutral command i-th synergy function of time, w is the constant weight of the i-th synergy referred to the m-muscle and K is the number of synergies.

The use of muscle synergies knowledge to rehabilitate postural control is still not clear. However, their role in functional movements and their importance have being already reported [10, 11]. This encourages Hyper scientific team to take them into account to promote the development of more efficient rehabilitation therapies by closely interacting with involved muscle synergies in balance control. In this way, FES can be used to develop and interact with synergies and muscle activation patterns.

3 Proposed System

In this section we describe the low-cost platform developed to perform static posturography tests and explore new therapies based of FES and muscle synergies. Balance control assessment platforms are usually not open and they are commercially available only as a posturography tool. Thus, a novel low cost and open posturography platform was developed to use it in this novel scenario. The main objective of this platform is to support the development of novel balance control rehabilitation therapies in the framework of the Spanish Hyper project.

Figure 1 shows the outline of the developed platform and a functional diagram of the different components that are further described in next sections. The platform is based on a centralized non real-time architecture, which includes several components: the Posturography Controller, the Neuroprosthetic Controller, a Wii Fit Balance Board, a Kinect camera and the TEREFES electrostimulation system [12].

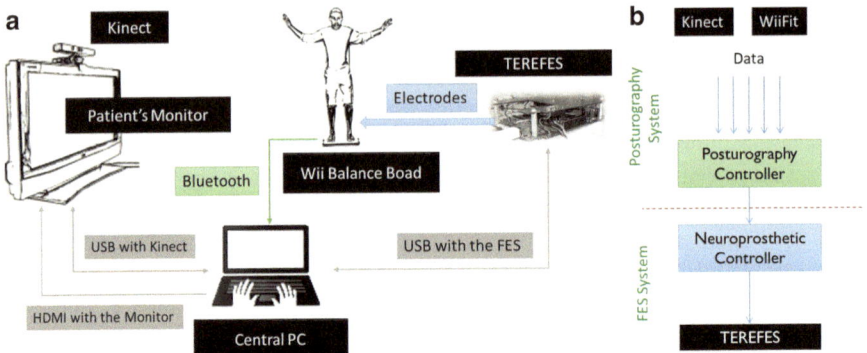

Fig. 1 (**a**) Proposed platform architectural and (**b**) the functional description diagram including the different components: the posturography system (balance control assessment) and the FES system (balance control training and rehabilitation)

3.1 Wii Fit Balance Board

Wii Fit balance board is an input device included in the Wii Fit from Nintendo®. It is a wireless device that uses Bluetooth technology to communicate with the Wii console. It is equipped with four resistive pressure sensors located in each corner of the table. In effect, it measures the displacement of the center of pressure and the weight of the user. It also gives an indication of the battery status.

Over the last years, the Wii Fit Balance Board have been used by the scientific community, specially as computer interface for disabled [13]. This device has two attractive features: it is wireless and low-cost. In our project, the Board will be used to measure the COP.

Data from Wii Fit Balance Board is accessed through a Microsoft Visual Studio C# application, using a library called WiimoteLib available at www.wiimotelib. codeplex.com. Visual C# was chosen because it is also compatible with the *Kinect* and its *Windows Software Development Kit* (SDK). Thus, the Wii Fit is connected as a HID interface device. Provided services by the Board are detected using the Service Discovery Protocol (SDP) of Bluetooth.

An important aspect to consider is the sampling frequency at which the Wii Fit sends the data to the PC, or more specifically, how often the data arrives, considering the nature of wireless transmission and the operating system behavior. To answer this question, we measured the time interval between samples using methods and public properties of the Microsoft Visual Studio C# class `System.Diagnostics.Stopwatch`. The program is executed in a almost dedicated HP Pavilion g6-1b70us Notebook (Intel Core i3 CPU M 370 @ 2.4 GHz, 4 GB of RAM) running 64-bit Windows 8. According to the obtained results, the average sampling frequency is 100 samples/s.

The aim of this analysis it to know how deterministic is the access to the data of the Board in terms of time. In other words, we want to know the probability that the

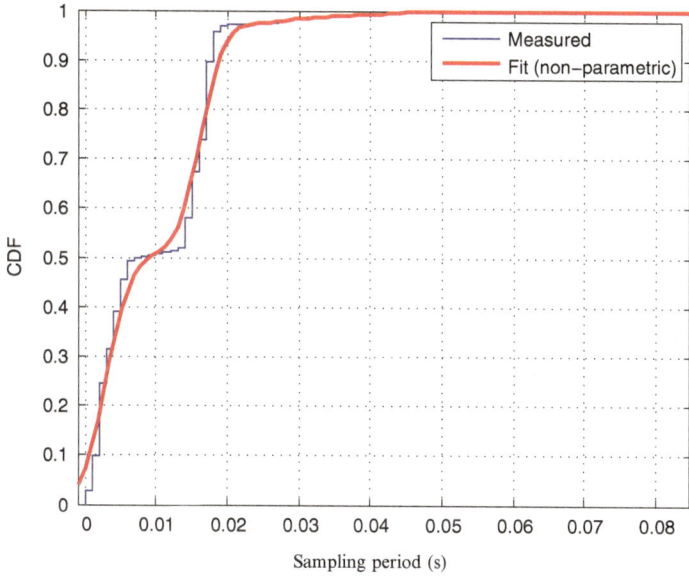

Fig. 2 Cumulative distribution function as a function of the sampling period (s)

sampling period of the Wii Fit is less than or equal to any given time. Cumulative distribution function for the Wii Fit data is depicted in Fig. 2. For example, the probability that the sampling period remains less than or equal to 0.02 s is 94.02 %.

3.2 Microsoft Kinect

The Kinect device is a natural user interface, which allows users to interact with games without physical contact. It was developed by PrimeSense Company. The user becomes the controller itself, having to rely on movements, natural gestures and voice commands to control game elements.

Kinect is equipped with an RGB-D camera that acquires images of 640 × 480 at 30 fps. It has a visual field range from 1.2 to 3.5 m, but can be reduced by optical coupling, as Niko Zoom Lenses®. Furthermore, its viewing angle is 57° horizontally, and 43° vertically. The vertical visual field can be expanded 27° with its servomotor. It is also equipped with an array of four microphones, each with a recording resolution of 16 bits sampled at 16 kHz. It also contains a stack of signal processing hardware that is able to handle all the data generated by cameras, infrared light, and microphones. By combining the output of these sensors, a program can track and recognize objects in front of it, determine the direction of the sound signals, and isolate them from the background noise.

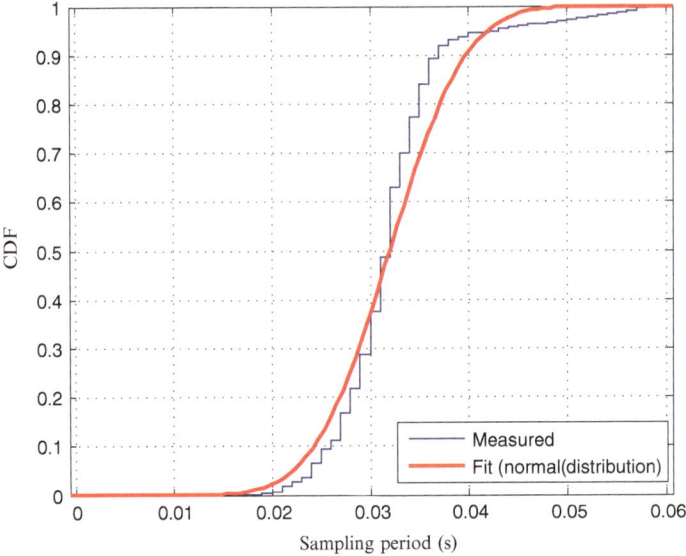

Fig. 3 Cumulative distribution function of the sampling period Kinect

The role of the Kinect in the platform is to enrich the visual feedback provided to the patient. Common posturography platforms are limited to provide users information about the center of pressure but the user does not know precisely his/her position and how good it is for the intended task. In this way, the Kinect provides kinematic information of full body segments, thus providing more complete information to users as well to the Neuroprosthetic Controller enabling better actuation commands.

Both, the Wii Fit as the Kinect, help to give visual feedback to the evaluated subject, and the information generated can be used by the Neuroprosthetic Controller to generate more precise and adequate stimulation patterns. A similar analysis done with the Wii Fit Board, was carried out with Kinect to evaluate the jitter effect when acquiring the frames.

The cumulative distribution function is shown in Fig. 3. For example, the probability that the sampling frequency is maintained below 35 fps is 75 %, approximately.

3.3 Posturography Controller

The Posturography Controller is implemented in a personal computer running Windows 8 operating system. The developed software includes traditional posturography tools and tests like Romberg's test, test of the stability limits and the rhythmic directional test.

The software was developed for easy use by medical personnel. It includes a database in which data of each patient is stored, allowing the physician to evaluate subject progress after several sessions. It also helps to diagnose potential diseases and program rehabilitation exercise routines for each evaluated subject. The application is also able to generate Matlab scripts containing the center of pressure points recorded during each rehabilitation session. In this way, the therapists can analyze data recorded in previous sessions.

The Posturography Controller receives all the data from the Kinect camera and the Wii Fit Balance Board. It fuses and displays the acquired data showing information like the center of pressure, the rigid body kinematic chain of the studied/analyzed subject, and information about current routines and tests. This controller and the Neuroprosthetic Controller are implemented in the same computer.

In next sections details on the parameters used to quantify the results of each test will be described [14].

- **Romberg's Test.** The subject is positioned on the Wii Fit Balance Board in an upright position with arms straight and close to the body trying not to move the head in neutral position facing forward, bare feet at an angle Opening of 30°. In this position is assessed for T seconds (configurable by the doctor) their ability to maintain balance in the following conditions:

  • Eyes Open (REO) and Eyes Closed (REC).
  • Foam on Wii Fit with Eyes Open (RGA) and Eyes Closed (RGC).

  The parameters evaluated in each test are:

  • *Shift angle (degrees).* The angle of the vector extending from the initial point to the subject portion to the end point of the trajectory.
  • *Swept area (mm².)* It estimates the area swept by the COP by mean of an ellipse whose axes correspond to the maximum mediolateral and anteroposterior displacement.
  • *Average speed (cm/s).* It estimates the average speed, which is the ratio between the displacement and time, T, that lasts the test.
  • *Maximum mediolateral and anteroposterior displacements (mm).* These parameters represent the longest displacement in the mediolateral and anteroposterior axes during the exercise.

  Figure 4 shows a screenshot during the execution of the application designed in this project. Specifically, this screenshot corresponds to a REO Test. The figure shows two visual feedbacks. The first one is the position of the center of pressure on the Wii Fit Balance Board, and the second, provided by Kinect, is the RGB image and a trace of its skeleton. The screen provides information about subject's skeleton (skeleton blue) and a given reference (red skeleton). The reference indicates correct estimated position during the tests.

  Some parameters are provided online by the application. For example, regarding the Kinect, it monitors the status of the tracking task, which can be *Tracking* (OK) or *not skeleton* (Subject not detected). Another parameter is the

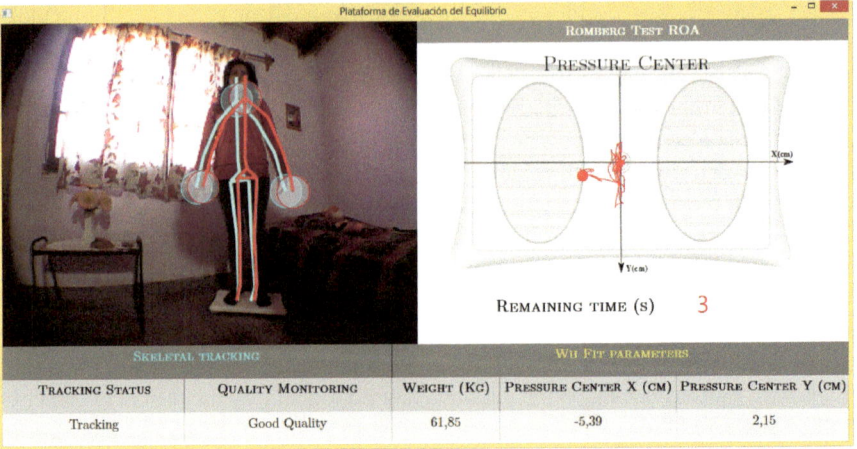

Fig. 4 Screenshot during the execution of REO Test. On the *right*, it is shown the center of pressure on the Wii Fit while on the *left* the subject with his/her skeleton (*blue lines*) and the given reference (*red lines*)

quality of the skeleton. This parameter indicates if the Kinect is showing the complete skeleton of the subject (*Good Quality*). This will help the therapist to point the camera in the correct position. Regarding the Wii Fit Balance Board, the parameters observed are subject's weight and the coordinates of the COP.

– **Stability Limit Test.** This test evaluates the following parameters:

- *LE max (mm).* It is the maximum value reached by the COP in the corresponding direction (8 targets separated of 45° and whose radial distance from the origin is configurable).
- *Stability zone (cm).* It is approximately the mean distance at which the patient is 90 % of time. It is calculated for each direction.

Figure 5 shows a screenshot during the execution of limit test. The displacement of the COP on the Wii Fit for the Limit Test is depicted on the right side. The red circle represents the current target to which the subject should direct his/her COP, while the green ones represent those already targeted. Traces of COPs in these directions have been deleted to not disturb the patient while reaching current target.

– **Rhythmic and Directional Test.** In this test, the patient is asked to follow the movements of a moving target (configurable frequency) in mediolateral (ML) and anteroposterior (AT) directions. The maximum excursion limit is calculated based on the parameters of normal stability limits (previously recorded with healthy subjects).

The following parameters are evaluated for each direction.

- *Reaction time (s).* It is defined as the time that the subject takes to bring his/her center of gravity closer than two centimeters from the reference target.

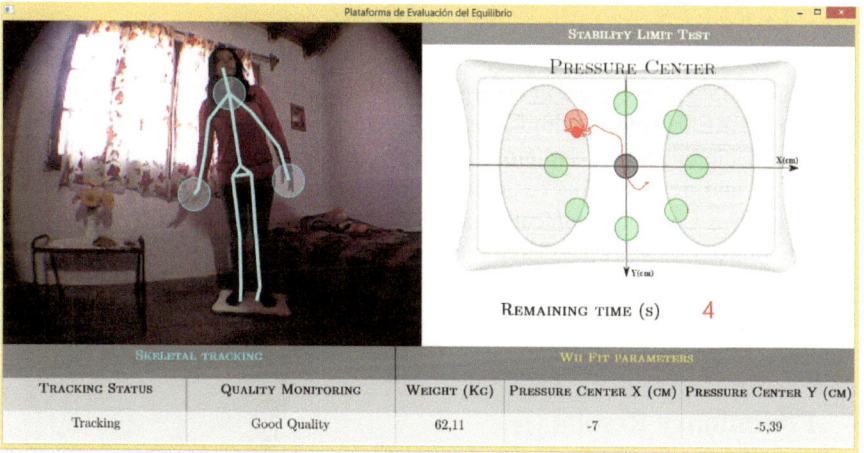

Fig. 5 Screenshot during the execution of the limit test. On the *right side* of the screen, the COP on the Wii Fit is shown in real time. On the *left side* is shown the patient with his/her stickman representation (*blue lines*)

- *Tracking capability (%)*. It quantifies subject's ability to follow the movement of the target in ML or AT directions. This parameter is calculated as the mean of error (*DesiredCOP − MeasuredCOP*), after the reaction time. If the error is lower than 2 cm (configurable), in other words the COP is inside the target circle, it is considered as zero for this sample.
- *Directional control (%)*. It quantifies the subject's ability to remain in the expected direction of the test. For example, if the target moves in the axis ML, the index is evaluated considering the AT axis error using the same process for calculating the tracking capability.

3.4   Neuroprosthetic Controller

The Neuroprosthetic Controller is responsible for the generation of muscle activation patterns and control of the actuation system: the TEREFES electrostimulator. It receives from the Posturography Controller all the kinematic data of the subject (acquired with the Kinect) and the coordinates of the center of pressure (COP). A driver decodes and converts the information into muscle activation patterns and specific TEREFES commands according to previously programmed synergies sets, theory and rehabilitation parameters. Full details of the proposed synergistic controller can be found in [15]. The functional stimulator TEREFES must act synchronously according with the exerted movements. Since Windows is not a real time operating system, a best effort approach is used in the Neuroprosthetic Controller. The TEREFES (real time system) is used to monitor possible delays in the Neuroprosthetic Controller. In next stages, system timing should be further analyzed.

3.5 TEREFES

The TEREFES was proposed in the framework of the TERERE and Hyper projects
[12]. The TEREFES electrostimulator provides up to 32 stimulation channels driven
by controllable and stable and close loop current sources. In addition, the system
is portable and flexible. This functional stimulator is powered by 4 AA batteries
and includes a USB communication interface that allows its online configuration
through an external software. Monophasic and biphasic stimulation signals can be
obtained in its 32 available channels. These channels are divided in two independent
groups of 16 channels each, that can be sequentially activated.

4 Preliminary Results

In this section preliminary results of posturography software are presented.
Described results were obtained with six healthy people, four men and two women.
The purpose of this functional validation is to technically verify the platform and to
compare result between different subjects. Unfortunately, at this stage of the work,
the system could not be tested with previously diagnosed pathological subjects,
and the results could not be compared with those obtained with other commercial
platforms like Neurocom.

The procedures for the tests were explained in previous sections. REO and REC
tests were conducted, as well as Stability Limit and Rhythmic tests. All of them
were realized a couple of times in order to make sure that the subjects understand the
test but without producing fatigue or previous learning/training [14]. The sampling
frequency was 30 frames/s, enough to detect any COP displacement [16].

4.1 Romberg's Test

Each Romberg's test lasted 30 s. The results of the six subjects are shown in Table 1.

All proposed parameters were calculated and presented in Table 1. Results
suggest a decrease of fine postural control in most subjects when they close their
eyes. For both men and women, the displacement angle is usually in the second
quadrant, and no significant differences are found among REO and REC tests. In
fact, according to [14], this parameter does not change significantly under these test
conditions.

Figure 6 shows the results of subject 4. Using similar data, proposed parameters
were calculated for each subject.

Regarding the swept area, calculated according to [17], it does not reflect
noticeable changes with the changing sensory conditions. Balaguer, in his work,
[14], has suggested that the calculation by fitting a geometric figure may not be
adequate to quantify this parameter.

Table 1 REO and REC test results

Subject	Gender	Years	Disp. angle (°)		S. area (cm²)		A. speed (cm/s)		Disp. ML (cm)		Displ. AT (cm)	
			REO	REC	REO	REC	REO	REC	REO	REC	REO	REC
1	M	23	108.22	114.45	18.28	17.21	1.94	2.28	2.42	3.16	1.82	3.00
2	M	26	90.77	91.45	2.71	6.06	1.25	1.65	0.817	1.552	1.01	1.22
3	M	34	95.89	113.77	12.91	12.52	1.69	2.15	2.46	2.71	2.01	2.31
4	M	47	76.55	75.71	9.05	4.88	1.23	1.34	1.81	1.55	1.85	1.21
Average			92.86	98.84	10.74	10.17	1.53	1.86	1.88	2.23	1.67	1.94
5	F	19	122.49	112.1	4.94	5.85	1.53	1.95	2.06	2.05	1.06	2.45
6	F	18	106.79	103.85	8.99	7.45	1.34	1.65	2.15	2.03	1.13	2.32
Average			114.64	107.96	6.97	6.65	1.44	1.80	2.11	2.04	1.095	2.385

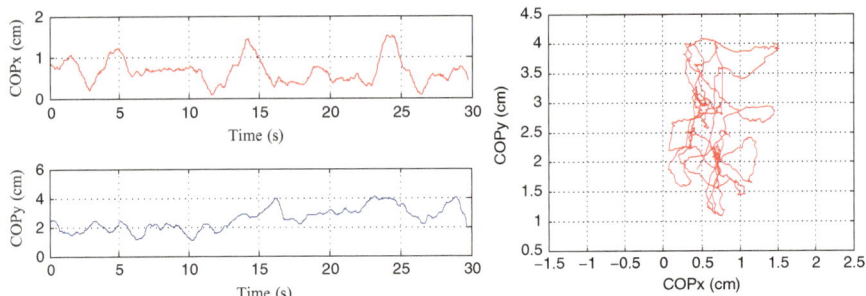

Fig. 6 Subject 4 (S4) REC Romberg's test plotted using Matlab. Parameters for S4 were calculated with these data

Table 2 Stability limit test results

	Average max. LE (cm)		Average stability zone (cm)	
	Gender			
Direction	Male	Female	Male	Female
Front	7.51	9.04	5.77	8.32
Front-right	9.68	9.695	8.05	8.73
Right	10.59	9.165	9.28	7.70
Rear-right	9.44	8.58	8.92	7.33
Rear	10.06	8.00	8.68	6.915
Rear-left	9.88	9.33	8.32	8.64
Left	11.01	10.05	9.00	8.35
Front-left	9.335	9.97	8.2675	8.55

Average values are shown. Similar data can be used to obtain normality patterns

Finally, the average speed of displacement, is found to increase without visual feedback. This same behavior is observed in the ML and AT displacement. Therefore, these parameters are used to differentiate visual system impact and potential dysfunction in balance control. Both men and women present larger variations in the ML direction, being even larger in men in these particular tests.

4.2 Stability Limit Test

The stability limit test lasted 10 s for each direction, and each target was located at a distance of 10 cm from the origin. The subject was asked to make his/her best effort to reach the targets.

Figure 7 shows the results of subject 1. Using similar data, proposed parameters were calculated for each subject (Table 2).

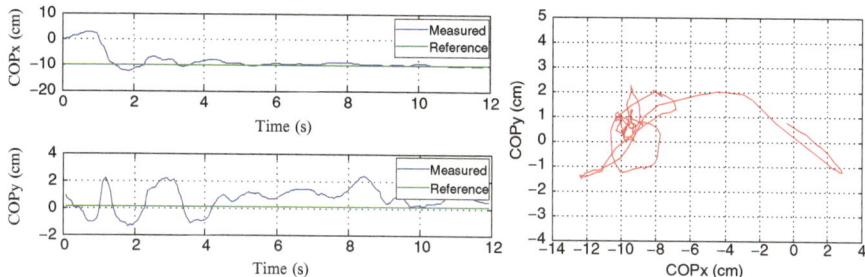

Fig. 7 Subject 1 (S1) limit test in left direction results plotted using Matlab. Parameters for S1 are calculated with these data

Table 3 Rhythmic control test results

Subject	Gender	Years	Reaction time (s)		Tracking capability (%)		Directional control (%)	
			ML	AT	ML	AT	ML	AT
1	M	23	0.037	0.50	81.86	85.7	81.7	99.9
2	M	26	0.119	0.039	80.55	70.44	90.17	99.05
3	M	34	0.12	1.059	87.7	63.8	78.22	99.83
4	M	47	0.198	1.046	79.6	57.8	77.51	99.92
Average			0.1185	0.66025	82.43	69.43	81.9	99.675
5	F	19	0.40	0.035	66.5	70.23	79.7	98.5
6	F	18	0.42	0.21	71.5	55.26	88.53	96.94
Average			0.4125	0.1225	69.0	62.74	84.11	97.72

According to these tests, areas of stability in both men and women vary with direction. In general terms, there are no significant differences. These results agree with [18]. However, a larger population is needed to obtain robust conclusions. Balaguer found that subject owns subjective perception (Previous Q&A about disability condition of the subject) of his/her skill or disabilities does not influence the stability limits [14].

4.3 Rhythmic and Directional Control Test

For the rhythmic tests, windows of 10 cm (configurable) long were defined directionally, first in the ML direction and then in the AT one. The subject was asked to follow a moving target traveling at a frequency of 0.25 Hz. Each test lasted 20 s. The results are shown in Table 3 for each patient.

Figure 8 shows the results of subject 5 (S5). Similar results were used to calculate all parameters for each subject.

According to these results, the reaction time increases with age. In addition, there is a shorter reaction time in women. In men, the reaction time is better in the ML

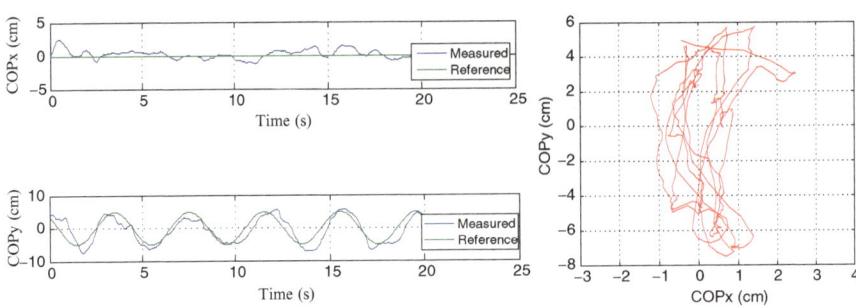

Fig. 8 Subject 5 (S5) rhythmic test results in the AT direction plotted using Matlab. Parameters for S5 are calculated with these data

direction with respect to the AT direction. Although this work has not made a study of subjects with specific pathologies [14] found that vestibular disorders can affect rhythmic and directional control in disagreement with the findings of Cortés [18].

5 Discussions

Posturography helps to assess the influence of any vestibular dysfunction in postural and balance control. However, a pathology that affects the balance in one patient, in other word the vestibular-spinal reflex, not necessarily will have the same effect it in another subject. In this case, tools like the one described in this work are not effective to help the diagnosis of the impairment.

Regarding the tool presented in this paper, it i not very clear in the literature the way how different assessment parameters are calculated. This lack of information makes more difficult to compare results. However, overall conclusions and trends obtained with this tool are similar to those reported in the literature and obtained with other platforms.

Nowadays, there is still a discrepancy between scientists regarding the results of each parameter and associated information. According to [14], this discrepancy exits because it is difficult to find clear relationships between functional assessment of balance and patient-perceived disability. Tests may be influenced by many factors like social, professional, technical, psychological, affective, and cognitive ones.

The current drawback of classical static posturography is limited only to study the subject during standing position, so it does not provide information on the dynamic aspects of postural control. To solve this shortcoming, we have followed the line proposed by García [14] and set dynamic tests, such as the rhythmic test.

Regarding the Neuroprosthetic Controller and the potential use of the system as a tool for training the balance control, it is ready to be used. However, specific scenarios and protocols should be defined to validate technically the tool in a better way.

6 Conclusions and Future Work

Postural rehabilitation boosts patient confidence and contribute to their self-improvement. In addition, knowledge of the particular deficit in postural control helps clinician and patient to develop prevention plans to avoid falls, and as mentioned before, it is the first step towards the rehabilitation of more complex processes like gait.

Current research projects in neuromotor rehabilitation like Hyper, are devoted to develop novel bioinspired rehabilitation treatments. The use of hybrid solutions including neurorobotics and neuroprosthetics devices has been shown as an efficient approach. However, the use and development of modern rehabilitation therapies based on novel knowledge need the support of non existing research tools.

We have seen how to make a low cost posturography system. It is based on a Wii Fit Balance Board, the Microsoft Kinect and the TEREFES electrostimulator. This tool can serve as a low cost balance control assessment tool and will allow the implementation of novel therapies that could improve current ones for the rehabilitation of balance control.

Future work includes the evaluation of the tool and developed system in clinical environments. After this validation, the final integration of the Neuroprosthetic Controller and the implementation of therapies based on muscle synergies will be done.

References

1. D'Avella A, Bizzi E. Shared and specific muscle synergies in natural motor behaviors. Proc Natl Acad Sci USA. 2005;102:3076–81.
2. Torricelli D, Aleixandre M, Alguacil IM, Cano R, Molina F, Carratalá M, Piazza S, Pons JL. Modular control of mediolateral postural sway. In: Proceedings of 34th annual international conference of the engineering in medicine and biology society (EMBC'12); 2012.
3. Piazza S, Torricelli D, Brunetti F, del-Ama AJ, Gil-Agudo A, Pons JL. A novel FES control paradigm based on muscle synergies for postural rehabilitation therapy with hybrid exoskeletons. In: Proceedings of 34th annual international conference of the engineering in medicine and biology society (EMBC'12); 2012.
4. Baydal J, Castelli A, Garrido J, Bermejo I, Broseta J, Amparo M, Perez J, Moya M. NedSVE/IBV v.5 a new system for postural control assessment in patients with visual conflict. Revista de Biomecánica (54). Valencia: Instituto de Biomecánica de Valencia; 2010.
5. Faraldo A. Registro postural en personas sanas: evaluación del equilibrio mediante el estudio comparativo entre la posturografía dinámica computerizada y el sistema sway star. Doctoral thesis, Universidad de Santiago de Compostela, Spain; 2009.
6. Khasnis A, Gokula R. Romberg's test. J Postgrad Med. 2003;2:169–72.
7. Del-Ama AJ, Koutsou AD, Moreno JC, Pons JL. Review of hybrid exoskeletons to restore gait following spinal cord injury. J Rehab Res Dev. 2012;49:497–514.
8. Cai LL, Fong AJ, Yongqiang L, Burdick J, Edgerton VR. Assist-as-needed training paradigms for robotic rehabilitation of spinal cord injuries. In: Proceedings of the 2006 IEEE international conference on robotics and automation (ICRA'06); 2006.

9. Torricelli D, Moreno JC, Pons JL. A new paradigm for neurorehabilitation based on muscle synergies. In: Proceedings of 34th annual international conference of the engineering in medicine and biology society (EMBC'12); 2011.

10. D'Avella A, Portone A, Fernandez L, Lacquaniti F. Control of fast-reaching movements by muscle synergy combinations. J Neurosci. 2006;26:7791–810.

11. Piazza S. Muscle synergies in postural sway movements: neurophysiological evidences and rehabilitation potentials. Master's thesis, University Carlos III of Madrid, Spain; 2013.

12. Brunetti F, Garay A, Moreno JC, Pons JL. Enhancing functional electrical stimulation for emerging rehabilitation robotics in the framework of hyper and project. In: 2011 IEEE international conference on rehabilitation robotics rehab week Zurich, ETH Zurich Science; 2011.

13. Martin A., Study and development of man-machine interface based on wireless sensor, Master Thesis, Escuela Técnica Superior de Ingeniería, Universidad Pontificia de Comillas; Madrid, Spain 2008.

14. García RB. Valoración de un método de posturografía estática con pruebas dinámicas para evaluar funcionalmente pacientes vestibulares en edad laboral y su relación con el índice de discapacidad. Doctoral thesis, Universidad Politécnica de Valencia, Spain; 2012.

15. Denis W, Brunetti F, Piazza S, Torricelli D, Pons JL. Functional electrical stimulation controller based on muscle synergies. In: Proceedings of the first international conference on neurorehabiltiation, converging clinical and engineering research on neurorehabilitation; 2012.

16. Enbom H, Magnusson M, Pyykko I, Schalen L. Presentation of a posturographic test with loading of the proprioceptive system. Acta Otolaryngol Suppl. 1988;455:58–61.

17. Black FO, Shupert CL, Peterka RJ, Nashner LM. Effects of unilateral loss of vestibular function on the vestíbulo-ocular reflex and postural control. Ann Otol Rhinol Laryngol. 1989;98:884–9.

18. Cortés O. Análisis clínico y posturográfico en ancianos con patología vestibular y su relación con las caídas. Doctoral thesis, Universidad Politécnica de Valencia, Spain; 2007.

A Pilot Study on the Feasibility of Hybrid Gait Training with Kinesis Overground Robot for Persons with Incomplete Spinal Cord Injury

Antonio J. del-Ama, Ángel Gil-Agudo, José Luis Pons Rovira, and Juan C. Moreno

Abstract Hybrid actuation and control have a considerable potential for walking rehabilitation of neurologically impaired people, but there is a need of novel hybrid control strategies that adequately manage the balance between functional electrical stimulation (FES) and robotic controllers. A case-study of a hybrid co-operative control strategy for overground gait training with a wearable robotic exoskeleton for persons with incomplete spinal cord injury (SCI) is presented. The feasibility of the control strategy to overcome muscular stimulation electro-mechanical delay, deterioration of muscle performance over time, and to balance muscular and robotic actuation cyclic overground walking is tested in one subject with incomplete spinal cord injury (L4, AIS grade D). The results show that the proposed hybrid cooperative control in Kinesis overground robot is able to autonomously compensate a bilateral pathologic walking pattern and the suitability of Kinesis hybrid gait training robot for conducting clinical experimentation.

Keywords Spinal cord injury • FES • Exoskeleton • Gait training

1 Introduction

Most therapies for rehabilitation of walking rely on the assumption that task-oriented practice promotes mechanisms of neural plasticity, muscle strength and learning of compensation strategies that increase walking ability of the person with SCI. Robotic technology holds a considerable potential to drive such interventions. Ambulatory robots, that have been developed mainly for functional compensation of walking, can offer a challenging and rich walking therapy. Furthermore, functional electrical stimulation can drive rehabilitation interventions of SCI providing several

A.J. del-Ama (✉) • Á. Gil-Agudo
Biomechanics and Technical Aids Unit, National Hospital for Spinal Cord Injury, Toledo, Spain
e-mail: ajdela@sescam.jccm.es

J.C. Moreno
Bioengineering Group, Spanish National Research Council, Madrid, Spain

J.L.P. Rovira
Neural Rehabilitation Group, Cajal Institute, Spanish Research Council (CSIC), Madrid, Spain

© Springer International Publishing Switzerland 2015   19
A.R. Londral et al. (eds.), *Neurotechnology, Electronics, and Informatics*,
Springer Series in Computational Neuroscience 13, DOI 10.1007/978-3-319-15997-3_2

physiological and psychological benefits to the user through artificial activation of paralyzed muscles. On the other hand, robotic exoskeletons can be used to manage the unavoidable loss of performance of FES-driven muscles. Hybrid exoskeletons are then regarded as a promising approach that blends complementary robotic and neuroprosthetic technologies. The overview of the state of the art on hybrid gait systems has demonstrated that such redundant actuated solutions can produce feasible systems for accurate control of joint movement [1]. Also, it has shown that diverse muscle fatigue management strategies could be applied for an effective closed-loop control of FES.

Under this hybrid scenario, the assist-as-needed control strategy has been proposed as a new redundant neuroprosthetic and robotic system that cooperates to optimize the outcome of the active control of the motion of the knee joint while providing assistance [2]. While various wearable exoskeletons were successful in achieving gait in subjects with incomplete spinal cord injury (SCI) [3–5] this has generally been proposed as a functional substitution. In this case study, we tested the feasibility of a novel hybrid gait training approach with an overground robot on a person with an incomplete SCI.

2   Material and Methods

2.1   Kinesis: Robotic Platform for Hybrid Therapy of Walking

Kinesis is a hybrid robotic device that has been developed for overground gait training in incomplete spinal cord injuries. Incomplete lesions were targeted because, first, prognosis of functional recovery of walking for these patients is that the they can walk short distances but depending on the wheelchair for community ambulation, therefore a successful hybrid walking therapy could provide benefits to this population. Second, motor-incomplete lesions have preservation of lower limb muscles innervation. This is a key feature when considering neuromuscular stimulation for therapy, because denervated muscles do not contract under artificially-delivered electric pulses. Among the different syndromes of incomplete spinal cord injury, target population is comprised by patients whose lesion is categorized as *Conus Medularis* [6], and those patients whose lesion met the functional characteristics of the *Conus Medularis*. These functional characteristics are intact upper body function and hip function preserved, while the knee and ankle muscles can vary from paralysis to healthy, depending on the specific characteristics of the injury.

The Kinesis system is a bilateral wearable knee-ankle-foot orthosis. Hip joint was not considered as the target population have preserved this function. Kinesis features active actuators at the knee hinges (Maxon DC flat motor, 90 W with Harmonic-Drive 100:1 gear) and a passive elastic actuator at the ankle. Force sensing resistors are employed for monitoring floor contact and custom force sensors are available to measure interaction torques. Kinesis has a PC-controlled stimulator (Rehastim,

Hasomed GmbH) which delivers biphasic current-controlled rectangular pulses. Rehastim can be pulse width and current controlled in real time.

The high-level control approach to achieve a cooperative behavior comprises four main components: (1) a knee joint robotic controller, (2) FES controller, (3) muscle fatigue estimator (MFE), and (4) a finite-state machine (FSM) that orchestrates the FES and joint controllers. Further details on the implementation of the high-level control can be found in [2, 7]. The cooperative behavior of Kinesis allows obtaining adequate and personalized stimulation patterns, estimating muscle fatigue and reducing robotic assistance during ambulatory walking. The ultimate goal is to give priority to the muscle-generated torque during gait training. A finite-state machine is employed to iteratively control the FES of knee muscles in a learning scheme for each leg while adapting torque field stiffness for a reference kinematic pattern. In this scheme, the resulting interaction torque (with added mass and inertia of the leg) is monitored and used towards convergence of stimulation parameters. The robot modulates its assistance by reducing joint stiffness, as shown in next section, ensuring the target flexion angle for effective swing of the leg. A muscle fatigue estimator is employed (based on the measurement of interaction torque) to trigger a fatigue compensation strategy (change stimulation firing rate). More details on the muscle fatigue monitoring and management strategy can be found in [8, 9]. A more detailed descriptions of the technique for hybrid cooperative control of Kinesis are discussed in [2, 7].

2.1.1   Robot Stiffness Modulation Strategy

The strategy to modulate the exoskeletal knee stiffness during cyclic walking is described in this section. The efficacy of the FES controller to generate the knee movement is inherently limited, due to the low efficiency of the force generated by the stimulated muscles and the electromechanical delay between the stimulus and the onset of joint movement. The goal of the hybrid control strategy was therefore to exploit the joint movement generated by the stimulation controller while supporting the movement through the joint controller. A controller was employed to provide compliant assistance to the knee, depending upon the parameter K_k, the stiffness of the force field applied around the trajectory.

Modulation of K_k was executed depending on the gait phase and the contribution of the FES to the knee trajectory. Thus, gait phase and muscle contribution were managed within a finite state machine (FSM), comprised of two FSMs operating in parallel: one FSM runs in the time domain (t-FSM) while the other operates in the cycle domain (c-FSM). The t-FSM detected the main walking states and the transitions among them. The c-FSM operated with discrete values, during each swing phase, that are related to performance of the stimulated muscle. Muscle fatigue results in a decrease of muscle performance thus increasing the interaction torque. This increase can be automatically compensated with the closed-loop FES controller reducing the interaction torque. This allows to uncouple the closed-loop control of stimulation from the muscle fatigue monitoring and management.

The t-FSM modulated the force field stiffness K_k and set the kinematic pattern, depending on the walking state. The compliant behavior of the exoskeleton was achieved by controlling knee trajectory through a first order torque field imposed around the joint trajectory. As a result, the joint torque imposed by the robot depended on the deviation of the knee trajectory from the kinematic pattern and the stiffness of the torque field K_k. The width of a virtual tunnel where the knee can actually move could be adjusted along time. During stance, a high stiffness torque field was imposed to provide support and avoid knee collapse. Conversely, the supportive actions of the exoskeleton must be reduced during the swing phase to allow for the contribution of stimulated muscles and passive dynamics. The former requirement was achieved by reducing the support of the robot through a wider virtual tunnel. At the end of each swing phase, prior to contacting the floor, the robot gradually increased its support to ensure full knee extension, through a progressive increment in the force field stiffness. However, this late stiffness for foot contact is insufficient for weight support and therefore, a quick transition to high stiffness required for stance was implemented.

2.2 Subject

A 43-year old male (75 kg and 1.77 m height) with a traumatic lesion at L4 (AIS [10] grade D) volunteered for participating in the experiment. He had preserved hip flexion ability, partial ability to generate voluntary knee extension and is in presence of mild spasticity. As consequence of the accident, the patient had a limited articular range of motion at both knees, which led to adaptation of the kinematic pattern of the left leg to meet these physical constraints (Table 1). The joint angle for the stance phase of the left leg was set at the maximum knee extension angle, and the kinematic pattern for the swing phase was consequently adapted. The subject provided written informed consent by signing a form that was approved by the Spinal Cord National Hospital Review Board.

2.3 Protocol

The patient participated in the hybrid gait training testing session (HGTT) to determine the feasibility of overground control of walking with Kinesis. Prior to the HGTT, the patient underwent a stimulation test and a training session in separate days. The stimulation test was employed to quantify the muscular response to the

Table 1 Articular range of knee joint		Left leg	Right leg
	Range of movement	20–100°	5–150°

Table 2 Stimulation test results: maximum force-time integral achieved (absolute and normalized by patient's leg weight)

Movement direction	Left leg	Right leg
Extension (N)	60 (134 %)	70 (156 %)
PA (mA)	40	48
Flexion (N)	10 (22 %)	25 (56 %)
PA (mA)	60	58

Stimulation PD was set to 450 μs, train frequency to 70 Hz, pulse train duration 14 s and duty cycle 43 %

muscle stimulation and also to get the patient used to the stimulation. Within this stimulation test, both flexor and extensor knee muscle groups of both legs were stimulated for 15 min. The results from the stimulation test are showed in Table 2. The only noticeable finding was a reduced performance of the left leg's flexor muscles.

After the stimulation test, a training session took place in which the subject carried out learning exercises with the Kinesis system. In this training session the basic walking technique was explained to the user (bend to the side to lift the heel prior to initiate a step and then pressing a manual button). Kinesis was adjusted to the patient anthropometry within this session. Total time walking in this training session did not exceed from 10 min. During HGTT kinesis hybrid-cooperative controller modulated both stimulation and robotic assistance during walking.

3 Results

The analysis of feasibility was performed at the biomechanical level. We assessed the actual knee joint kinematics and stiffness during overground hybrid control of gait. In Fig. 1, the kinematic pattern designed to meet patient's left knee angular limitation is showed (blue curves, light blue curves are actual knee angle). Looking at the right leg, high interaction forces towards flexion during stance were appreciated (Fig. 1c, d red curve, leg over exoskeleton). These flexor forces reflected on the kinematics, where the right knee is flexed during stance. In contrast, the interaction forces of the left leg during stance were lower, and the actual knee angle during stance remained close to the reference. Transition from stance to swing phase of the right leg resulted also on a transition of the interaction forces from flexor direction for stance, to markedly extensor direction for swing. Stimulation of flexor muscles for both legs exhibited in general a saturated pattern for the swing phase (black curves).

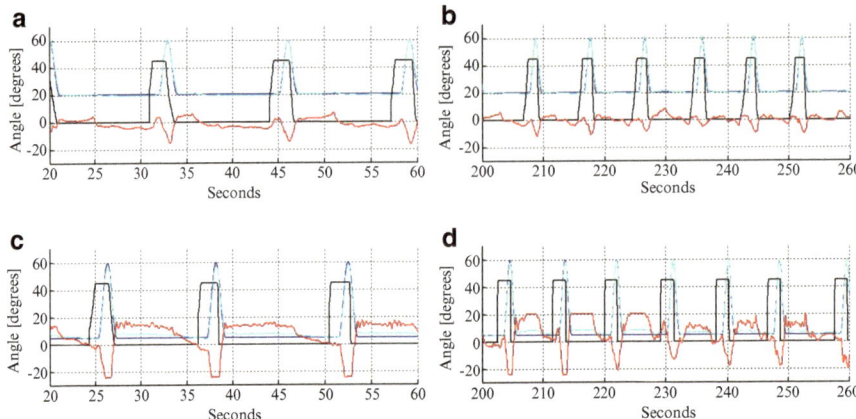

Fig. 1 Pattern (*blue curve*, deg) and actual (*light blue curve*, deg) knee kinematics, interaction torque (*red curve*, N·m/deg) and stimulation controller output (*black curve*, μs, only stimulation of flexor muscles is showed for representation purposes). (**a**) First steps of left leg; (**b**) last steps of left leg; (**c**) first steps of right leg; (**d**) last steps of right leg

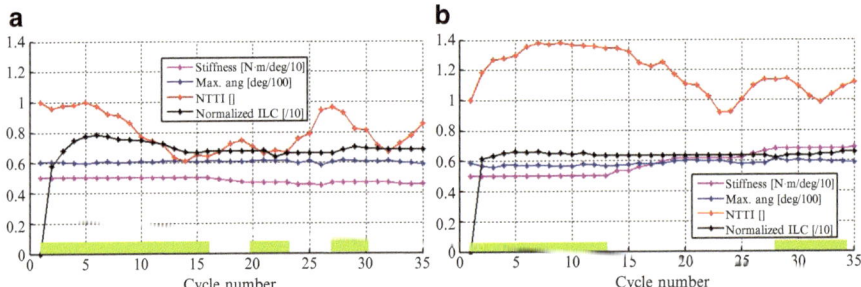

Fig. 2 Experiment results in cycle domain. Stiffness of the robotic knee joint (*magenta curve*, N·m/deg), maximum angle achieved during flexion (*blue curve*, deg) normalized torque-time integral (NTTI, *red curve*), and normalized stimulation controller output for knee flexor muscles (NILC, *black curve*) of both legs for the entire walking experiment. *Green boxes* shows *learning* state active. No box means *monitoring* state active. Controller stiffness, maximum angle and normalized stimulation curves are scaled for representation purposes. (**a**) Left leg; (**b**) right leg

Maximum flexion angle of the right leg for the first learning period was slightly lower than 60° (Fig. 1c,[1] blue and light blue curves). This was further compensated during the first monitoring period by an increase in the stiffness (Fig. 2b). It was also noticeable the difference in step cadence between the beginning and the end of the walking experiment, which indicates that the patient increased expertise and confidence on the use of Kinesis after a number of cycles (Fig. 1a–d respectively.)

---

[1]This is actually difficult to observe in the figures. During the first learning period, the maximum flexion angle was 58° in average.

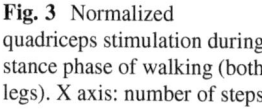

Fig. 3 Normalized quadriceps stimulation during stance phase of walking (both legs). X axis: number of steps

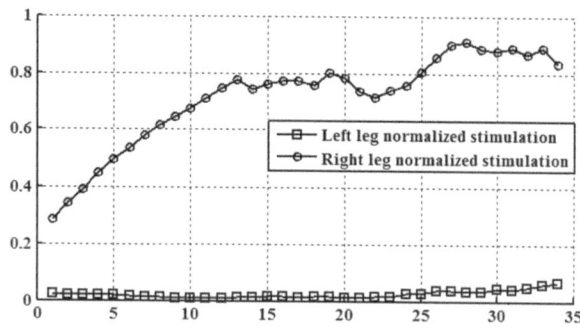

Figure 2a, b shows the operation of Kinesis within cycles, thus the c-FSM operation. The averaged interaction torque during the swing phase was obtained, and divided by the first averaged interaction torque. This procedure allowed for monitoring the cycle-to-cycle evolution of the physical interaction between the leg and the exoskeleton (NTTI, red curve in the figure). Furthermore, the stimulation controller output for knee flexor muscles was averaged for the entire swing phase and divided by a saturated stimulation for the entire swing phase. The value obtained reflect an scalar value that is representative of the stimulation intensity during the swing phase (NILC, black curve in the figure).

NILC for both legs achieved the 70 % of the maximum achievable stimulation intensity for the swing phase. After the first learning period of the right leg, a decrease on NTTI is observed along with an increase on the stiffness. In cycle 27 fatigue was detected and a new learning period took place (Fig. 2b). Regarding the left leg, a more moderated increase on stimulation intensity (Fig. 2, black curve) for the first learning period was observed, reaching similar stimulation intensity than for the right leg, corresponding with a decrease on NTTI. After the learning period, the stiffness was slightly reduced.

Figure 3 shows the averaged quadriceps stimulation intensity for the stance phases, for both legs, during the HGTT. These value was obtained similarly to the NILC procedure. It is noticeable the high stimulation intensity applied to the right leg, in response to the flexion posture exhibited by the patient during stance phases. Furthermore, as the experiment progressed and more steps were taken, the stimulation intensity progressively augmented until reaching a plateau (step 13), although after step 22, stimulation intensity increased across cycles.

4 Discussion

During and after the HGTT, no adverse effects were observed, and the patient reported neither uncomfortable nor excessive physical demand. He was able to learn the use of the system after 1 day of practice, although increases in the number of steps were observed in the experiment, which could be related to gaining expertise in the use of Kinesis.

The specific functional deficit of the subject lead to a limitation on the maximum (left) knee extension, which caused the patient to exert compensating actions. These compensations were effectively counteracted by the hybrid gait control. The compensating actions differed for both side and stance and swing phases of gait. During stance, on average, the patient flexed the right knee in an attempt to compensate for the flexion angle of the left knee. The robotic actuation compensated the stimulation as needed: the displayed stiffness was sufficient to provide compliant but adequate support during stance phases. We also noticed that during pre-swing phases, the subject consistently changed from flexion to extension, probably as a response to the limited range of movement of the right knee.

The results from the case study presented here supports performing a more detailed clinical study in order to investigate the impact of the HGTT in the walking ability of an adequate sample of spinal cord injured patients.

5 Conclusion

The HGTT delivered by Kinesis was tolerated by the patient, completing the 6 min of walking test. Mutual adaptations were observed between the patient and Kinesis that were assessed through the analysis of the physical interaction. The hybrid cooperative control in Kinesis is able to compensate a bilateral pathologic walking pattern by autonomously increasing the stimulation of the flexor muscles and increasing the displayed stiffness of the robotic actuator.

Acknowledgements This work was supported by grant CSD2009-00067 CONSOLIDER INGE-NIO 2010. The authors thanks to all the participants that volunteered for the experiments, and the clinical staff from National Hospital for Spinal Cord injury.

References

1. del-Ama AJ, Koutsou AD, Moreno JC, de los Reyes-Guzmán A, Gil-Agudo A, Pons JL. Review of hybrid exoskeletons to restore gait following spinal cord injury. J Rehab Res Dev. 2012;49(4):497–514.
2. del-Ama AJ, Gil-Agudo Á, Pons JL, Moreno JC. Hybrid FES-Robot cooperative control of ambulatory gait rehabilitation exoskeleton. J Neuroeng Rehab. 2014;11:27. doi: 10.1186/1743-0003-11-27.
3. Dollar AM, Herr H. Active orthoses for the lower-limbs: challenges and state of the art. IEEE Int Conf Rehab Robot. 2007;1:968–77.
4. Dollar AM, Herr H. Lower extremity exoskeletons and active orthoses: challenges and state-of-the-art. IEEE Trans Robot. 2008;24(1):144–58.
5. Hesse S, Waldner A, Tomelleri C. Innovative gait robot for the repetitive practice of floor walking and stair climbing up and down in stroke patients. J Neuroeng Rehab. 2010;7:30.

6. Hayes K, Hsieh JTC, Wolfe D, Potter P, Delaney G. Classifying incomplete spinal cord injury syndromes: algorithms based on the international standards for neurological and functional classification of spinal cord injury patients. Arch Phys Med Rehab. 2000;81(5):644–52.
7. del-Ama AJ, Moreno JC, Gil-agudo Á, Pons JL. Hybrid FES-robot cooperative control of ambulatory gait rehabilitation exoskeleton for spinal cord injured users. In: 2012 international conference on neurorehabilitation (ICNR2012): converging clinical and engineering research on neurorehabilitation; 2012. p. 155–9.

del-Ama AJ, Gil-Agudo A, Pons JL, Moreno JC. Hybrid FES-robot cooperative control of ambulatory gait rehabilitation exoskeleton. J Neuroeng Rehabil. 2014;11:27. doi:10.1186/1743-0003-11-27.
8. del-Ama AJ, Koutsou AD, Bravo-Esteban E, Gómez-Soriano J, Gil-agudo Á, Pons JL, Moreno JC. A comparison of customized strategies to manage muscle fatigue in isometric artificially elicited muscle contractions for incomplete SCI subjects. J Autom Control. 2013;21(1):19–25.
9. del-Ama AJ, Moreno JC, Gil-Agudo A, De-los-Reyes A, Pons JL. Online assessment of human-robot interaction for hybrid control of walking. Sensors (Basel). 2012;12(1):215–25.
10. Maynard FM Jr, Bracken MB, Creasey G, Ditunno JF Jr, Donovan WH, Ducker TB, Garber SL, Marino RJ, Stover SL, Tator CH, Waters RL, Wilberger JE, Young W. International standards for neurological and functional classification of spinal cord injury. American Spinal Injury Association. Spinal Cord. 1997;35(5):266–74.

Human-Like Sensor Fusion Implemented in the Posture Control of a Bipedal Robot

Georg Hettich, Vittorio Lippi, and Thomas Mergner

Abstract Posture control represents the basis for many human sensorimotor activities such as standing, walking or reaching. It involves inputs from joint angle, joint torque, vestibular and visual sensors as well as fusions of the sensor data. Roboticists may draw inspirations from the human posture control methods when building devices that interact with humans such as prostheses or exoskeletons. This study describes multisensory fusion mechanisms that were derived from human perception of ego-motion. They were implemented in a posture control model that describes human balancing of biped stance during external disturbances. The fusions are used for estimating the disturbances and the estimates, in turn, command joint servo controls to compensate them (disturbance estimation and compensation, DEC, concept). An emergent property of the network of sensory estimators is an automatic adaptation to changes in disturbance type and magnitude and in sensor availability. Previously, the model described human and robot balancing about the ankle joints in the sagittal plane. Here, the approach is extended to include the hip joints. The extended human-derived model is again re-embodied in a biped posture control robot constructed with human anthropometrics. The robot is tested in direct comparison with human subjects. Results on hip and ankle sway responses to support surface rotation are described. Basic resemblance of the results suggests that the robot's DEC controls capture important aspects of the human balancing.

Keywords Sensor fusion • Postural control • Sensory feedback • Humanoid robot

1 Introduction

Sensors and sensor fusion play a fundamental role in the sensorimotor behavior of animals and humans. Their use offloads computational burdens to the periphery and early processing stages of the central nervous system (CNS; e.g. [1]). Furthermore,

G. Hettich (✉) • V. Lippi • T. Mergner
Neurocenter, Neurological University Clinic, Freiburg, Germany
e-mail: georg.hettich@uniklinik-freiburg.de; vittorio.lippi@uniklinik-freiburg.de; thomas.mergner@uniklinik-freiburg.de

© Springer International Publishing Switzerland 2015
A.R. Londral et al. (eds.), *Neurotechnology, Electronics, and Informatics*,
Springer Series in Computational Neuroscience 13, DOI 10.1007/978-3-319-15997-3_3

sensor data fusions represent the basis for the perceptual reconstruction of the external world and the interaction with it. Current understanding of the involved mechanisms in humans owes mainly to sensory physiology and to psychophysics, a research method that relates the perception to the physical stimuli it evokes, allowing inferences on the underlying information processing. The founders were, more than a century ago, Fechner and Weber (see [2]) and major contributions dealt with visual and vestibular mechanisms. Cybernetics then introduced engineering methods of describing information processing and control into biomedical research [3]. The present study uses psychophysical findings on human ego-motion perception and their model-based descriptions for the sensorimotor control of a humanoid robot. This represents a neurorobotics approach where neuroscientists apply engineering methods to unveil human neural control and roboticists draw inspirations from the human control methods [4].

Human sensorimotor control involves not only movement planning and movement commanding, but also posture control. Posture control is an instrumental constituent of skeletal motor activity. It copes with inter-segmental coupling torques and movement coordination, adequate buttressing of movements (e.g. push off), maintaining balance, and automatizing the compensation of external disturbances. Posture control functions may be selectively impaired in neurological patients as witnessed by disabling consequences. Both, sensory loss and cerebellar lesions cause ataxia with jerkiness of movements, dysmetria (inappropriate metrics), falls, and motor timing problems [5]. In basal ganglia diseases such as Parkinson's disease, the posture control impairment causes falls, akinesia (difficulties in movement execution), movement freezing, impaired motor adaptability to external disturbances, and muscular stiffness ('rigor') [6].

Modeling the role of sensors and sensor fusions in human posture control has been successful only recently. The problem to overcome was how humans manage to deal with sensory feedback despite long neural time delays (see [7]). Before, it was often held that passive joint stiffness and viscosity, stemming from intrinsic musculoskeletal properties and acting virtually without time delay, play a major role, for example in stabilizing biped stance [8]. Later work showed, however, that this owes primarily to the neural reflexes (ankle joint: [9, 10]; ankle, knee and hip joint: [11, 12]). Several types of reflexes appear to be involved, some with short time delay (40–80 ms) and others with long time delay (>100 ms), and this applies not only to proprioceptive reflexes, but also to the vestibular reflexes [13].

The total time delay of the reflexive feedback mechanisms in biped balancing is approximately 180 ms (e.g. [10]). Yet the neural control of biped balancing in the ankle joints is stable, owing mainly to the fact that the loop gain is very low, hardly exceeding the minimum required for the balancing [10, 14]. The sensory feedback stems primarily from joint angle and torque proprioception, the vestibular system and vision (see [15]). The underlying neural sensor fusions, often referred to as 'multi-sensory integration', allow humans to adapt their posture control to changes in the environmental conditions and to the availability of sensory information. They do so mainly by changing sensory weights, which has been called 'sensory reweighting' [10, 14, 16–18]. The sensory integration and reweighting mechanisms are still a topic of on-going research.

This paper presents a concept of human-derived sensor fusion mechanisms for use in the posture control of a humanoid robot that balances biped stance. In the following, basic aspects of the multi-sensory fusions are explained, before their use in the human posture control model is described and the model is implemented in a humanoid robot for balancing biped stance in the ankle joints. The model is then extended to include the hip joints in the balancing and is again re-embodied into a robot for direct robot-human comparisons. Finally, an outlook is given on how the control concept can further be extended in a modular control architecture for humanoid robots that we expect to show human-like characteristics when interacting behaviorally with humans or in the form of prostheses or exoskeletons.

2   Sensor Fusion and Posture Control Mechanisms

Sensor fusion is an important technical issue. Position tracking design technologies rely heavily on the integration of several sensors: e.g. inertial measuring units (IMUs) integrates gyros and accelerometers, and IMUs output itself is often fused with global positioning system (GPS) data. Published work on sensor fusion for postural control in robots typically used Kalman filters [19–22]. Simulation models for human posture control [23, 24] also implemented Kalman filters, combining in 'sensory integration centers' multiple sensory signals with centrally generated information (motor command) to find the most accurate sensory representation for a given environmental situation. Drawbacks of these approaches are high demands on computational power in multi degree of freedom (DoF) systems and problems of control stability if the plant is not accurately reflected in the model.

A different disturbance estimation method was used in the posture control model considered here. It proceeded from psychophysical work that investigated (i) which sensory information are humans using for their ego-motion perception during passive motion of the body or parts of it (e.g. head, trunk, legs, feet with respect to each or in space), (ii) how humans fuse sensor data to obtain information that is not directly available from their sensors (e.g. trunk motion in space), and (iii) how they obtain estimates of external disturbances that may affect the ego-motion. The approach was model-based and originally aimed to formally describe the experimentally obtained human responses in the form of time series and performance data.

The psychophysical studies showed, for example, that humans involve joint proprioceptive information when using the vestibular information arising in the head for estimating the kinematic state of the trunk and legs in space as well as of the haptically experienced body support. From this information, humans internally reconstruct the external disturbances, which in the experiments consisted of support surface rotation and translation, and experienced their self-motion as a consequence of these external physical stimuli (see [25, 26]).

The concept of external disturbance estimation was extended to include field forces such as gravity or Coriolis forces (e.g. [27]) and to contact forces such as a push against, or a pull on the body [17, 28]. Neural correlates of some of the observed sensor fusions were found in neuron recordings in the vestibular nuclei [29, 30] and in cortical vestibular centers [31]. Furthermore, down and up channeling of vestibular signals in pathways of the spinal cord and their convergence with proprioceptive signals, have been described [32]. Also, representations of processed sensory signals in terms of kinematic variables have been observed in spino-cerebellar pathways [33–35].

It was hypothesized that humans use the same or similar sensory information as observed in, or inferred from the psychophysical studies also for their sensori-motor control, at least as concerns re-active (sensor-driven) responses to external disturbances. On this basis, human posture control experiments were performed and modeled, leading to a disturbance estimation and compensation, DEC, concept.

2.1   Sensor Fusion in the DEC Concept

The DEC concept involves essentially two steps of sensor fusion, schematically illustrated in Fig. 1. In the first step, information from several sensory transducers is fused to obtain measures of kinematic and kinetic variables. In the second step, these physical variables are combined to yield estimates of the external disturbances.

2.1.1   Fusion of Sensory Transducer Data

An example of the first step is the human sense of joint angle proprioception. It combines information from several sensory transducers such as muscle spindles, Golgi tendon organs and cutaneous receptors [36]. This also applies to the human perception of head on trunk rotation, which in addition is complicated by the fact that rotations between several segments of the cervical vertebral column are

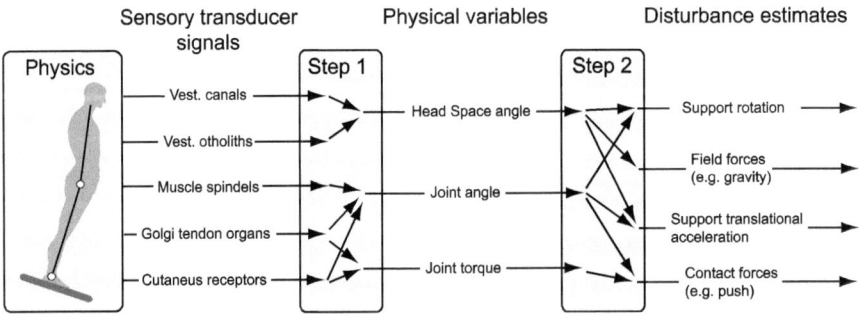

Fig. 1 Schematic illustration of the sensor fusion mechanisms. Information of sensory transducer signals is fused in the first step to yield physical variables. These variables are used in the second step to reconstruct external disturbances

involved. Yet, the result is a sense of angular head-on-trunk velocity and position, as if an angular rate sensor and a goniometer in a single joint were measuring head-trunk speed and rotation, respectively [37, 38].

Another example for the first step, well known to engineers who work with IMUs, is the fusion of angular and linear accelerometer signals. A problem with linear accelerometers is that they do not distinguish between inertial and gravitational forces (i.e. between linear acceleration and tilt of the sensor). There exists also a problem with the angular accelerometers, often used in the form of gyros that measure angular velocity. They show low frequency signal variations over time ('drifts'). Both problems can be solved for the earth vertical planes by fusing the inputs from the two sensors in an appropriate way. This has an analogy in the human vestibular system that is located in the inner ears. Its otolith organs and canal systems represent biological equivalents of linear and angular accelerometers, respectively [39]. The solutions for both, the technical system and its biological equivalent involve information of the gravitational vector. In the horizontal translational and rotational planes, however, there is no such information available, so that further sources of information are required. In technical systems, often the GPS is used. Humans usually use the visual system for this purpose.

In the following we will speak of joint angle and angular velocity sensors and by this we mean virtual sensors that result from step one. The same applies when we refer to the vestibular sensor and its three output measures, i.e. 3D angular velocity and linear acceleration in space and 2D orientation with respect to the gravitational vertical. These measures of the physical variables represent the inputs to the second step of the sensor fusions.

2.1.2   Disturbance Estimation

In the second step of Fig. 1, the signals of the variables resulting from step one are combined to reconstruct external disturbances that have impact on the body. In the DEC concept, it is assumed that four physical quantities suffice to define the external disturbances that affect human balancing in moderate stimulus conditions (body sway amplitudes and velocities, $<8°$ and $<80°/s$; frequencies, <3 Hz). The four types of external disturbances are: (1) Support surface rotation, (2) support surface translational acceleration, (3) field forces such as gravity, and (4) contact forces such as a pull on, or push against the body.

The second step in Fig. 1 was originally motivated by reports of the subjects in the aforementioned psychophysical experiments. When asked to report their percepts during passive rotations on a rotation chair, subjects typically started the report with the rotation of the chair, even though the percept primarily stems from the vestibular system in the head. Thus, without being aware of it, the subjects reconstructed the physical cause of their body rotation, i.e. the chair rotation in space, by internally reversing the linkages from the vestibular signal 'head rotation in space' via the proprioceptive signal 'trunk rotation relative to the head' to the haptical information of 'sitting on the chair'. This can formally be described in terms of a transformation by which the trunk and chair kinematics are referenced to the vestibular derived notion of inertial space [25]. The concept applies to both, the vestibular-able subjects' estimation of 'support rotation' and 'support translational acceleration' in Fig. 1 (formal description in Sect. 2.2).

Vestibular-able subjects furthermore use vestibular information for estimating body lean with respect to the earth vertical when balancing stance in the sagittal plane. From lean of the whole-body's center of mass (COM_B) above the ankle joints and knowledge about body mass and COM height they to estimate the required ankle joint torque to compensate for the gravity effect. For field forces in general, it is known that subjects, when presented with a new aspect of a field force, they perceive it and readily learn to counteract its impact on the body. Thereafter, they no longer perceive it consciously, as has been shown in Coriolis force experiments by Lackner and DiZio [27]. The subconscious estimation and compensation of field forces makes it difficult to study them psychophysically.

Estimation of contact force effects on the ankle joint balancing requires internal measurement of the overall ankle torque (or related measures such as the center of pressure, COP, shift) and the distinct contributions to the ankle torque such as active torque and the gravitational torque. Details have been described before [40] for sagittal plane balancing of moderate disturbances, where the balancing is performed predominantly in the ankle joints ('ankle strategy'; [41, 42]). In such situations, a single inverted pendulum, SIP, can approximately mimic human biomechanics.

2.1.3 Feedback Control Model

The two steps of sensor fusion are used for feedback control of one joint (Fig. 2). Its lower half represents a servo control consisting of a negative joint angle proprioceptive feedback and a controller with a proportional and a derivative factor (PD controller). The controller provides the motor command that is transformed by the muscles into joint torque (not shown in Fig. 2). Given appropriate parameters

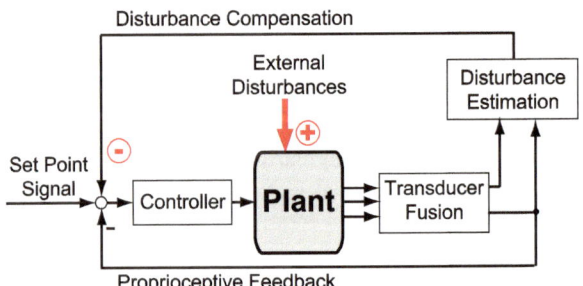

Fig. 2 Simplified feedback control scheme of the Disturbance Estimation and Compensation (DEC) concept. The *Proprioceptive Feedback* loop yields a servo control, by which actual joint angle approximately equals the desired joint angle. Signals from the *Disturbance Estimation* part command the servo to compensate the disturbances

of the servo control, actual joint angle approximately equals the desired joint angle without requiring a feed forward of plant dynamics. Feedback from passive stiffness and viscosity with virtual zero delay is assumed to amount to 10 % of the proprioceptive feedback (not shown in Fig. 2).

Noticeably, in the SIP scenario, the P and D factors identified in human stance control are surprisingly low [10, 14, 43]. They appear to be geared to the pendulum mass m, the height h of the COM, and gravitational acceleration g (mgh; $P \approx mgh$; $D \approx mgh/4$). The values that humans use for balancing are only slightly higher. A consequence is that the servo alone is insufficient to cope with external disturbances such as gravity or a push against the body.

The upper half of Fig. 2 shows schematically the loop that carries the estimates of the external disturbances and compensates for them. To insure control stability in face of the neural time delays, the field and contact force estimates are not used directly, but in the form of body-space angle equivalents. For example, the estimate of body lean commands the servo to compensate for the gravitational torque it produces. Then, the loop gain (at the level of the controller) is raised accordingly. Noticeably, the increase occurs only at the time of, and to the extent that the disturbance has impact. Note furthermore that disturbance compensation applies even with superposition of several disturbances as well as with superposition of disturbances and voluntary movements [39].

The DEC loops are not simply representing additional sensory feedback loops, but are thought to represent long-latency loops through basal ganglia and cerebral cortex [40]. They contain central detection thresholds and allow for voluntary scaling the disturbance compensations and for predictions of the disturbance estimates (e.g. self-produced disturbances during voluntary movements).

It has been shown by comparing human data with model simulations that the DEC concept describes the human ankle joint balancing in a variety of disturbance scenarios. Furthermore, the control automatically adapts to changes in disturbance scenario and magnitude as well as sensor availability. This also applied when the model was implemented in a humanoid robot with ankle joint actuation, and tested in the human experimental setup (PostuRob I; overview [39, 40]). These experiments demonstrated that the DEC concept is robust against real world problems such as inaccurate and noisy sensors and mechanical dead zones.

The following describes an extension of the DEC concept to include the hip joints in the balancing. The hip joints contribute considerably when strong transient disturbances are applied ('hip strategy'; [41, 42]). Then humans may use hip joint accelerations to produce shear forces under the feet to counteract body COM excursions. Another, more common involvement of the hip joints deals with adding to the task of body COM balancing a secondary task of keeping the orientation of the upper body upright. This 'head stabilization in space' task is thought to improve under dynamic conditions such as walking the sensory feedback from the vestibular and visual cues arising in the head [44, 45].

2.2   Extended DEC Concept: Sensor Fusion in Ankle and Hip Joint

The extension of the DEC concept for including the hip joints entails that double inverted pendulum (DIP) rather than SIP biomechanics are considered, and with this the occurrence of inter-segmental coupling torques [46]. In an extended DEC concept for DIP biomechanics, we postulated two DEC controls, one for the hip joint and the other for the ankle joint. This approach allowed to use again the above described sensor fusion principles for disturbance estimation.

2.2.1   DIP Biomechanics

The DIP biomechanical model is shown in Fig. 3. In Fig. 3a, COM_T, COM_L and COM_B stand for the COM of the trunk (including head and arms), leg and whole body, respectively. Leg length is given by l_L, the trunk and leg COM heights are given by h_T and h_L, respectively. Figure 3b shows the angular excursion of the trunk and leg segments with respect to earth vertical (trunk-space angle α_{TS}, leg-space angle α_{LS}). Angular excursion of COM_B is defined as body-space angle α_{BS}. The foot has firm contact with the support surface, therefore platform tilt angle equals foot angle with respect to earth horizontal (foot-space angle α_{FS}). The trunk-leg joint angle is α_{TL} and the leg-foot joint angle is α_{LF}. In perfectly upright body position, all angles are 0°. Angular speed during reactive human balancing can be assumed to be slow enough such that the Coriolis and centrifugal forces can be neglected; the model can be linearized using small angle approximation, assuming that the subject is maintaining his upright position close to the vertical.

Fig. 3 DIP biomechanics

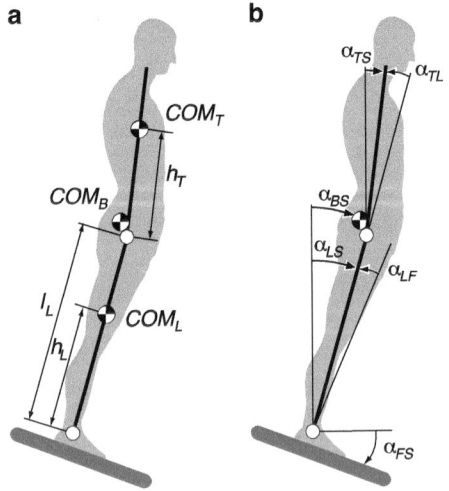

Maintaining upright stance in the situation of a support surface tilt in the sagittal plane requires corrective joint torque in the ankle and hip joints. This torque can be expressed by the following equations for hip torque T_H

$$T_H = \left(J_T + m_T h_T^2 + m_T l_L h_T\right) \ddot{\alpha}_{LS} + \left(J_T + m_T h_T^2\right) \ddot{\alpha}_{TL} - (m_T g h_T)\alpha_{LS}$$
$$- (m_T g h_T)\alpha_{TL} \tag{1}$$

and for ankle torque T_A

$$T_A = \left(J_L + J_T + m_L h_L^2 + m_T (l_L^2 + h_T^2 + 2l_L h_T)\right) \ddot{\alpha}_{LS}$$
$$+ \left(J_T + m_T h_T^2 + m_T l_L h_T\right) \ddot{\alpha}_{TL} - (m_L g h_L + m_T g l_L + m_T g h_T) \alpha_{LS}$$
$$- (m_T g h_T)\alpha_{TL} \tag{2}$$

where $\ddot{\alpha}_{LS}$, and $\ddot{\alpha}_{TL}$ represent angular accelerations, m_L and m_T are the segment masses, and J_L and J_T the segment moments of inertia (details in Al Bakri [47]).

In the extended DEC concept for the DIP, the hip joint is used for orienting and balancing the trunk segment and the ankle joint for balancing the whole-body using two separate controls. The vestibular-derived signals used for the controls are: the trunk-space angle α_{ts}, trunk-space angular velocity $\dot{\alpha}_{ts}$, and head translational acceleration $\ddot{x}_{Head}$. The proprioceptive signals are: the trunk-leg angle α_{tl} and the trunk-leg angular velocity $\dot{\alpha}_{tl}$; the leg-foot angle α_{lf} and the leg-foot angular velocity $\dot{\alpha}_{lf}$. Uppercase letters in the angle subscripts indicate physical angles, lowercase letters the sensory derived representations of these angles.

2.2.2 Hip Joint Control

The DEC control of the trunk reflects the principles described already above for the SIP biomechanics. Considering the support surface tilt scenario in the sagittal plane shown in Fig. 3, the legs tend to rotate somewhat with the platform, due to passive ankle joint stiffness and a imperfect tilt compensation that is typical in humans with eyes closed. Since the legs represent the support base for the trunk, an eccentric hip rotation represents:

(a) A support base tilt disturbance for the trunk, evoked by the leg rotation, α_{LS}.
(b) A hip translational acceleration $\ddot{x}_{Hip}$. It produces a hip torque (T_{H_in}) in relation to m_T, h_T and J_T. This torque is treated here as if it were an external disturbance rather than an inter-segmental coupling effect.

Furthermore, trunk lean is associated with a gravitational hip torque disturbance (T_{H_grav}).

These three disturbances are estimated in the DEC control of the hip joint control in the following form:

(i) Estimation of leg tilt, $\widehat{\alpha}_{LS}$. This estimate is derived from fusing the vestibular velocity signal $\dot{\alpha}_{ts}$ with the proprioceptive velocity signal $\dot{\alpha}_{tl}$ by $\dot{\alpha}_{ls} = \dot{\alpha}_{ts} - \dot{\alpha}_{tl}$ (Assumption: these transformations are performed as vector summations of co-planar rotations, separately for the three body planes). $\widehat{\alpha}_{LS}$ is obtained by applying to the signal a detection threshold and a mathematical integration.

(ii) Estimation of hip translational acceleration $\widehat{\ddot{x}}_{Hip}$. The estimate is derived from fusing the vestibular signals $\dot{\alpha}_{ts}$ and $\ddot{x}_{Head}$ in the form

$$\widehat{\ddot{x}}_{Hip} = \ddot{x}_{Head} - \frac{d\,(\dot{\alpha}_{ts})}{dt} l_T,\tag{3}$$

where the trunk length l_T gives the height of the vestibular system above the hip. $\widehat{\ddot{x}}_{Hip}$ is, in turn, used to estimate the inertial disturbance torque in the form of

$$\widehat{T}_{H\_in} = \widehat{\ddot{x}}_{Hip} m_T h_T.\tag{4}$$

(iii) Estimation of gravitational hip torque $\widehat{T}_{H\_grav}$. Using the vestibular signal α_{ts}, the third and fourth term of Eq. (1) becomes

$$\widehat{T}_{H\_grav} = m_T g h_T \alpha_{ts}.\tag{5}$$

2.2.3 Ankle Joint Control

The DEC control of the ankle joints is used to balance the whole body above the ankle joint. To this end, it combines the leg and trunk angular excursions in the form of COM_B excursions in space, α_{BS}. In this respect, also the DEC control of the ankle deals with a SIP. The following three disturbances that have impact on the ankle torque during support surface tilts are:

(a) The support surface tilt, α_{FS}.
(b) The gravitational ankle torque, T_{A_grav}. It results from α_{BS}.
(c) Inter-segmental coupling torque in the ankle joint, T_{A_coup}. It arises with angular acceleration of the trunk segment.

For the estimation of these disturbances, the DEC control of the ankle fuses sensory signals from the vestibular system and the hip *and* ankle joint proprioception. To this end, sensory signals from the hip DEC control are transmitted ("down-channeled") to the ankle joint DEC control. The estimates are:

(i) Estimation of foot-space rotation, $\widehat{\alpha}_{FS}$. This estimate uses a down-channeled version of $\dot{\alpha}_{ls}$ and combines it with the ankle joint angular velocity signal $\dot{\alpha}_{lf}$ in the form

$$\dot{\alpha}_{fs} = \dot{\alpha}_{ls} - \dot{\alpha}_{lf}.\tag{6}$$

Analogous to $\widehat{\alpha}_{LS}$, the estimate $\widehat{\alpha}_{FS}$ contains a detection threshold and a mathematical integration.

(ii) Estimation of gravitational ankle torque, $\widehat{T}_{A\_grav}$. This estimate relates to the third and fourth term of Eq. (2), which are mathematically combined in the COM_B excursion α_{bs}. From this, the gravitational torque is obtained in the form

$$\widehat{T}_{A\_grav} = m_B g h_B \alpha_{bs} \qquad (7)$$

where m_B represents whole-body mass and h_B represents COM_B height. Small angular excursions allow approximating h_B by a constant value.

(iii) Estimation of the inter-segmental coupling torque, $\widehat{T}_{A\_coup}$. This torque arises upon trunk rotational acceleration and tends to evoke a leg counter-rotation. In view of the DEC concept, the trunk acceleration exerts a 'push' against the hip like a contact force disturbance (compare external torque estimate in [40]). This disturbance is expressed by the second component of Eq. (2). Since its implementation was not critical for the stability of the DIP control in the present context (compare [48]), it is omitted in the following.

The hip and the ankle DEC controls can be viewed as separate control modules that are interconnected by 'down-channeling' of sensory information from the hip DEC control to the ankle DEC control. Recent experimental evidence suggests in addition an "up-channeling" of information between them (details in [49]). A schematic illustration of the DIP control is given in Fig. 4.

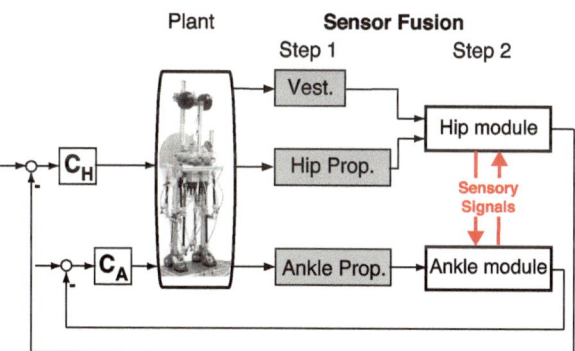

Fig. 4 Basic aspects of the DIP control concept used for PostuRob II. C_H and C_A are the hip and ankle controllers, Vest. is the vestibular input while Hip Prop. and Ankle Prop. are the proprioceptive inputs

3 Human and Robot Experiments

The extended DEC concept was tested experimentally by comparing sway to support surface tilt in the sagittal plane with sway of a bipedal robot (PostuRob II) in a human posturography laboratory.

3.1 Bipedal Robot PostuRob II

PostuRob II consists of mechanical, mechatronic, and computer control parts. The mechanical part comprises one trunk segment, two legs and two feet, with a total mass of 59 kg and a total height of 1.78 m. Two hip joints and two ankle joints connect the segments (4 DOF in the sagittal plane; Fig. 5). The mechatronic part comprises an artificial vestibular sensor [39] that is fixed to the trunk segment. Artificial pneumatic 'muscles' (FESTO, Esslingen, Germany; Typ MAS20) connected with serial springs (spring rate 25 N/mm) are used for actuation. An electronic inner torque control loop ensures that actual torque equals approximately desired torque. Sensory signals are sampled at 200 Hz by an acquisition board. Computer control is performed through a real time PC that executes a compiled Simulink model using Real-Time Windows Target (The Math Works Inc., Natick, USA).

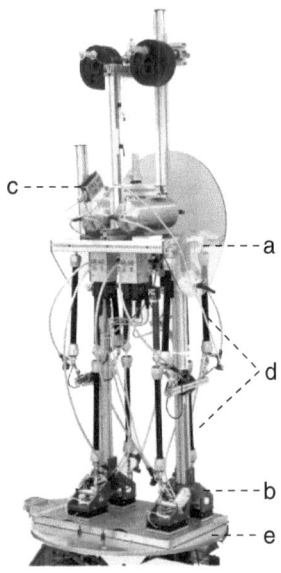

Fig. 5 PostuRob II. The robot consists of trunk, leg, and foot segments interconnected by the hip joints (*a*) and ankle joints (*b*). Sensory information stems from artificial vestibular system (*c*) and ankle and hip joint angle and angular velocity sensors. Actuation is through pneumatic 'muscles' (*d*). PostuRob II stands freely on a motion platform (*e*)

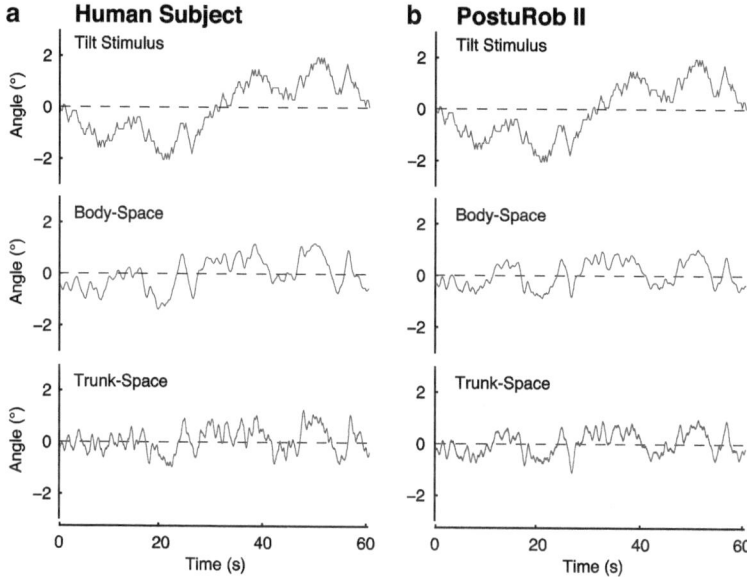

Fig. 6 Tilt stimulus and angular excursion responses of body in space and trunk in space from one representative subject (**a**) and of PostuRob II (**b**)

3.2 Experimental Methods

Seven healthy human subjects (3 female; mean age, 28 ± 3 years) participated after giving their informed consent. The subjects (eyes closed) and the robot stood freely on a motion platform (see Fig. 5), while six successive pseudorandom ternary tilt sequences, each 60.5 s long, with peak-to-peak amplitude of 4° were applied (PRTS stimulus; frequency range 0.017–2.2 Hz; [10]). The first rows in Fig. 6a, b show one 60.5 s long tilt stimulus sequence.

Trunk, leg, and COM_B angular excursions in space were calculated on the basis of opto-electronically measured marker data (Optotrak 3020®; Waterloo, Canada) that were recorded with a sampling frequency of 100 Hz. Data analysis took into account human anthropometric measures [50] and was performed using custom-made software programmed in Matlab (The MathWorks, Natick, USA). The responses were expressed as gain and phase from the frequency response function [10] in a form where zero gain means no body excursion and unity gain means that body angular excursion equals platform tilt. Phase represents the temporal relationship between stimulus and response. Variability of averaged values was expressed as 95 % confidence limits [51].

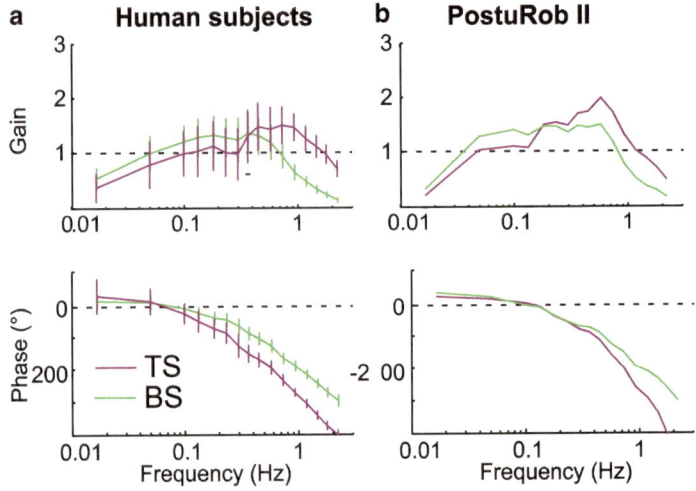

Fig. 7 Tilt responses in terms of gain, phase and coherence curves of human subjects (**a**; 7 subjects, medians ±95 confidence intervals) and PostuRob II (**b**)

3.3 Results

Subjects and PostuRob II balanced the tilts in similar ways. Time series of the responses of one subject and the robot are shown in Fig. 6. Note that the responses resemble each other, both for the body-space and the trunk-space responses. As shown in Fig. 7, the resemblance also holds for the mean gain and phase curves of the human subjects and the robot. In both, the gain values of trunk-space (TS) and of body-space (BS) vary similarly with stimulus frequency. In the low frequency range (<0.3 Hz), TS gain is lower than BS gain. In contrast, in the high frequency range (>0.3 Hz), TS gain exceeds BS gain, while the phase shows a larger phase lag.

4 Conclusions

The here proposed feedback control system of a bipedal robot takes advantage of the sensor fusion and posture control mechanisms that were derived from findings in human experiments. The disturbance estimators that were used are non-iterative and remarkably simpler than estimators that were used in Kalman filters. Furthermore, the multi-sensory feedback control is performed without integrating any dynamic model of the whole body in the control architecture. Filtering the estimates through a nonlinear operation provided by a deadband threshold tends to reduce noise, which appears to stem mainly from vestibular signals [39]. The noise shows 1/f properties and therefore overlaps with the bandwidth of human sensorimotor behavior. The threshold shuts off any estimator if there is no corresponding disturbance, which in a multi-DoF system may help to prevent accumulation of noise. The threshold also explains a non-linear behavior in the human disturbance responses that were observed with increase in stimulus magnitude [14]. Due to the non-linearity, small stimuli yield relatively smaller responses than larger stimuli. This is an aspect of the automatic sensory re-weightings, which emerged from the sensory network of estimators. Other important aspects of it are that the control automatically adjusts to changes in disturbance type and to sensor availability (for SIP, see [40]).

The here obtained good match of the data between the human subjects and PostuRob II suggests that the proposed sensor fusion and posture control mechanism capture important constituents of the human balancing system. The application of the extended DEC concept to the balancing of upright stance using hip and ankle joints in terms of a DIP required the integration of sensory signals from almost the whole body. In a recent study that used this approach, coordination between hip and ankle joint emerged from the multi-sensory feedback control [49]. These experiences with the extended DEC concept led us explore its usefulness with further DoF in a modular control architecture. In the generalized description, each DoF is controlled by one DEC control, which stabilizes a SIP (defined by the COM and moment of inertia of the segments above) on a moving support (given by the upper end of the segment below). Adjoining DEC controls are synergistically interconnected to exchange sensory information and disturbance estimates [52].

Taken together, although optimizing the DEC concept and its control parameters is still under research, the concept proved to have several promising features. These include: (i) a computationally very simple implementation, since almost all sensor fusions are based on algebraic operations; (ii) the control complexity scales linearly with the number of joints, since every joint is controlled as if a SIP and the signals are exchanged only between adjoining modules; (iii) noise rejection makes it possible to fuse the input of an high number of sensors; and (iv) the system, originally proposed for its predictive power of human behavior, can be employed to control actuated prostheses and exoskeletons to provide users with a human-like feeling.

Acknowledgements Supported by the European Commission (FP7-ICT-600698 H2R). GH is supported by the Konrad-Adenauer-Stiftung.

References

1. Wehner R. Matched filters - neural models of the external world. J Comp Physiol A. 1987;161:511–31.
2. Stevens SS. Psychophysics. Introduction to its perceptual, neural, and social prospects. New York, NY: John Wiley & Sons; 1975.
3. Wiener N. Cybernetics. Sci Am. 1948;179:14–8.
4. Mergner T, Tahboub KA. Neurorobotics approaches to human and humanoid sensorimotor control. J Physiol Paris. 2009;103:115–8.
5. Bastian AJ. Mechanisms of ataxia. Phys Ther. 1997;77:672–5.

6. Visser JE, Bloem BR. Role of the basal ganglia in balance control. Neural Plast. 2005;12: 161–74.
7. Rack PMH. Limitations of somatosensory feedback in control of posture and movement. In: Brooks VB, editor. Handbook of physiology: the nervous system, vol. 2. Baltimore, MD: Williams and Wilkins; 1981.
8. Winter DA, Patla AE, Prince F, Ishac MG, Gielo-Perczak K. Stiffness control of balance in quiet standing. J Neurophysiol. 1998;80:1211–21.
9. Morasso PG, Schieppati M. Can muscle stiffness alone stabilize upright standing? J Neurophysiol. 1999;82:1622–6.
10. Peterka RJ. Sensorimotor integration in human postural control. J Neurophysiol. 2002;88: 1097–118.
11. van Soest AJ, Haenen WP, Rozendaal LA. Stability of bipedal stance: the contribution of cocontraction and spindle feedback. Biol Cybern. 2003;88:293–301.
12. Edwards W. Effect of joint stiffness on standing stability. Gait Posture. 2007;25:432–9.
13. Britton TC, Day BL, Brown P, Rothwell JC, Thompson PD, Marsden CD. Postural electromyographic responses in the arm and leg following galvanic vestibular stimulation in man. Exp Brain Res. 1993;94:143–51.
14. Maurer C, Mergner T, Peterka RJ. Multisensory control of human upright stance. Exp Brain Res. 2006;171:231–50.
15. Horak FB, Macpherson JM. Postural orientation and equilibrium. In: Rowell L, Shepherd J, editors. Handbook of physiology. New York, NY: Oxford University Press; 1996.
16. Nashner LM, Berthoz A. Visual contribution to rapid responses during postural control. Exp Brain Res. 1978;150:403–7.
17. Mergner T, Maurer C, Peterka RJ. A multisensory posture control model of human upright stance. Prog Brain Res. 2003;142:189–201.
18. van der Kooij H, Peterka RJ. Non-linear stimulus-response behavior of the human stance control system is predicted by optimization of a system with sensory and motor noise. J Comput Neurosci. 2011;30(3):759–78.
19. Tahboub K, Mergner T. Biological and engineering approaches to human postural control. Integ Comput Aid Eng. 2007;13:1–17.
20. Mahboobin A, Loughlin PJ, Redfern MS, Anderson SO, Atkeson CG, Hodgkins JK. Sensory adaptation in balance control: lessons for biomimetic robotic bipeds. Neural Netw. 2008;21(4):621–7.
21. Tahboub K. Biologically-inspired humanoid postural control. J Physiol Paris. 2009;103: 195–210.
22. Klein TJ, Jeka J, Kiemel T, Lewis MA. Navigating sensory conflict in dynamic environments using adaptive state estimation. Biol Cybern. 2011;105:291–304.
23. van der Kooij H, Jacobs R, Koopman B, Grootenboer H. A multisensory integration model of human stance control. Biol Cybern. 1999;80:299–308.
24. Kuo AD. An optimal state estimation model of sensory integration in human postural balance. J Neural Eng. 2005;2:235–49.
25. Mergner T, Huber W, Becker W. Vestibular-neck interaction and transformations of sensory coordinates. J Vestibul Res Equil. 1997;7:119–35.
26. Mergner T, Rosemeier T. Interaction of vestibular, somatosensory and visual signals for postural control and motion perception under terrestrial and microgravity conditions – a conceptual model. Brain Res Rev. 1998;28:118–35.
27. Lackner JR, DiZio P. Rapid adaptation to Coriolis force perturbations of arm trajectories. J Neurophysiol. 1994;72:299–313.
28. Mergner T. The Matryoshka Dolls principle in human dynamic behavior in space – a theory of linked references for multisensory perception and control of action. Cah Psychol Cogn. 2002;21:129–212.
29. Boyle R, Pompeiano O. Convergence and interaction of neck and macular vestibular inputs on vestibulospinal neurons. J Neurophysiol. 1981;45:852–68.

30. Anastasopoulos D, Mergner T. Canal-neck interaction in vestibular nuclear neurons of the cat. Exp Brain Res. 1982;46:269–80.
31. Mergner T, Becker W, Deecke L. Canal-neck interaction in vestibular neurons of the cat's cerebral cortex. Exp Brain Res. 1985;61:94–108.
32. Coulter JD, Mergner T, Pompeiano O. Effects of static tilt on cervical spinoreticular tract neurons. J Neurophysiol. 1976;39:45–62.
33. Bosco G, Poppele RE. Representation of multiple kinematic parameters of the cat hindlimb in spinocerebellar activity. J Neurophysiol. 1997;78:1421–32.
34. Poppele RE, Bosco G, Rankin AM. Independent representations of limb axis length and orientation in spinocerebellar response components. J Neurophysiol. 2002;87:409–22.
35. Casabona A, Valle MS, Bosco G, Perciavalle V. Cerebellar encoding of limb position. Cerebellum. 2004;3:172–7.
36. Gandevia SC, Refshauge KM, Collins DF. Proprioception: peripheral inputs and perceptual interactions. Adv Exp Med Biol. 2002;508:61–8.
37. Mergner T, Nardi GL, Becker W, Deecke L. The role of canal-neck interaction for the perception of horizontal trunk and head rotation. Exp Brain Res. 1983;49:198–208.
38. Mergner T, Siebold C, Schweigart G, Becker W. Human perception of horizontal trunk and head rotation in space during vestibular and neck stimulation. Exp Brain Res. 1991;85: 389–404.
39. Mergner T, Schweigart G, Fennell L. Vestibular humanoid postural control. J Physiol Paris. 2009;103:178–94.
40. Mergner T. A neurological view on reactive human stance control. Annu Rev Control. 2010;34:177–98.
41. Nashner L, McCollum G. The organization of human postural movements: a formal basis and experimental synthesis. Behav Brain Sci. 1985;8:135–72.
42. Horak FB, Nashner LM. Central programming of postural movements: adaptation to altered support-surface configurations. J Neurophysiol. 1986;55:1369–81.
43. Alexandrov AV, Frolov AA, Horak FB, Carlson-Kuhta P, Park S. Feedback equilibrium control during human standing. Biol Cybern. 2005;93:309–22.
44. Bronstein AM. Evidence for a vestibular input contributing to dynamic head stabilization in man. Acta Otolaryngol. 1988;105:1–6.
45. Pozzo T, Berthoz A, Lefort L, Vitte E. Head stabilization during various locomotor tasks in humans. II. Patients with bilateral peripheral vestibular deficits. Exp Brain Res. 1991;85: 208–17.
46. Zajac FE, Gordon ME. Determining muscle's force and action in multi-articular movement. Excerc Sport Sci Rev. 1989;17:187–230.
47. Al Bakri M. Development of a mathematical model and simulation environment for the postural robot (PostuRob II). 2008. Available from http://www.posturob.uniklinik-freiburg.de
48. Hettich G, Fennell L, Mergner T. Double inverted pendulum model of reactive stance control. Multibody Dynamics Conference 2011. Available from http://www.posturob.uniklinik-freiburg.de
49. Hettich G, Assländer L, Gollhofer A, Mergner T. Human hip-ankle coordination emerging from multisensory feedback control. Hum Mov Sci. 2014;37:173.
50. Winter DA. Biomechanics and motor control of human movement. 2nd ed. New York, NY: Wiley; 1990.
51. Otnes RK, Enochson LD. Digital time series analysis. New York, NY: Wiley; 1972.
52. Lippi V, Mergner T, Hettich G. A bio-inspired modular system for humanoid posture control. In: Ugur E, Oztop E, Morimoto J, Ishii S (editors), Proceedings of IROS 2013 Workshop on Neuroscience and Robotics. "Towards a robot-enabled, neuroscience-guided healthy society", November 3, 2013, Tokyo, Japan

Microchannel Scaffolds for Neural Signal Acquisition and Analysis

Rouhollah Habibey, Asiyeh Golabchi, and Axel Blau

Abstract Replica-casting finds wide application in soft lithography and microfluidics. Most commonly, structures are molded with micro- to nano-patterned photoresists as master casts into polydimethylsiloxane (PDMS). PDMS features many favorable properties. It reproduces geometric details with nanometer fidelity, has low cytotoxicity and is transparent in the visible spectrum. It is furthermore biostable both *in vitro* and *in vivo*, can be plasma-bonded to itself, has low water permeability and is easy to handle and process. After curing, the PDMS can be peeled from the master and latter usually be reused if patterns are not undercut. Here, we describe the straightforward replica-molding process for devices that can be exploited either as perforated microchannel scaffolds for the *in vitro* use in axonal guidance and regeneration studies on microelectrode arrays (MEAs) or for the production of tissue-conformal *in vivo* MEAs for neuroprosthetic applications.

Keywords Neuroengineering • Microchannel confinements • Polymer composite neuroprosthetics • Multi-level photolithography • Polydimethylsiloxane (PDMS)

1 Introduction

In microfluidics, PDMS microchannel systems [21, 23, 24] inspired by Taylor *et al.*, allow the separation of somata from their axons in two different fluidic environments [1]. Over the last decade, devices like these have been extensively utilized for studying axonal injury and regeneration, myelination, protein and mRNA synthesis, as well as transport phenomena [2–6]. Inexpensive microlithography techniques [22] for the generation of microtunnels have been used recently to structure networks in variable spatial designs. They allow microwell-confined populations of neural networks to connect through microchannels [7]. Microchannels can also be utilized in electrophysiological studies of neural networks *in vitro* by placing cells in specific substrate locations including the electrodes [8, 9]. By combining microchannels

R. Habibey • A. Golabchi • A. Blau (✉)
Department of Neuroscience and Brain Technologies (NBT), Neurotechnologies (NTech),
Fondazione Istituto Italiano di Tecnologia (IIT), Via Morego 30, 16163 Genoa, Italy
e-mail: axel.blau@iit.it

© Springer International Publishing Switzerland 2015
A.R. Londral et al. (eds.), *Neurotechnology, Electronics, and Informatics*,
Springer Series in Computational Neuroscience 13, DOI 10.1007/978-3-319-15997-3_4

with microelectrode arrays (MEAs), activity can be selectively recorded from axons. In addition, amplitudes of extracellularly recorded signals from neurites are usually amplified by two orders of magnitude (millivolts instead of tens of microvolts) [10, 11]. In a non-conventional fabrication scheme, microchannel devices can also be transformed into MEAs. Filling the cavities of multi-level microchannel scaffolds with electrically conductive polymers or composite materials is a rapid prototyping approach for the design of various *in vitro* and *in vivo* recording and stimulation devices [12]. This opens the door for both application-specific prototyping as well as mass production of more tissue-mimetic neuroprosthetic recording and stimulation devices. After a brief methodological description of the general device fabrication steps, we present exemplary results on device exploitation as physical cell culture guidance cues and as MEAs for neuroprosthetics.

2 Methods

2.1 Fabrication of Multi-level PDMS Microchannel Devices

Bi-level patterns can be casted into high-aspect ratio negative photoresist (*e.g.*, SU-8) in a two-step photolithography technique using separate masks (Expert, Silvaco). To generate both microchannels and perforating vias, the first mask defines all features, whereas the second mask delineates the through-holes only. The principle fabrication steps and overall process flow are depicted in Fig. 1: A clean silicon wafer (a) is spin-coated with the first photo resin layer (<150 μm) (b) and photo cross-linked through a first photomask to define both the channels and the sockets for vias (c). The procedure is repeated in (d) and (e) for the second photoresist layer (<150 μm) to define the via through-holes. After removing the uncured photoresin in a developing step (f), the bi-level microstructure (g) can be covered with PDMS pre-polymer (h), cured after its leveling, and be peeled off to result in a microchannel scaffold with via holes (i). Such microchannels can either be used as physical guidance cues for axons and dendrites or be filled with conductive polymers (*e.g.*, PEDOT:PSS, carbon-polymer composites) (j) and backside-insulated (k) to yield all-polymer MEAs (*polyMEAs*) with electrodes and contact pads at the via holes for their application *in vitro* or *in vivo* (l).

2.2 Conductive Materials

If not stated differently, all materials were purchased from Sigma-Aldrich. A composite of carbon, graphite, and poly (3,4-ethylenedioxythiophene):polystyrene sodium sulfonate (PEDOT:PSS) (p Jet 700, Clevios) with up to 5 % of conductivity enhancers was used as a rubber-like electrode-, conductive track- and connection pad-filling. The carbon/graphite/PEDOT:PSS composite was mixed into pre-mixed

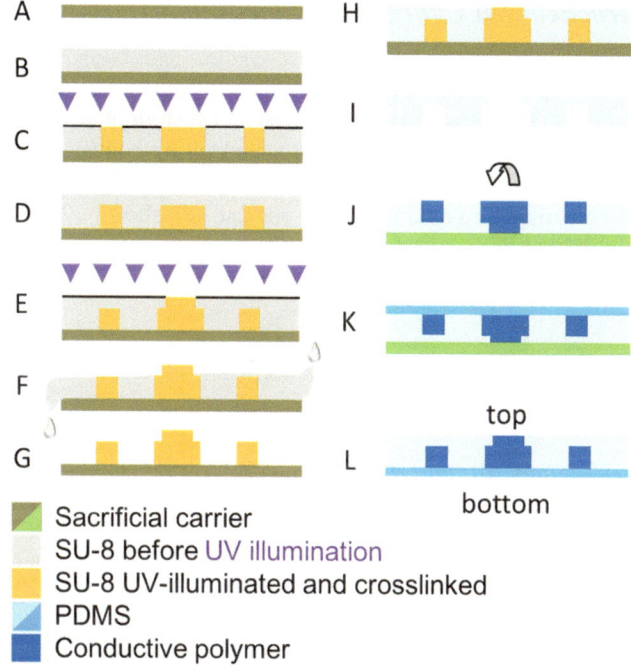

Fig. 1 Process flow of master generation (**a–f**) and microchannel replica molding from master in PDMS (**g–i**). The scaffold can either be used directly in neural guidance studies or be functionalized *e.g.*, with conductive polymers (**j–l**)

PDMS prepolymer and curing agent (Sylgard 184, Dow Corning) in a 1:1:3 ratio until its DC resistivity dropped below 10 kΩ over a distance of 5 mm. The track-, contact pad- and electrode-cavities were filled with the conductive material by spreading it onto the *polyMEA* scaffold. Excess material was removed by first swiping the surface with the edge of a ruler, then with a cotton swab and filter paper. This procedure removed the shortcuts between pads and tracks. In a final step, the surface of the *polyMEA* was cleaned by rolling a lint remover over it. To ensure that no shortcuts had remained, adjacent contact pads were probed by an ohmmeter. The conductive composite was cured at 120 °C for one hour. The *polyMEAs* were then backside-insulated by a second layer of PDMS. The contact pads of *in vivo* *polyMEA* probes were slid between the pins of Omnetics connectors. A folded laser transparency between the folded *polyMEA* helped in pressing the pads against the connector pins, thereby establishing electrical contact. A thin PDMS coat can be used for pad/pin insulation thereafter. Device biocompatibility was tested in cell culture. The electrical characteristics were determined by electrical impedance spectroscopy. *PolyMEA* recording functionality was validated with cortical cultures after 7 days *in vitro* (DIV).

2.3 Electrochemical Characterization of polyMEA Electrodes

The electrodes of *polyMEAs* were characterized by electrochemical impedance spectroscopy (EIS). The *polyMEAs* were placed in a custom-made Faraday box and their pads sequentially connected by a gold-coated spring-contact probe. The electrodes were immersed in saturated KCl at room temperature. A silver/silver chloride wire (∅1 mm) was used as a reference electrode and a Pt sheet (∼2 × 3 mm^2) as the counter electrode. Impedance spectroscopy was performed at frequencies between 1 Hz and 100 kHz (Perstat 2273, Princeton Applied Research).

2.4 Cell Culture

If not stated differently, all cell culture chemicals were purchased from Invitrogen. PDMS *in vitro polyMEAs* and commercial MEAs (Multi Channel Systems), PDMS caps [13] and PDMS microchannel tiles were autoclaved at 120 °C for 20 min before being moved to the sterile hood. To increase their surface hydrophilicity, MEAs and *polyMEAs* were exposed to oxygen plasma (0.5–2 min, 50 W, 2.45 GHz, 0.3 mbar O_2) [femto, Diener]. Their central surface was then coated with 20 μL of 0.1 mg/mL poly-D-lysine (PDL) and 0.05 mg/mL laminin, which was allowed to dry in vacuum. To remove soluble PDL, MEAs were rinsed twice with sterile ultrapure water and dried. After complete water evaporation, the PDMS microchannels were aligned on the surface of the commercial MEAs using a sterile water droplet. The crossing points of the microchannels were placed on the electrodes. Therefore, every microchannel included eight electrodes in one row or column (*e.g.*, 68 to 61 in Fig. 2). Cell suspensions (rat E18) were prepared following standard protocols [2]. For microchannel devices on commercial MEAs, a small drop (∼5 μL; 10,600 cortical neurons/μL, ∼50,000 cells per device) was placed into just one out of four reservoirs. On *polyMEAs*, about 100,000 cells were plated onto their central electrode area. Cells were allowed to settle for 30 min in a cell culture incubator (5 % CO_2, 37 °C, 95 % RH) before adding 1 mL of serum-free medium (Neurobasal medium, B27 2 %, Glutamax 1 %, penicillin/streptomycin 1 %). Cultures were protected by a PDMS cap against evaporation and contamination [13] and stored in the incubator. Every week, 450 μL of media was exchanged with fresh warm media. Cultures were imaged once a week on an inverted microscope (DMIL, Leica Microsystems) equipped with a 5 MP camera (DFC420C, Leica Microsystems).

2.5 Electrophysiology

Extracellular signals were recorded with a 60-channel filter-amplifier (MEA60-Up, Multi Channel Systems), featuring a built-in thermal sensor and heating element.

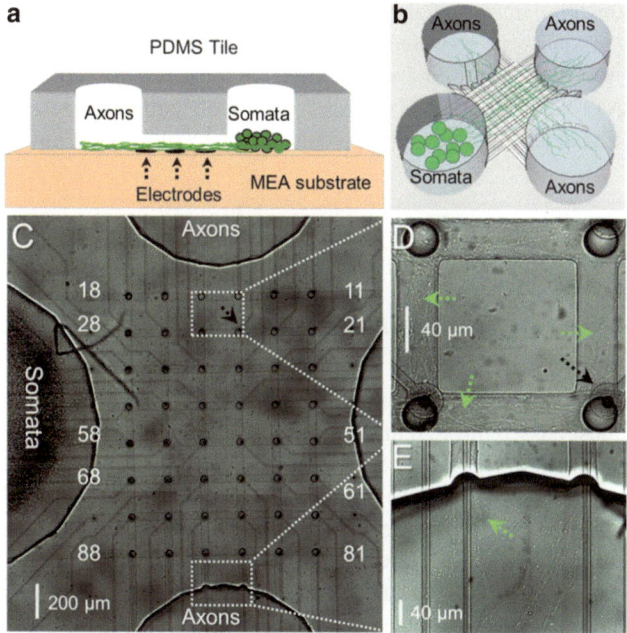

Fig. 2 PDMS microchannel tiles on MEAs for neural network compartmentalization. (**a**) A schematic cross-section of a microchannel and the reservoirs shows how microchannels selectively let axons grow on top of an electrode row while preventing cell bodies to enter. (**b**) PDMS microstructure including four big reservoirs interconnected by an 8 × 8 matrix of channels. (**c**) Cells seeded in a somal compartment (*left*) had grown their axons after 9 DIV through the entire length of a microchannel into the empty axonal compartment. (**d**) Magnified view of the axons inside the channels between electrodes 15-14 and 25-24. (**e**) Magnified view of axons entering the axonal compartment. *Black arrows* indicate an electrode and *green arrows* point at axons

For microchannel tiles, acquired activity (MC_Rack, Multi Channel Systems) was filtered and analyzed offline. The low frequency noise, which was augmented by the microchannels, was removed by a second-order Bessel high-pass filter (cut-off at 200 Hz). Spikes were detected in the filtered data stream by passing a negative threshold set to −4.5 StDev of the peak-to-peak noise. Only downward threshold-crossings were analyzed.

2.6 Statistical Analysis

Spike trains from microchannel tile recordings were transformed to timestamps (NeuroExplorer, Nex Technologies) for mean frequency and burst analysis. Individual units on each electrode were detected by k-means clustering (Offline Sorter, Plexon). Thereafter, sorted units were checked visually to merge the same units or exclude unsorted rare signals from analysis. Numerical results were further analyzed by a one-way repeated measures analysis of variance (ANOVA) or a two-way ANOVA, followed by a Bonferroni post-test for all groups (Prism, GraphPad). A probability of $p < 0.05$ was considered significant. Data are expressed as mean $\pm$ SEM or as quartiles (Whisker-bars) with maximum and minimum values.

3   Results

3.1   Recording Activity from Microchannel-Confined Axons

Cells started to grow their neurites into the microchannels after 3 DIV. After 9 DIV, axons had almost passed through the entire channel and entered into one of the three axonal compartments (Fig. 2). The neuronal tissue mass inside the channels was increasing up to 27 DIV. Afterward, neural tissue inside the channels and the axonal compartments started to degenerate and thick axonal bundles appeared (Fig. 8c).

Because the microchannels were aligned with the electrodes, axons were forced to grow over the electrodes. The small microchannel cross sections (40 μm × 5 μm) increased the electrical resistance to ground, thereby amplifying the weak extracellular axonal signals. At 9 DIV, first signals could be recorded. They were similar in shape to random networks, but had higher amplitudes (>100 μV; Fig. 3a). The

Fig. 3 Sample recording from electrodes in channel 2 at 30 DIV. The *right panel* shows the magnified propagating signal along the full length of the channel from electrode 28 toward electrode 21 (see Fig. 2c)

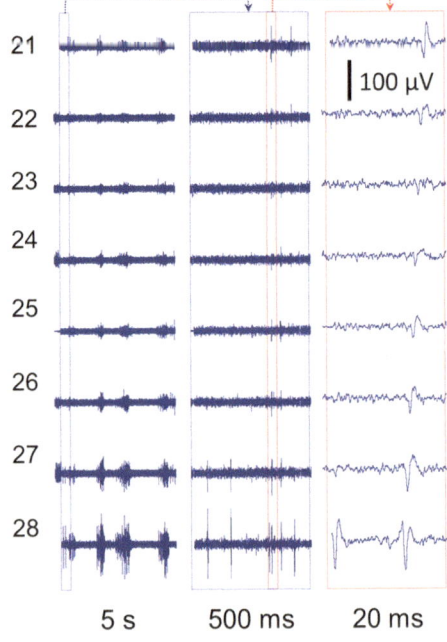

signal amplitudes increased up to 600 µV after 20 DIV (Fig. 3b). The microchannels not only allowed to record from axons, but also forced the same axon to pass over or nearby all electrodes in one row of the electrode array (Fig. 2c; *e.g.*, electrodes 28 to 21). This made it possible to record from the same axon at its different lengths. Therefore, the same signal appeared with short delay on subsequent electrodes as it propagated along the axon (Fig. 3).

Compared to the diameter of an axon ($\sim$1 µm), the width of a microchannel was sufficiently large (40 µm) to let different axonal branches from the same network enter into it. Therefore, every electrode inside the microchannel could simultaneously record from different axons. Signals were sorted by shape to distinguish between the different axons (sources) (Fig. 4a–c). In general, two main signal categories were detected inside the microchannels; monophasic signals (mainly with negative wave) and biphasic signals (mainly with a negative followed by a positive wave). In contrast to quickly decaying biphasic signals, monophasic

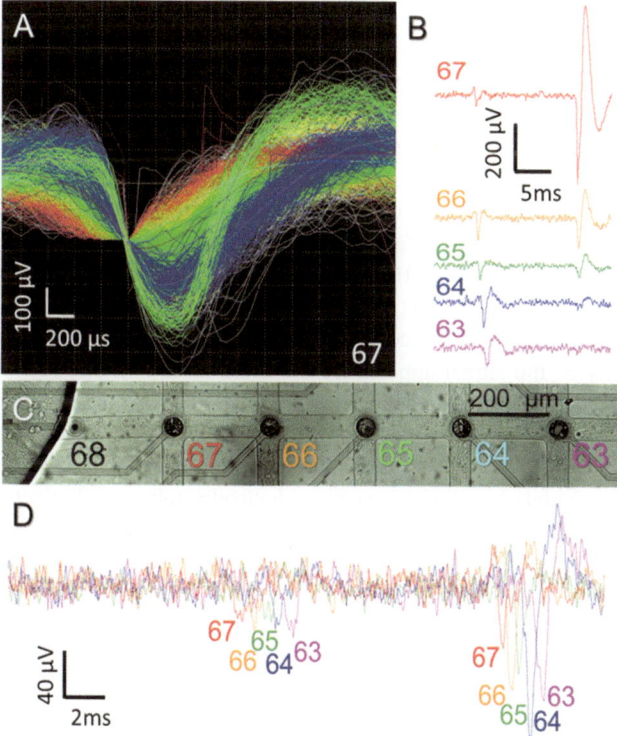

Fig. 4 Signals with different amplitudes and shapes were propagating in channels. (**a**) Signal sorting for electrode 67 shows five different waveforms that were recorded by this electrode. (**b**) Shape and propagation length of monophasic and biphasic signals in channel 6 from electrode 67 to 63. (**c, d**) Overlaid signals recorded from electrodes 67 to 63 show the propagation delays for a signal traveling within an axon over sequential electrodes

signals could propagate to distant electrodes inside a microchannel. This feature was amplitude-independent (Fig. 4b). Overlaying the signals recorded from subsequent electrodes inside the same channel showed the propagation delay and how the signal shape varied with location (Fig. 4c).

To analyze the network activity evolution and the propagation velocity along the axons inside the microchannels over time, the electrical activity at 10, 20, 30 and 53 DIV was compared. To evaluate how activity levels change over time or along channels, two subsequent electrodes in a channel were considered as one segment, thereby dividing every channel into the following four segments: Seg 1 (0–250 μm), Seg 2 (250–650 μm), Seg 3 (650–1,050 μm) and Seg 4 (1,050–1,300 μm) (Fig. 5a). Because the amplifiers for electrodes 27, 52 and 62 were switched off at 10 and 20 DIV, data for the respective segments was collected on these two days from one electrode only. A comparison of the overall number of spikes per minute in the proximal segments of all considered channels with that of their distal segments showed that the signal frequency decreases along the channel for all recording days ($p < 0.001$ at 10 and 30 DIV, $p < 0.01$ at 20 DIV; Fig. 5b). Monitoring the mean signal frequency in each segment over time (Fig. 4c) showed a significant decrease at 53 DIV compared to 10 DIV in segments 1, 2 and 3 ($p < 0.05$).

Signal frequencies and propagation velocities were evaluated in detail for three selected channels (channel 2 (electrodes 21–28), 5 (electrodes 51–58) and 6 (electrodes 61–68)). The mean signal frequency for each channel was calculated by averaging all recorded signals from any of the electrodes inside a single channel. The mean signal frequency for each channel tended to decrease over time. This decrease was significant between 10 DIV and the subsequent DIVs for channel 5 ($p < 0.05$ vs. 20 and 30 DIV, and $p < 0.001$ vs. 53 DIV; Fig. 6a) and channel 6 ($p < 0.05$ vs. 30 DIV and $p < 0.01$ vs. 53 DIV; Fig. 5a). The propagation velocity was calculated by dividing the constant distance between a pair of electrodes (200 μm) by the temporal delay of the signal appearance on two subsequent electrodes. The mean propagation velocity for each channel was calculated by averaging the propagation velocities of all subsequent electrode pairs in a channel. For electrodes from which no signals were acquired in 10 and 20 DIV (electrodes 27, 52 and 62), the average velocity was calculated for the two nearest electrodes (channel 2; electrodes 28 and 26, and channel 6; electrodes 63 and 61). The mean propagation velocity tended to increase with culture age, which was contrary to the observed decrease in the spike frequency (Fig. 6a, b). Propagation velocity clearly increased with respect to 10 DIV in channel 2 ($p < 0.001$ vs. 20 DIV and $p < 0.01$ vs. 53 DIV; Fig. 6b), channel 5 ($p < 0.05$, $p < 0.001$ and $p < 0.001$ vs. 20, 30 and 53 DIV, respectively; Fig. 6b) and channel 6 ($p < 0.001$ vs. 20, 30 and 53 DIV; Fig. 6b). This increase was also significant when compared between 20 DIV and the subsequent recording days at 30 and 53 DIV in all channels ($p < 0.001$; Fig. 6b).

In addition to determining the mean propagation velocity, the changes in the propagation velocity along the axon was calculated by comparing the velocity propagation between subsequent electrode pairs (Fig. 7). In channel 5, only few signals propagated along the entire length of the channel at 10, 20 and 53 DIV. Therefore, these DIVs were excluded from the analysis. Despite variations in the

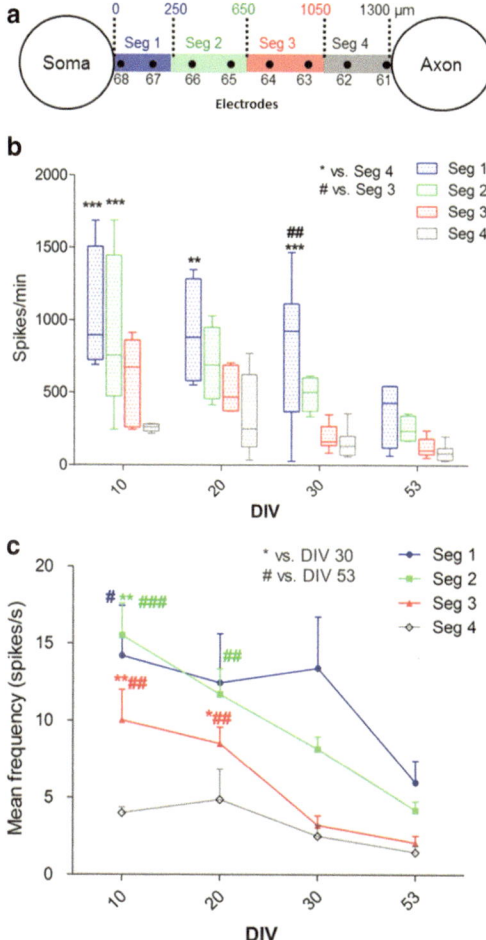

Fig. 5 The spike frequency decreased along a channel and over time. (**a**) Each microchannel was divided into four sections for a spike frequency analysis in the following way: segment 1 (Seg 1 = 250 μm from channel entrance), Seg 2 (250–650 μm), Seg 3 (650–1,050 μm), Seg 4 (1,050–1,300 μm). Each segment represents data that was collected from two adjacent electrodes. (**b**) Spike frequency along the channel. Each *bar* represents the spike frequency range in one segment (*color*) on the mentioned day. The mean spike frequency was calculated by averaging the number of spikes per minute in a selected segment of channels 2, 5 and 6. A two-way ANOVA was applied for comparing the mean values between different segments on the mentioned day. * vs. Seg 4 and # vs. Seg 3 at the same DIV. (**c**) Spike frequency evolution over time. Each line represents the changes in the mean spike frequency of the same segment (averaged over three channels) at different days. A one-way repeated measures ANOVA was applied for analyzing the mean frequency between different DIVs in each segment. * vs. DIV 30 and # vs. DIV 53 of the same segment. * or #$p < 0.05$, ** or ##$p < 0.01$ and *** or ###$p < 0.001$

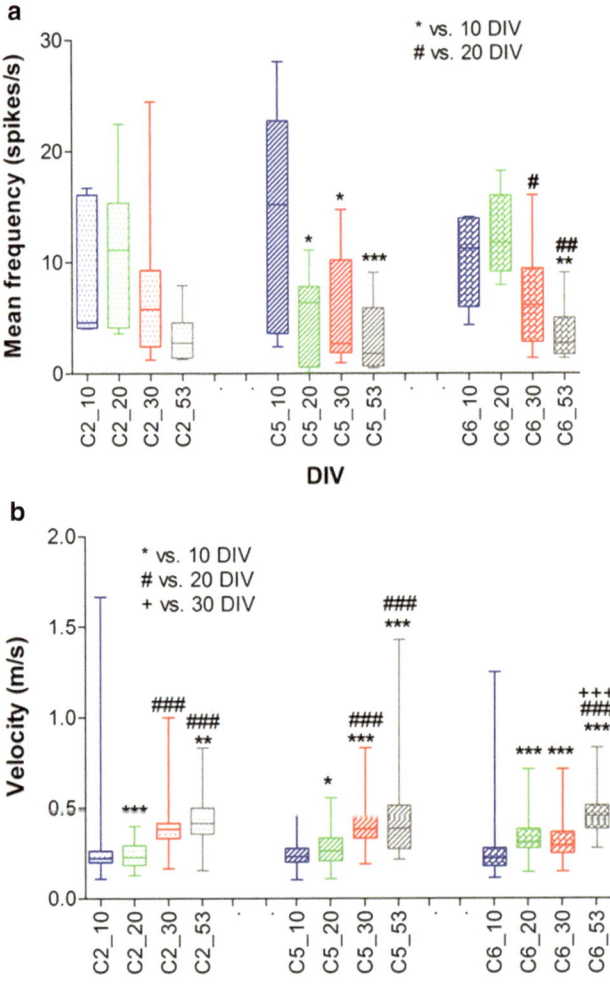

Fig. 6 Different profiles for spike frequency and propagation velocity in three evaluated channels over time. (**a**) Each *bar* represents the range of spike frequencies in one channel on the mentioned day. *x*-axis denominators code for the channel number followed by the DIV on which the recording was performed. One-way repeated measures ANOVA applied for analyzing the mean frequency between different DIVs in each channel. * vs. DIV 10 and # vs. DIV 20 in same channel. (**b**) Mean propagation speed in each channel for a given DIV. After calculating the velocity between each electrode pair in a channel, the mean propagation speed was determined by averaging the velocities of all pairs. Each *bar* represents the range of propagation speed in one channel at the mentioned day. One-way repeated measures ANOVA applied for analyzing the mean velocity between different DIVs in each channel. * vs. DIV 10, # vs. DIV 20 and + vs. DIV 30 in same channel. * or #$p < 0.05$, ** or ##$p < 0.01$ and ***, ### and +++$p < 0.001$. Electrodes 27, 52 and 62 did not record any spikes at 10 and 20 DIV. Therefore, the velocity was estimated by dividing the time delay between two nearby electrodes by a distance of 400 μm

Fig. 7 Changes in the signal propagation velocity along the axon in different channels. The propagation velocity was calculated by dividing the constant distance between each electrode pair by the temporal delay between time stamps. (**a, b**) In channels 2 and 5: no significant differences in the propagation velocity between two subsequent electrode pairs could be detected. (**c**) In channel 6, however, the velocity decreased or increased significantly along the axon at different days. One-way repeated measures ANOVA was applied for analyzing the mean propagation velocity between two subsequent electrode pairs at the mentioned day. $*p < 0.05$, $**p < 0.01$ and $***p < 0.001$ vs. mean propagation speed between previous electrode pairs. For all channels, the propagation speed was calculated for 30 randomly selected propagating signals at different time points during the recording window (1 min)

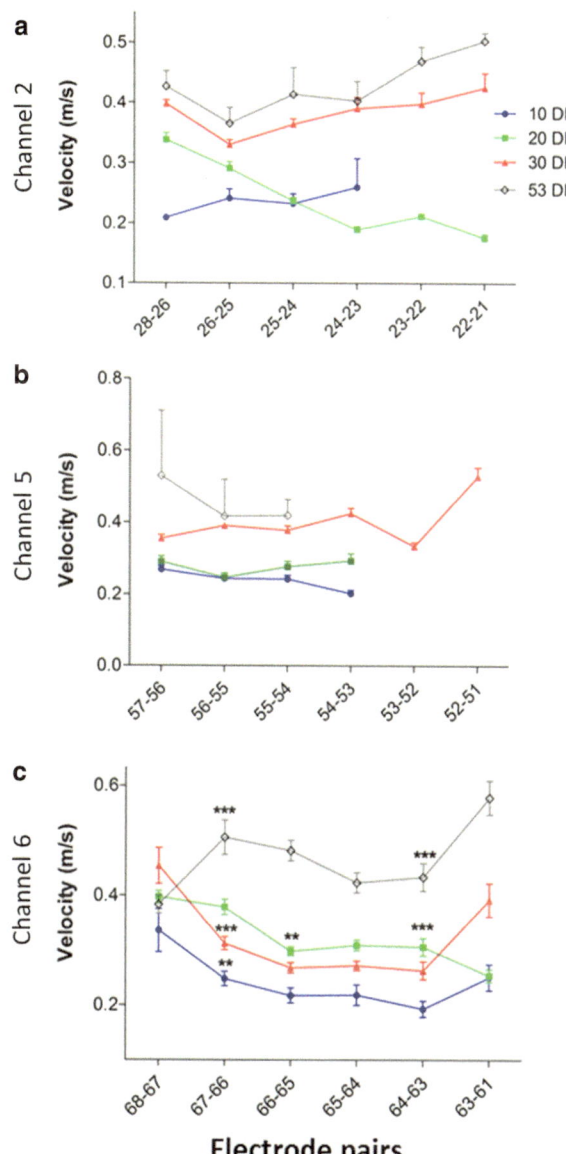

propagation velocities along channels 2 and 5, they were not significant between two adjacent segments on any of the recording DIVs (Fig. 7a, b). However, in channel 6, significant changes in the propagation velocity were observed between different segments, which could either decrease (*e.g.*, $p < 0.001$ from 68-67 to 67-66 at 30 DIV; Fig. 7c) or increase (*e.g.*, $p < 0.001$ from 68-67 to 67-66 at 53 DIV; Fig. 7c). Besides the local fluctuations in velocity, an increase in velocity over time is clearly evident for all channels in Fig. 7.

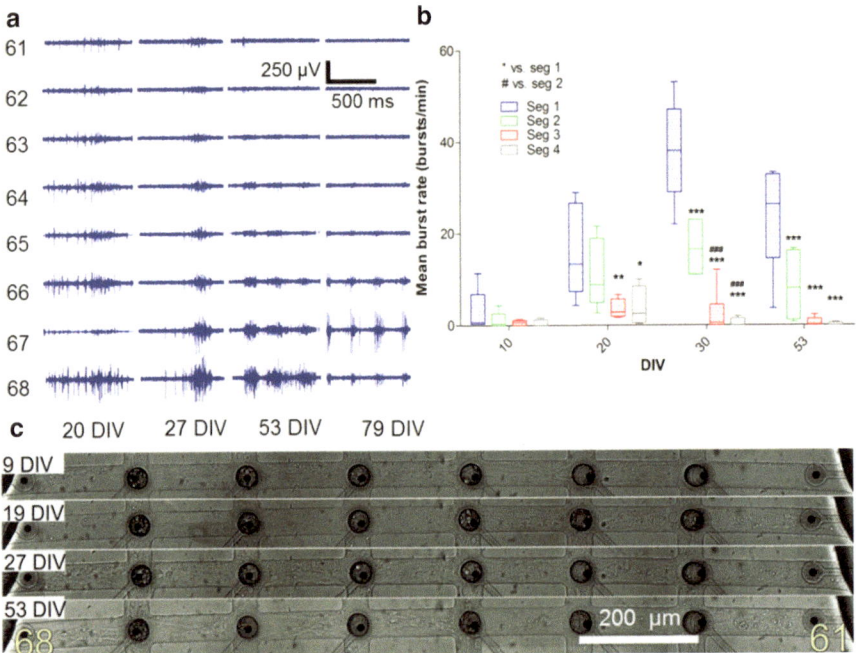

Fig. 8 Burst propagation along channels. (**a**) Raw signals from channel 6 (electrodes 68-61) at different DIVs. (**b**) Mean burst frequency in different segments of three channels. After burst detection, the burst frequency was calculated for each channel segment of channels 2, 5 and 6, and then averaged for all three channels. Each *bar* represents the burst frequency range in one segment of all three channels. Two-way ANOVA applied for analyzing the mean burst rate between two segments of the same channel at the mentioned day. * vs. seg 1 and # vs. seg 2. *$p < 0.05$, **$p < 0.01$ and *** or ###$p < 0.001$. (**c**) Changes in morphology along channel 6 (electrodes 68–61) over 53 DIV

Bursts propagation was analyzed for channels 2, 5 and 6 using the following criteria in NeuroExplorer: maximal interval to start burst = 0.02 s, maximal interval to end burst = 0.01 s, minimal interval between bursts = 0.01 s, minimal duration of burst = 0.02 s, minimal number of spikes in burst = 4, and bin size = 1 s). Figure 8 summarizes the mean burst rate in different segments of channels 2, 5 and 6. At 10 DIV, only a few bursts were detected (Fig. 8b). The burst frequency (bursts/min) increased over time and reached its maximum value after 30 DIV (Fig. 8b). A burst rate analysis for each segment of these three channels showed that bursts faded after 400-600 μm propagation length within a channel (Fig. 8b). The burst rate in segments 3 and 4 was significantly lower than in segment 1 ($p < 0.01$ and $p < 0.05$ at 20 DIV, $p < 0.001$ at 30 DIV, and $p < 0.001$ at 53 DIV; Fig. 8b).

3.2 polyMEA Features and Device Characteristics

Most deep brain implants are designed to record from biological tissue at near cellular resolution. This requires the electrode diameters to be on a similar scale. The design considerations for the implantable polymer microelectrode arrays included: (1) compatibility to commercial or custom-made signal processing electronics; (2) preferably a seamless, flexible and stable connection between implanted electrodes and the electronic platform; (3) an insertion depth control that may be defined by the probe geometry and guidance features (e.g., shaft edges, stoppers, ...).

Once geometries and boundary conditions are defined, photomasks can be designed by any microelectromechanical systems (MEMS) layout editor (*e.g.*, Expert, Silvaco; CleWin, WieWeb). Exemplarily, five different *polyMEA* designs for different brain areas are depicted in Fig. 9. Overall device features and the connecting scheme for a low-density electrode *polyMEA* are shown in Fig. 10. Its feature sizes are summarized in Table 1.

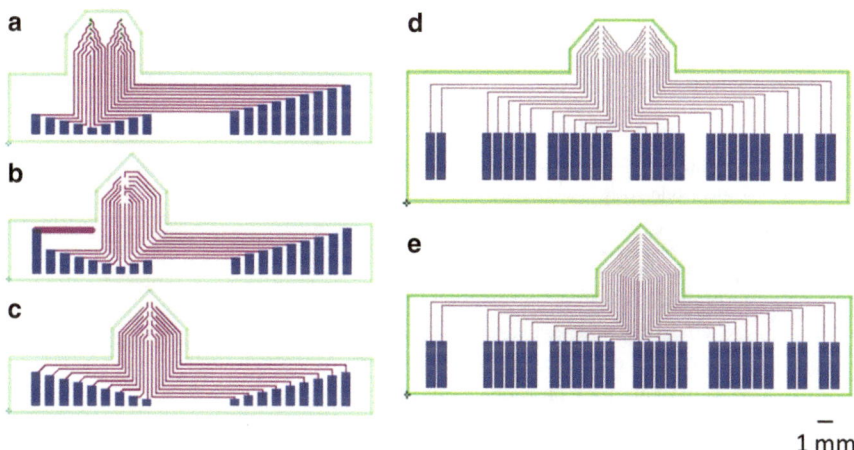

Fig. 9 Five *polyMEA* design examples for recording from various brain areas. The number of electrodes is the same in **a–c** (18) and in **d, e** (28). Design B with two separate electrode fields is suitable for the local recording at two specific depths while designs **a, c, d** and **e** are universal probes with equal electrode pitches along their insertion shafts. The sharpened probe tips in designs **b, c** and **e** may facilitate probe insertion into the brain. Electrode pitch and diameter, wire widths and pitches are different in these designs

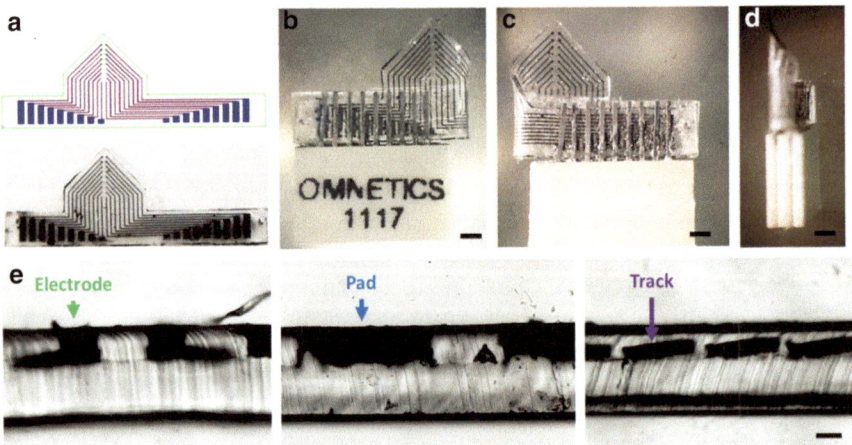

Fig. 10 Overall device layout and connecting scheme for an *in vivo polyMEA* with 2 × 9 electrode array. (**a**) CAD design and resulting device with tracks and electrodes made from a conductive polymer composite; pad width: 414 μm. (**b–d**) *polyMEA* squeeze-clamped between an Omnetics (A79006-001) 0.757 mm pitch, double-row pin connector (*front, back* and *side*). In this case, no extra PDMS sealing coat was applied to the connector and the *polyMEA* yet. (**e**) Cross section views of the *polyMEA* electrodes (*left*), pads (*middle*) and buried tracks (*right*). Scale bars: 1 mm (**b–d**), 100 μm (**e**)

Table 1 Feature sizes of a universal *in vivo polyMEA* with 2 × 9 electrode array

Parameter	Size (μm)
Maximum electrode site diameter	130
Minimum electrode site diameter	50
Vertical electrode center-to-center pitch	300
Horizontal electrode center-to-center pitch	300
Vertical electrode array dimension (center-to-center)	2,400
Shaft width	5,370
Shaft thickness	280.6
Shaft length	3,920
Track width	50

3.3 Electrical Characteristics of Carbon-, Graphite-, PEDOT:PSS-in-PDMS Composite Electrodes

In the *in vivo polyMEA* featuring a 2 × 9 electrode array, all 18 microelectrodes were functional. The average and extreme electrode impedance distribution between 1 Hz and 100 kHz is depicted in Fig. 11. Impedances ranged between 5 MΩ at low frequencies and several hundred kΩ at 1 kHz. Electrodes had almost resistive character at 1 kHz.

Fig. 11 Impedance characteristics of a *polyMEA* with a 2 × 9 electrode array. Electrode diameters ranged from 50 to 130 μm by 10 μm increments. Comparison of the average (Avg.) and extreme (Min., Max.) electrode impedances in a Bode magnitude plot (*top*) and of their capacitive-resistive characteristics in a phase plot (*bottom*)

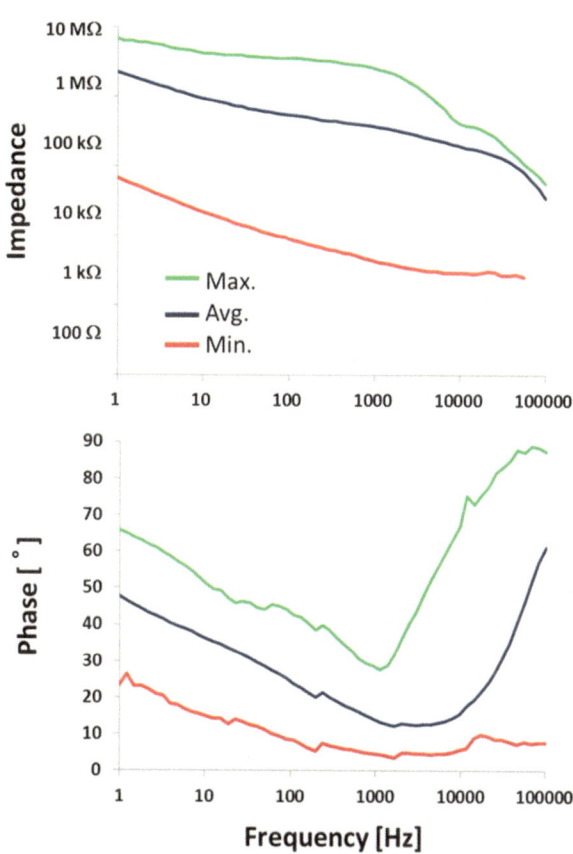

4 Discussion

The presented examples demonstrated how a straightforward replica-molding process for PDMS sheets generates perforated microchannel scaffolds for diverse applications. We exemplarily discussed their *in vitro* use in axonal guidance and regeneration studies on microelectrode arrays (MEAs) and the production of tissue-conformal *in vivo* MEAs for neuroprosthetic applications.

4.1 PDMS Microchannels for Axon Guidance and Electrophysiology

We showed that axons can be easily guided in microchannels that were aligned on top of the recording electrodes of a MEA. This allowed the detailed evaluation of axonal morphology, the different types of activity and their propagation velocity along channels over time.

We found that the signal frequency decreased between proximal and distal segments of a microchannel. This could have two reasons: the signal could be fading along the channel as it was evident in most cases (Fig. 3b). Furthermore, axons may leave the straight path by changing their growth direction at the channel crossing points (Fig. 2b).

The effect of microchannel geometry on signal quality has been evaluated by Wang et al. [11]. They showed that in microchannels with lengths between 200 and 3,000 μm the signal amplitude decreases along the channel length. Therefore, signals could become too small to be detected by electrodes in distal sections. In accordance with Fig. 3, large amplitude signals in proximal sections (electrode 68) decreased as they propagated within the channel. In line with the findings by Wang et al. [11], the high microchannel resistance combined with its stray capacitance acted like a first-order low-pass filter that attenuate high frequency signal amplitudes (i.e., the negative peaks of an action potential). Equally, the channel cross-section can affect signal properties [11]. Although all microchannels had the same cross-section, the number of axons and the thickness of the axon bundles inside the microchannel varied with channel depth and over time, thereby changing the total cross-section of a microchannel at different channel locations. In 2009, Dworak and Wheeler showed that the growth of neurites inside microchannels ($750 \times 10 \times 3$ μm^3) increased the resistivity from 75 Ω cm of empty channels to 300 Ω cm for microchannels filled with neural tissue due to an effective decrease of the channel cross-section [10].

The increase of neural tissue mass inside the microchannels between 20 and 30 DIVs (Fig. 8c) could be correlated with an increased spike frequency for almost all channels and all channel segments (Figs. 5 and 6). Equally, a decrease in neural tissue density after 53 DIV led to a decrease in the firing frequency in all channels (Figs. 5 and 6). A significant decrease of the signal frequency after 30 DIV in all segments could be related to the axonal degeneration and the formation of axonal bundles (Fig. 8c).

Recording an action potential from subsequent electrodes along a microchannel enabled us to determine the propagation velocity and direction. The propagation velocities varied between 0.1 to 2.5 m/s, which is in same range as previously reported by Pan et al. (0.18–1.14 m s^{-1} in unmyelinated axons) [14]. In this study, we also evaluated changes in the propagation velocities in different segments of the same channel as well as in the same segment or channel over time. Changes in action potential velocity along the axon could be related to changes in thickness and curvature along the axon. We also observed an increased propagation velocity in all channels between 30 and 53 DIVs, which could possibly be related to increased axonal diameters or the fasciculation of axons. Such spatial and temporal changes in stimulus-driven action potential propagation velocity have been reported recently by Bakkum et al., however in random cultures on dense microelectrode arrays [15]. In the present study, we evaluated the propagation velocity of signals derived from spontaneous network activity over time.

Bursts are spike flares within a short time window. Bursts are usually recorded simultaneously from different MEA electrodes, which shows that burst activity involves and propagates within large parts of the network [16]. We also evaluated how bursts propagate inside the microchannels. While individual monophasic spikes are able to propagate over long distances within a channel, bursts failed in reaching the microchannel endings in most cases (Fig. 8a). A significant difference in burst frequency on electrodes in proximal and distal parts of the microchannels confirmed the observed burst propagation fading along the microchannel (Fig. 8b). Because bursts are composed of individual spikes with different shapes, each of them will have its specific half-life for traveling along the channel. Fading spikes inside a burst will cause the burst to disintegrate along the channel. In consequence, a spike sequence that was categorized as a burst at the proximal end of a channel may not be recognized as a burst anymore at the distal end of a channel. Another mechanism for burst fading could be based on a phase cancelation effect. Signals with different shapes and phases can cancel each other out while traveling along the channel [11].

4.2 PDMS Microchannels for the Production of Tissue-Conformal in vivo MEAs for Neuroprosthetic Applications

Compared to stiff implants, flexible probes will more easily relieve the strain caused by micromotion forces that result from the relative displacement between the implant and the brain tissue. This may minimize chronic tissue damage or inflammation due to a better match in the stiffness of the probe and the brain microenvironment [17]. In this context, PDMS has potential as a scaffold material for neural interfaces [18, 19]. As thin sheets it can follow the curvature of a tissue and provide a uniform and tight contact [20]. Here, we exemplarily depicted the design and fabrication of a flexible *polyMEA* with 18 recording sites that can be implanted into the brain tissue. Impedance characteristics indicated that the electrodes performed well even in the low frequency range. At 1 kHz, their average impedance stayed at several hundred $k\Omega$ with an almost resistive behavior.

Acknowledgements Many thanks to Marina Nanni, Francesca Succol and Claudia Chiabrera for their excellent assistance in cell culture preparation. Thanks to Francesco Difato and Mattia Pesce for their advice on imaging techniques. Intramural funding is highly appreciated.

References

1. Taylor AM, Blurton-Jones M, Rhee SW, et al. A microfluidic culture platform for CNS axonal injury, regeneration and transport. Nat Methods. 2005;2:599–605.
2. Banker G, Goslin K. Culturing nerve cells. Cambridge: MIT Press; 1998.
3. Kim HJ, Park JW, Byun JH, et al. Quantitative analysis of axonal transport by using compartmentalized and surface micropatterned culture of neurons. ACS Chem Neurosci. 2012;3:433–8.

4. Kim YT, Karthikeyan K, Chirvi S, et al. Neuro-optical microfluidic platform to study injury and regeneration of single axons. Lab Chip. 2009;9:2576–81.
5. Park J, Koito H, Li J, et al. Microfluidic compartmentalized co-culture platform for CNS axon myelination research. Biomed Microdevices. 2009;11:1145–53.
6. Yang IH, Gary D, Malone M, et al. Axon myelination and electrical stimulation in a microfluidic, compartmentalized cell culture platform. Neuromolecular Med. 2012;14:112–8.
7. Wheeler BC, Brewer GJ. Designing neural networks in culture. Proc IEEE. 2010;98:398–406.
8. Claverol-Tinture E, Ghirardi M, Fiumara F, et al. Multielectrode arrays with elastomeric microstructured overlays for extracellular recordings from patterned neurons. J Neural Eng. 2005;2:L1–7.
9. Ravula SK, Mcclain MA, Wang MS, et al. A multielectrode microcompartment culture platform for studying signal transduction in the nervous system. Lab Chip. 2006;6:1530–6.
10. Dworak BJ, Wheeler BC. Novel MEA platform with PDMS microtunnels enables the detection of action potential propagation from isolated axons in culture. Lab Chip. 2009;9:404–10.
11. Wang L, Riss M, Buitrago JO, et al. Biophysics of microchannel-enabled neuron-electrode interfaces. J Neural Eng. 2012;9:026010.
12. Blau A, Murr A, Wolff S, et al. Flexible, all-polymer microelectrode arrays for the capture of cardiac and neuronal signals. Biomaterials. 2011;32:1778–86.
13. Blau A, Neumann T, Ziegler C, et al. Replica-moulded polydimethylsiloxane culture vessel lids attenuate osmotic drift in long-term cell cultures. J Biosci. 2009;34:59–69.
14. Pan LB, Alagapan S, Franca E, et al. Propagation of action potential activity in a predefined microtunnel neural network. J Neural Eng. 2011;8.
15. Bakkum DJ, Frey U, Radivojevic M, et al. Tracking axonal action potential propagation on a high-density microelectrode array across hundreds of sites. Nat Commun. 2013;4:2181.
16. Maeda E, Robinson HP, Kawana A. The mechanisms of generation and propagation of synchronized bursting in developing networks of cortical neurons. J Neurosci. 1995;15:6834–45.
17. Subbaroyan J, Kipke DR. The role of flexible polymer interconnects in chronic tissue response induced by intracortical microelectrodes—a modeling and an in vivo study. Conf Proc IEEE Eng Med Biol Soc. 2006;1:3588–91.
18. Lacour S, Benmerah S, Tarte E, et al. Flexible and stretchable micro-electrodes for in vitro and in vivo neural interfaces. Med Biol Eng Comput. 2010;48:945–54.
19. Maghribi M, Hamilton J, Polla D et al. (2002) Stretchable micro-electrode array. In: Microtechnologies in medicine & biology 2nd annual international IEEE-EMB special topic conference, p 80–83
20. Guo L, Meacham KW, Hochman S, et al. A PDMS-based conical-well microelectrode array for surface stimulation and recording of neural tissues. IEEE Trans Biomed Eng. 2010;57:2485–94.
21. Anderson JR, Chiu DT, Jackman RJ, et al. Fabrication of topologically complex three-dimensional microfluidic systems in PDMS by rapid prototyping. Anal Chem. 2000;72:3158–64.
22. Bettinger CJ, Borenstein JT. Biomaterials-based microfluidics for engineered tissue constructs. Soft Matter. 2010;6:4999–5015.
23. Qin D, Xia Y, Whitesides GM. Soft lithography for micro- and nanoscale patterning. Nat Protoc. 2010;5:491–502.
24. Yun K-S, Yoon E. Fabrication of complex multilevel microchannels in PDMS by using three-dimensional photoresist masters. Lab Chip. 2008;8:245–50.

Brain-Controlled Selection of Objects Combined with Autonomous Robotic Grasping

Christoph Reichert, Matthias Kennel, Rudolf Kruse, Hans-Jochen Heinze, Ulrich Schmucker, Hermann Hinrichs, and Jochem W. Rieger

Abstract A Brain–Computer Interface (BCI) could help to restore mobility of severely paralyzed patients, for instance by prosthesis control. However, the currently achievable information transfer rate of noninvasive BCIs is insufficient to control complex prostheses continuously in many degrees of freedom. In this paper we present an autonomous system for grasping natural objects that compensates the low information flow from noninvasive BCIs. Using this system, one out of several objects can be grasped without any muscle activity. Rather, the grasp is initiated by decoded voluntary brain wave modulations. Object selection and grasping are performed in a virtual reality environment. A universal grasp planning algorithm calculates the trajectory of a gripper online. The system can be controlled after less

C. Reichert
Department of Neurology, University Medical Center A.ö.R., Magdeburg, Germany

Department of Knowledge and Language Processing, Otto-von-Guericke
University, Magdeburg, Germany
e-mail: christoph.reichert@med.ovgu.de

M. Kennel • U. Schmucker
Fraunhofer Institute for Factory Operation and Automation IFF, Magdeburg, Germany
e-mail: matthias.kennel@iff.fraunhofer.de; klaus-ulrich.schmucker@iff.fraunhofer.de

R. Kruse
Department of Knowledge and Language Processing, Otto-von-Guericke
University, Magdeburg, Germany
e-mail: kruse@iws.cs.uni-magdeburg.de

H.-J. Heinze • H. Hinrichs
Department of Neurology, University Medical Center A.ö.R., Magdeburg, Germany

Leibniz Institute for Neurobiology, Magdeburg, Germany

German Center for Neurodegenerative Diseases (DZNE), Magdeburg, Germany
e-mail: hans-jochen.heinze@med.ovgu.de; hermann.hinrichs@med.ovgu.de

J.W. Rieger (✉)
Department of Applied Neurocognitive Psychology, Carl-von-Ossietzky
University, Oldenburg, Germany
e-mail: jochem.rieger@uni-oldenburg.de

© Springer International Publishing Switzerland 2015
A.R. Londral et al. (eds.), *Neurotechnology, Electronics, and Informatics*,
Springer Series in Computational Neuroscience 13, DOI 10.1007/978-3-319-15997-3_5

than 10 min of training. We found that decoding accuracy increases over time and that an increased sense of agency achieved by permitting free selections renders the system to work most reliably.

Keywords Brain–computer interface (BCI) • P300 • Oddball paradigm • Grasp robot • Magnetoencephalogram (MEG) • Virtual reality (VR) • Gaze independence

1 Introduction

Brain–Computer Interfaces (BCI) can serve as a communication channel between the human brain and a computational or robotic device by translating human brain activity to machine commands [1]. Due to the independence from muscles and peripheral nerves they are in the focus of research to replace motor functions of severely paralyzed patients. In the recent years, highly invasive techniques were applied to control prosthetic devices by voluntary modulation of brain activity [2, 3]. In humans, the use of noninvasive techniques, like the electroencephalogram (EEG), is preferable over invasive recordings. However, only a small number of commands can be discriminated with noninvasively assessed brain activity and, as a consequence, noninvasive systems do not allow for full control of complex manipulators with many degrees of freedom. Here we report progress in our development of a noninvasive BCI that enables users to grasp natural objects with a complex manipulator. Our approach combines the development of both efficient brain decoding techniques and autonomous actuator control to overcome the limited information transfer from noninvasive BCIs.

A commonly used control signal for movement related BCIs is the so-called μ-rhythm [4], which can be measured during motor imagery. However, a considerable percentage of people are unable to control motor imagery BCIs [5, 6]. In contrast, it was shown that a larger fraction of people is able to select items in speller paradigms using an oddball task [7]. The widely used matrix speller, which relies on the oddball task, was first introduced by Farwell and Donchin [8]. In this paradigm a P300, a positive EEG deflection, is evoked when a rare target stimulus appears in a series of irrelevant stimuli. While it is often assumed that the accuracy of visually stimulated P300 speller is independent of gaze direction, it has recently been shown that the performance of the matrix speller drops significantly if the eyes are not moved toward the target [9]. The reason is that two EEG components, the P300 and the N200, contribute information when the centre of regard is moved to the target [10], whereas only the P300 is present if the eyes don't move. This could render the P300 paradigm less useful for patients who cannot move their eyes.

In this work we demonstrate that the visual oddball paradigm can be successfully applied to initiate targeted grasps in a visually complex virtual environment with multiple realistic objects. Importantly, we show that the paradigm developed here works independent of the user's ability to direct gaze towards the target object. This is of high relevance for the targeted user group.

The development of an efficient noninvasive brain decoder is complemented by the development of a robotic manipulator with the ability to manipulate in an intelligent and autonomous way predefined objects. Here we propose a new analytical grasp planning algorithm to achieve autonomous grasping of arbitrary objects. In contrast to other motion planning algorithms, our algorithm is not based on Learning by Demonstration (for a review see [11]) and involves, but is not limited to, the robot's kinematics.

2   Materials and Methods

2.1   General Procedure

In this study we decoded in real-time the magnetoencephalogram (MEG) of 17 subjects (9 male, 8 female, mean age: 26.6 years) to determine their intention to select one of six selectable realistic objects for grasping. We used the decoding results to initiate a grasp of a robotic gripper. All subjects gave written informed consent. The study was approved by the ethics committee of the Medical Faculty of the Otto-von-Guericke University of Magdeburg.

2.2   Virtual Environment

We presented six objects placed at fixed positions in a virtual reality environment as shown in Fig. 1. The visual angle between outmost left and right objects was 8.5°. We used circular regions around the objects on the table to provide cue flashes for the P300 paradigm and to provide feedback by coloring the region's shape. A photo transistor placed on the screen was used to synchronize the ongoing MEG with the events displayed on the screen. To provide realistic feedback, the model of a robot (Mitsubishi RV E2) equipped with a three finger gripper (Schunk SDH) was part of the scene. The robot in the virtual reality scene visualized the autonomously calculated grasp to the selected object mimicking the movements of the real robot.

2.3   Paradigm

The oddball paradigm we employed is designed to elicit a P300 potential which is a deflection of sensor values evoked approximately 300 ms after a rare target stimulus occurs in a series of irrelevant stimuli. The goal for the brain decoder is to reliably detect the P300 to generate a control signal for the robot. In our variant of the paradigm, we marked objects by flashing their background for 100 ms. Objects

Fig. 1 VR scenario used for visual stimulation. This snapshot shows one flash event of an object

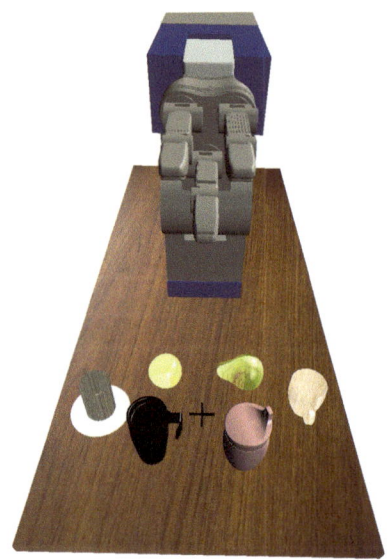

Table 1 Number of runs the subjects performed in different selection modes

	Instructed			
Subject #	Training	Decoder	Free	Grasp
1	2	5	–	–
2	4	4	1	–
3	3	4	–	–
4–6	3	4	1	1
7	2	4	2	1
8	3	3	2	1
9	2	4	2	1
10	2	4	2	–
11	2	4	2	1
12	2	5	2	–
13–17	2	4	2	1

were marked in random order with an interstimulus interval of 300 ms. Each object was marked five times per selection trial resulting in a stimulation interval length of 10 s.

Subjects were instructed to fixate the black cross centred to the objects and to count how often the target object was marked. The counting ensured that attention was maintained on the stimulus stream. In addition, subjects were instructed to avoid eye movements and blinking during the stimulation interval.

Each subject performed a minimum of seven runs with 18 selection trials per run. The runs were performed in three different modes that served different purposes. The number of runs each subject performed in each mode is listed in Table 1. We started with the *instructed selection* mode. In this mode, the target object was cued by a light grey circle at the beginning of a trial and subjects were instructed to attend the cued object. Instructed selection was used in the initial training runs in which we collected data to train the classifier. In this mode true classifier labels are available which are required to train the classifier. We provided random feedback during training runs because no classifier was available in these initial runs. After the training runs, each subject performed several instructed runs with feedback. We denote the second selection mode *free selection*. In this mode, subjects were free to choose the target object. In instructed selection mode and in free selection mode, a green circle was presented at the end of the trial on the decoded object as feedback. All other objects were marked by red circles. Free selection runs were performed after the instructed selection runs. In the third mode, the *grasp selection* mode, the virtual robot grasped and lifted the decoded target for feedback. Grasp selection runs were performed after free selection runs. In both modes, the free selection and grasp selection mode, the subject said "no" to signal that the classifier decoded the wrong object and did not respond otherwise.

The results reported in this paper arise from online experiments. We did not exclude subjects participating in early sessions, causing slight changes in the experimental protocol during the study (Table 1). The number of runs performed in the different modes depended on cross validated classifier performance estimation and the development of detection accuracy. In total, five subjects performed three, one subject four and the remaining 11 subjects two initial training runs. Two subjects performed only instructed selections. Twelve of the subjects performed one run in the grasp selection mode. Here, only six instead of 18 trials were performed, due to the longer feedback duration.

2.4 Data Acquisition and Processing

The MEG was recorded with a whole-head BTi Magnes 248-sensors system (4D-Neuroimaging, San Diego, CA, USA) at a sampling rate of 678.17 Hz. Simultaneously, the electrooculogram (EOG) was recorded for subsequent inspection of eye movements. MEG data and event channels were instantaneously forwarded to a second workstation capable of processing the data in real-time. The data stream was cut into intervals including only the stimulation sequence. The MEG data were then band-pass filtered between 1 and 12 Hz and down sampled to 32 Hz sampling rate. Then, the stimulation interval was cut in overlapping 1,000 ms segments starting at each flash event. In instructed selection mode, the segments were labelled as target or nontarget segments depending on whether the target or a nontarget object was marked.

We used a linear support vector machine (SVM) as classifier because it proved to reliably provide high performance in single trial MEG discrimination [12, 13]. These previous studies showed that linear SVM is capable of selecting appropriate features in high dimensional MEG feature spaces. We performed classification in the time domain, meaning that we used the magnetic flux measured in 32 time steps as classifier input. To reduce the dimensionality of the feature space, we excluded 96 sensors located farthest from the vertex (the midline sensor at the position halfway between inion and nasion) which is the expected site of the P300 response. We further reduced the number of sensors by selecting the 64 sensors providing the highest sum of weights per channel in an initial SVM training on all preselected 152 sensors of the training run data. The selected feature set (64 sensors × 32 samples = 2,048 features) was then used to train the classifier again and retrain the classifier after each run conducted in instructed selection mode.

2.5 Grasping Algorithm

In this section we describe the general procedure of our grasp planning algorithm, whereas we present the mathematical details in the Appendix. The algorithm was developed to physically drive a robot arm, but in this experiment it was used to provide virtual reality feedback. Importantly, in this strategy the robot serves as an intelligent, autonomous actuator and does not drive predefined trajectories. The algorithm assumes that object position and shape coordinates relative to the manipulator are known to the system. In this experiment, coordinates of CAD-modelled objects were used. However, coordinates could as well be generated by a 3D object recognition system.

Central to our approach is that the contact surfaces of the gripper's fingers and the surfaces of the objects were rasterized with virtual point poles. We assumed an imaginary force field between the poles on the manipulator and the poles on the target object (see Appendix for details). The goal of the algorithm is to initially generate a manipulator posture that ensures a force closure grasp. The following grasp is organized by closing the hand in a real world scenario and by locking the object coordinates relative to the finger surface coordinates in the virtual scenario.

3 Results

3.1 Decoder Accuracy

We determined the decoding accuracy as the ratio of correctly decoded objects divided by the total number of object selections. All subjects performed the task reliably above guessing level which was 16.7 % according to the six objects on the table. On average, the intended object selections were correctly decoded from the MEG data in 77.7 % of all trials performed. Single subject accuracies ranged from 55.6 to 92.1 %. In the instructed selection mode the average accuracy was 73.9 and 85.9 % in the free selection mode. A Wilcoxon rank sum test revealed a p-value of 0.03 which indicates that the performance difference between the instructed and

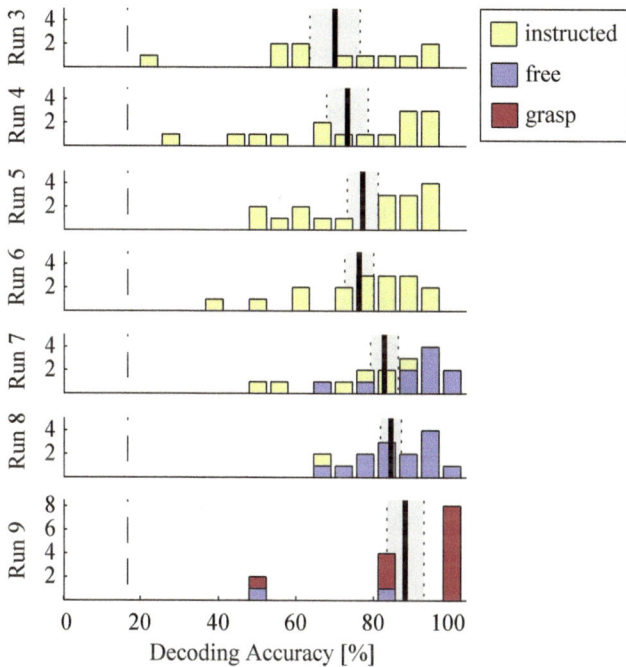

Fig. 2 Performance histograms. The ordinate indicates the number of subjects who achieved the respective decoding performance out of 19 possible percentage bins, equally spaced from 0 to 100 %. The histograms show data from different runs, chronologically ordered from top to bottom. The run modes are coded by color. *Vertical dashed lines* indicate the guessing level and *thick solid lines* indicate the average decoding accuracies over subjects. Standard error is marked *grey*

the free selection mode is statistically significant with higher performance in the free selection mode. When subjects received feedback by moving the virtual robot to the grasp target, the average accuracy was even higher and reached 91.2 %. Figure 2 depicts the evolution of decoding accuracies over runs. The height of the bars indicates the number of subjects (y-axis) who achieved the respective decoding performance out of 19 possible percentage bins. Each histogram shows the results from one run and the performance bins are equally spaced from 0 to 100 %. The histograms are chronologically ordered from top to bottom. Yellow bars indicate results from instructed selection runs, blue bars indicate free selection run results and red bars indicate results in runs with grasp feedback. Vertical dashed lines indicate the guessing level and thick solid lines indicate the average decoding accuracies over subjects whereas the standard error is marked grey. The average decoding accuracy increases gradually over the course of the experiment. Moreover, the histograms show that the highest accuracy over subjects was achieved in free selection runs. Note that our system achieved perfect detection in 8 of the 12 subjects who received virtual grasp feedback. However, only six selections were performed by each subject in these grasp selection runs.

An established measure for the comparison of BCIs is the information transfer rate (ITR) which combines decoding accuracy and number of alternatives to a unique measure. We calculated the ITR according to the method of Wolpaw et al. [14] at 3.4–12.0 bit/min for single subjects and 8.1 bit/min on average. Note that the maximum achievable bit rate with the applied stimulation scheme is 15.5 bit/min.

For online eye movement control, we observed the subjects' eyes on a video screen. In addition, we inspected the EOG measurements offline. Both methods confirmed that subjects followed the instruction to keep fixation.

3.2   Grasping Performance

We evaluated the execution duration of the online grasp calculation for different setups and objects. We implemented our grasping algorithm with the ability to distribute force computations to several parallel threads. Here, we permitted five threads employing a 2.8 GHz AMD Opteron 8220 SE processor. We calculated grasps of the six objects shown in Fig. 1. To assess effects of object position, we arranged the objects at different positions within the limits of our demonstrating robot's work space. Each object was placed once at each of the positions depicted in Fig. 1. The time needed to plan the trajectory and execute the grasp until reaching force closure is listed in Table 2 for each object/position combination. Calculation times ranged from 11 to 72.6 s depending on the object and the position. The diagonal of Table 2 represents the actual object/position setup during our experiment.

The results indicate that the duration of grasp planning depends on many parameters. The most important determinant of execution time is the number of point poles ($O(n^2)$) which depends on the level of detail of the object surface. This is reflected in the high variation of the average calculation time of objects 1–6 over positions (18, 50.3, 12.9, 15.3, 26.1, 39.4 s). The execution times also depend on the physical constraints of the robot and object position. We observed that even minimal differences in object arrangement appear to have strong influence on force closure termination. This is also indicated by different execution times of identical objects at symmetric positions (e.g. left/right). We consider it likely that these differences

Table 2 Duration of grasp planning calculation for all object/position combinations in seconds

	Object #					
Object position	#1	#2	#3	#4	#5	#6
Left	**33.0**	68.5	11.0	14.1	16.0	24.5
Upper left	25.5	**34.0**	16.5	13.5	39.0	46.7
Lower left	11.0	72.6	**11.0**	18.0	22.5	65.0
Upper right	15.5	42.5	15.0	**12.5**	37.0	46.0
Lower right	12.0	48.5	11.0	19.0	**22.5**	24.5
Right	11.0	35.6	13.0	14.5	19.5	**29.5**

Bold numbers indicate the object positions used in the online experiment are caused by numerical precision issues due to the high number of summations in Eqs. (3) and (4) (see Appendix).

4 Discussion

In the present work we demonstrated that the oddball paradigm is well suited for use in a BCI to reliably select one of several objects for grasping. In a previous study we showed that other selection paradigms also are suitable to select virtual reality objects [15]. The advantage of the present study over the previous study is that the selection of objects was achieved independent from eye movements to the target and realistic grasp feedback was presented. Importantly, our results suggest that performance improves with training and the improvement is even higher when subjects obtain more control by permitting free selections and when realistic visual feedback was presented. This suggests that BCI control in our P300 paradigm is improved with an increase of the subject's sense of agency. A gaze independent BCI based on directing covert attention is a fundamental requirement for patients who cannot easily orient gaze to the target object.

Earlier reports suggested that eye movements greatly improve performance in a P300 speller [9, 10, 16], due to contribution from visual areas to brain wave classification [17]. We extend these previous studies and show that the P300 paradigm is well suited for a gaze independent object grasping BCI. We achieved independence from visual components by instructing our subjects to fixate and by excluding occipital sensors from the analysis. This approach simulates a realistic setting with patients who cannot move their eyes and are therefore dependent on covert attention shift based activation for control. To date, only a small number of studies successfully implemented such a more restrictive covert attention P300 approach [18–20].

We observed increasing decoder accuracy during the course of the experiment. This suggests that the increasing amount of training is beneficial for performance in our BCI paradigm. However, due to classifier updates performed in the course of the experiment, the learning process is likely bilateral and involves both the subjects and the classifier [21]. Importantly, when subjects were free to select the target object, the decoding success was significantly higher compared to the instructed selections. This suggests a strong role for task involvement and the sense of agency in our paradigm. When subjects performed runs receiving grasp feedback, most of them achieved perfect decoding accuracy. We expect the reliability of the system to be further increased by extending the stimulation interval [18, 22]. Note that system reliability is often more important for the user than a rapid but error prone detection of intention. Furthermore, the system presented here is efficient for use with nearly no training. Most subjects performed less than 10 min of training in order to provide data to the decoding algorithm. This is a very small effort compared to motor imagery based systems aiming to control movement in a few degrees of freedom [2, 23].

In order to reduce the burden of controlling a complex manipulator with many degrees of freedom by voluntary modulation of brain activity, we combined a P300 BCI with a grasping system that autonomously executes the grasp requiring only a very low input bit rate, namely the command to grasp an object known to the system. To execute the grasp intended by the BCI user, we developed an algorithm for autonomous grasp planning that can place a reliable grasp on natural objects. The execution times we achieved were practical for the proposed task but can be further reduced in future work. We should note that we did not focus on timing optimization in this work but on reliable placement of the grasp. Moreover, the proposed algorithm is universal in the sense that it is not restricted to a specific manipulator. Consequently, this algorithm should also be easily transferable to arbitrary prosthetic devices suitable for grasping potential target objects with a force closure grasp.

In online closed-loop BCI studies the decoding algorithm has to be fixed before the start of the experiment. We decided to use SVM classification because this is an established classifier for high dimensional feature spaces that provides high and robust generalization by upweighting informative and downweighting uninformative features [24]. Furthermore, it was shown that linear SVM was equally accurate for P300 detection compared to Fisher's linear discriminant and stepwise linear discriminant analysis [25]. Several existing studies make use of extended linear discriminant analysis algorithms applied to EEG data [19, 20]. However, because MEG data are based on a much larger number of sensors, these approaches are not applicable in a suitable time.

As input brain signal for the BCI, we used the MEG. This noninvasive technique measures magnetic fields of cortical dipoles. While the dynamic signal characteristics are comparable to those in EEG, MEG tends to provide higher spatial resolution [26]. We are aware that this modality is not suitable for daily use and particularly not for use of a prosthetic device. In fact, we consider our study basic research, and to our knowledge, this is the first implementation of a MEG based P300 closed loop BCI. In this study, we tested our system with healthy subjects. A confirmation of our results with patients is an open issue for future work.

5   Conclusions

In this paper we showed that noninvasive BCI in combination with an intelligent actuator can be used in real world settings to grasp and manipulate objects. This is an important step towards the development of assistive systems for severely impaired patients.

Acknowledgements This work has been supported by the EU project ECHORD number 231143 from the 7th Framework Programme and by Land-Sachsen-Anhalt Grant MK48-2009/003.

Appendix

In Sect. 2.4 we stated the rasterizing of the object and gripper surfaces with virtual point poles. Here we describe the algorithm in more detail.

Our grasp planning algorithm is organized by simulating the action of forces between target object and manipulator in consecutive time frames. While the object poles P^O are defined as positive, the manipulator poles P^M are defined as negative. In accordance with Khatib [27], we assume that opposite poles attract each other while like poles do not interact. The magnitude of the force between two poles P_i^O and P_j^M we calculated as

$$\vec{F}\left(P_i^O, P_j^M\right) = e^{-\overline{P_i^O P_j^M}} \tag{1}$$

where $\overline{P_i^O P_j^M}$ is the distance between the poles, and the unit of F is arbitrary. The exponential function limits F to a maximum of 1 unit. This avoids infinite forces at collision scenarios and provides a suitable scaling to instantiate both propulsive forces between manipulator and object and repulsive forces to reject manipulator poles that penetrate the object's boundary.

The total propulsive force $\vec{F}\left(P_i^M\right)$ affecting one point pole P_i^M on the manipulator is calculated from a set of object point poles A_O where

$$A_O\left(P_i^M\right) := \left\{ P_j^O \middle| P_j^O \in P^O \wedge \vec{n}_j^O \cdot \vec{n}_i^M < 0 \right\} \tag{2}$$

which indicates that only pairwise point poles with an angle between the surface normal $\vec{n}_i^M$ and $\vec{n}_j^O$ greater than $\pi/4$ are involved. We included this constraint to restrict interactions to opposing surface force vectors. The force $\vec{F}\left(P_i^M\right)$ that moves the manipulator is then calculated as

$$\vec{F}\left(P_i^M\right) = \sum_{P_j^O \in A_O\left(P_i^M\right)} \vec{F}\left(P_i^O, P_j^M\right) \tag{3}$$

The manipulator's effective joint torque $\vec{\tau}$ can be calculated by means of the Jacobian J generated from the joint angles $\vec{q}$ and the point poles P^M [28] by

$$\vec{\tau} = \sum_i J\left(P_i^M, \vec{q}\right)^T \left[\begin{matrix} \vec{F}\left(P_i^M\right) \\ \vec{M} \end{matrix} \right] \tag{4}$$

where external moments are considered $\vec{M} = \vec{0}$. In order to simulate the manipulator movement, we calculated the new joint angle $q_k(t)$ of an axis k by solving the equation system

$$\dot{q}_k(t) = \dot{q}_k(t - \Delta t) + \Delta t * \frac{\tau_k}{\vec{a}_k^T I\left(\vec{q}\right) \vec{a}_k} \tag{5}$$

$$q_k(t) = q_k(t - \Delta t) + \Delta t * \dot{q}_k(t) \tag{6}$$

where $I\left(\vec{q}\right)$ is the inertia tensor of the robot's solid elements and $\vec{a}_k$ defines one of the manipulator axes. We chose a heuristically dynamic calculation of the time frame length Δt which is proportional to the mean distance between the set of point poles P^M and P^O.

Collision detection was performed for the new posture before a new time frame was assigned to be valid and the position update was sent to the manipulator. We used standard techniques [29] to detect surface intersections. If intersections were detected, repulsive forces were calculated for the affected point poles directing to their position of the last valid time frame and satisfying Eq. (1). If no intersections were detected, the robot moved to the new coordinates. This procedure was repeated until the force closure condition [30] was satisfied.

References

1. Wolpaw JR. Brain-computer interfaces. In: Barnes MP, Good DC, editors. Neurological rehabilitation, Handbook of clinical neurology, vol. 110. Amsterdam: Elsevier; 2013. p. 6774.
2. Hochberg LR, Bacher D, Jarosiewicz B, Masse NY, Simeral JD, Vogel J, Haddadin S, Liu J, Cash SS, van der Smagt P, Donoghue JP. Reach and grasp by people with tetraplegia using a neurally controlled robotic arm. Nature. 2012,485(7398):372 5.
3. Velliste M, Perel S, Spalding MC, Whitford AS, Schwartz AB. Cortical control of a prosthetic arm for self-feeding. Nature. 2008;453(7198):1098–101.
4. Pfurtscheller G, Neuper C, Guger C, Harkam W, Ramoser H, Schlögl A, Obermaier B, Pregenzer M. Current trends in Graz Brain-Computer Interface (BCI) research. IEEE Trans Rehabil Eng. 2000;8(2):216–9.
5. Guger C, Edlinger G, Harkam W, Niedermayer I, Pfurtscheller G. How many people are able to operate an EEG-based brain-computer interface (BCI)? IEEE Trans Neural Syst Rehabil Eng. 2003;11(2):145–7.
6. Vidaurre C, Blankertz B. Towards a cure for BCI illiteracy. Brain Topogr. 2010;23(2):194–8.
7. Guger C, Daban S, Sellers E, Holzner C, Krausz G, Carabalona R, Gramatica F, Edlinger G. How many people are able to control a P300-based brain-computer interface (BCI)? Neurosci Lett. 2009;462(1):94–8.
8. Farwell LA, Donchin E. Talking off the top of your head: toward a mental prosthesis utilizing event-related brain potentials. Electroencephalogr Clin Neurophysiol. 1988;70(6):510–23.
9. Brunner P, Joshi S, Briskin S, Wolpaw JR, Bischof H, Schalk G. Does the 'P300' speller depend on eye gaze? J Neural Eng. 2010;7(5):056013.
10. Frenzel S, Neubert E, Bandt C. Two communication lines in a 3×3 matrix speller. J Neural Eng. 2011;8(3):036021.
11. Sahbani A, El-Khoury S, Bidaud P. An overview of 3D object grasp synthesis algorithms. Robot Auton Syst. 2012;60(3):326–36.

12. Quandt F, Reichert C, Hinrichs H, Heinze HJ, Knight RT, Rieger JW. Single trial discrimination of individual finger movements on one hand: a combined MEG and EEG study. Neuroimage. 2012;59(4):3316–24.

13. Rieger JW, Reichert C, Gegenfurtner KR, Noesselt T, Braun C, Heinze H-J, Kruse R, Hinrichs H. Predicting the recognition of natural scenes from single trial MEG recordings of brain activity. Neuroimage. 2008;42(3):1056–68.

14. Wolpaw JR, Birbaumer N, Heetderks WJ, McFarland DJ, Peckham PH, Schalk G, Donchin E, Quatrano LA, Robinson CJ, Vaughan TM. Brain-computer interface technology: a review of the first international meeting. IEEE Trans Rehabil Eng. 2000;8(2):164–73.

15. Reichert C, Kennel M, Kruse R, Hinrichs H, Rieger JW. Efficiency of SSVEF recognition from the magnetoencephalogram - a comparison of spectral feature classification and CCA-based prediction. In: NEUROTECHNIX 2013 - Proceedings of the International Congress on Neurotechnology, Electronics and Informatics, pp. 233–237. Vilamoura: SciTePress (2013)

16. Treder MS, Blankertz B. (C)overt attention and visual speller design in an ERP-based brain-computer interface. Behav Brain Funct. 2010;6:28.

17. Bianchi L, Sami S, Hillebrand A, Fawcett IP, Quitadamo LR, Seri S. Which physiological components are more suitable for visual ERP based brain-computer interface? A preliminary MEG/EEG study. Brain Topogr. 2010;23(2):180–5.

18. Aloise F, Schettini F, Aricò P, Salinari S, Babiloni F, Cincotti F. A comparison of classification techniques for a gaze-independent P300-based brain-computer interface. J Neural Eng. 2012;9(4):045012.

19. Liu Y, Zhou Z, Hu D. Gaze independent brain–computer speller with covert visual search tasks. Clin Neurophysiol. 2011;122(6):1127–36.

20. Treder MS, Schmidt NM, Blankertz B. Gaze-independent brain-computer interfaces based on covert attention and feature attention. J Neural Eng. 2011;8(6):066003.

21. Curran EA, Stokes MJ. Learning to control brain activity: a review of the production and control of EEG components for driving brain-computer interface (BCI) systems. Brain Cogn. 2003;51(3):326–36.

22. Hoffmann U, Vesin J-M, Ebrahimi T, Diserens K. An efficient P300-based brain-computer interface for disabled subjects. J Neurosci Methods. 2008;167(1):115–25.

23. Wolpaw JR, McFarland DJ. Control of a two-dimensional movement signal by a noninvasive brain–computer interface in humans. Proc Natl Acad Sci U S A. 2004;101(51):17849–54.

24. Cherkassky V, Mulier FM. Learning from data: concepts, theory, and methods. New York, NY: John Wiley & Sons; 1998.

25. Krusienski DJ, Sellers EW, Cabestaing F, Bayoudh S, McFarland DJ, Vaughan TM, Wolpaw JR. A comparison of classification techniques for the P300 Speller. J Neural Eng. 2006;3(4):299–305.

26. Bradshaw LA, Wijesinghe RS, Wikswo Jr JP. Spatial filter approach for comparison of the forward and inverse problems of electroencephalography and magnetoencephalography. Ann Biomed Eng. 2001;29(3):214–26.

27. Khatib O. Real-time obstacle avoidance for manipulators and mobile robots. Int J Robot Res. 1986;5(1):90–8.

28. Siciliano B, Villani L. Robot force control. Norwell, MA: Kluwer Academic Publishers; 1999.

29. Ericson C. Real-time collision detection. San Francisco, CA: Elsevier, Morgan Kaufmann Publishers; 2005.

30. Prattichizzo D, Trinkle JC. Grasping. In: Siciliano B, Khatib O, editors. Springer handbook of robotics. Heidelberg: Springer; 2008. p. 671–700.

The Merging of Humans and Machines

Kevin Warwick

Abstract In this article a look is taken at some ways in which humans and machines are merging together to become one entity. A look is taken at culturing biological neurons and embodying them within a robot body, the use of implants to link a human nervous system with the internet and recent results from Turing's Imitation Game which concentrates on differences in human communication. In each case the background is described, practical results are discussed and implications and future directions are considered.

Keywords Cyborgs • Implant technology • Bio-tech hybrids • Human enhancement • Turing test

1 Introduction

As technology improves and human dependence on that technology increases so humans and machines are rapidly merging into one. The focus of attention has hence been placed on interfaces between technology and the human brain. This is done from a practical perspective with applications in mind, however some of the implications are also considered. Results from experiments are considered here in terms of their meaning and application possibilities. The article is written from the perspective of scientific experimentation opening up realistic possibilities to be faced in the future rather than providing conclusive comments. Human implantation and the merger of biology and technology are important elements, in the next two sections at least.

In this article different experiments in linking biology and technology together in a cybernetic fashion, essentially ultimately combining humans and machines in a relatively permanent merger, are considered. However a look is also taken, by means of the Turing test, at conversational abilities and how easy or difficult it is to tell the

K. Warwick (✉)
School of Systems Engineering, University of Reading, Reading, UK
e-mail: k.warwick@reading.ac.uk

© Springer International Publishing Switzerland 2015   79
A.R. Londral et al. (eds.), *Neurotechnology, Electronics, and Informatics*,
Springer Series in Computational Neuroscience 13, DOI 10.1007/978-3-319-15997-3_6

difference between humans and machines. Each of the sections involves practical experiments as something that have been actually realised, i.e. we are looking here at actual real world experiments as opposed to mere philosophical speculations.

Each different experiment is described in its own section. Whilst there is distinct overlap between the sections, they all throw up individual considerations. Following a description of each investigation some pertinent issues on the topic are therefore discussed.

2 Biological Brains in a Robot Body

When one thinks of linking a brain with technology then it is probably in terms of a brain already functioning within its own body. Here however we consider the possibility of a fresh merger where a brain, consisting of biological neurons, is grown and then given its own body in which to operate.

An experimental control platform, a robot body, can move around in a defined area purely under the control of such a network and the effects of the brain, controlling the body, can be witnessed. Investigations can therefore be performed into memory formation, habituation and reward/punishment scenarios—elements that underpin the functioning and growth mechanisms of a brain.

Growing (culturing) networks of brain cells (up to 150,000) in vitro begins by using enzymes to separate neurons obtained from foetal rodent cortical tissue. They are then grown in a specialised chamber, in which they can be provided with controlled environmental conditions (e.g. appropriate temperature) and nutrients [1, 2]. An array of electrodes embedded in the base of the chamber (a Multi Electrode Array; MEA—see Fig. 1) acts as a bi-directional electrical interface with which to provide signals to the culture and to monitor signals from the culture. This allows for electrical signals to be supplied both for input stimulation and also for recordings to be taken as outputs from the culture. The neurons in such cultures spontaneously connect, communicate and develop, within a few weeks.

With the MEA it is possible to separate the firings of small groups of neurons by monitoring the output signals on the electrodes. Thereby a picture of the global activity of the brain network can be formed. It is also possible to electrically stimulate the culture via any of the electrodes to induce neural activity. The multi-electrode array therefore forms a bi-directional interface with the cultured neurons [3, 4].

The cultured brain is, typically after 7–10 days, coupled to its physical robot body [5]. Sensory data fed back from the robot is delivered to the culture, thereby closing the robot-culture loop. The processing of signals can be broken down into two discrete sections (a) 'culture to robot', in which live neuronal activity is used as the decision making mechanism for robot control, and (b) 'robot to culture', which involves an input mapping process, from robot sensor to stimulate the culture.

The actual number of neurons in a brain depends on natural density variations in seeding the culture in the first place. The electrochemical activity of the culture is

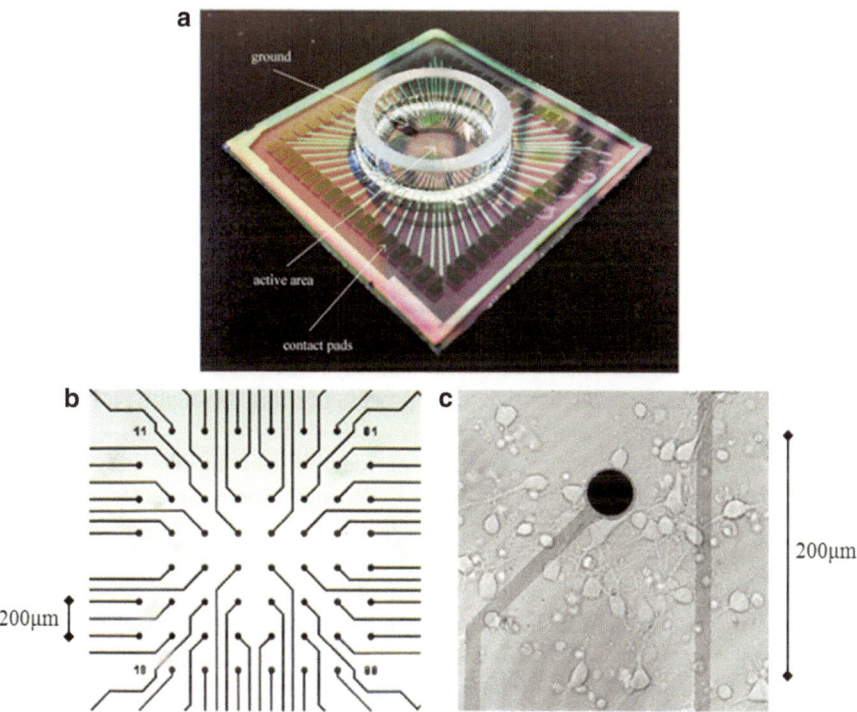

Fig. 1 (**a**) A Multi Electrode Array (MEA) showing the electrodes. (**b**) Electrodes in the centre of the MEA seen under an optical microscope (**c**) An MEA at × 40 magnification, showing neuronal cells in close proximity to an electrode

sampled and is used as input to the robot's wheels. The robot's (ultrasonic) sensor readings are converted into stimulation signals received by the culture, closing the feedback loop.

Once the brain has grown for several days, an existing neuronal pathway through the culture is identified by searching for strong relationships between (input–output) pairs of electrodes. A rough input–output response map of the culture can be created by cycling through the electrodes in turn. In this way, a suitable input/output electrode pair can be chosen in order to provide an initial decision making pathway for the robot. This is then employed to control the robot body—for example if the ultrasonic sensor is active and we wish the response to cause the robot to turn away from the object being located ultrasonically (possibly a wall) in order to keep moving.

For experimentation purposes at this time, the robot is required to follow a forward path until it nears a wall, at which point the front sonar value decreases below a threshold, triggering a stimulating pulse. If the responding/output electrode registers activity within a few milliseconds then the robot turns to avoid the wall. The most relevant result is the occurrence of the chain of events: wall detection–stimulation–response. However from a neurological perspective it is also interesting to speculate why there is activity on the response electrode when no stimulating pulse has been applied.

The cultured brain acts as the sole decision making entity within the overall feedback loop, any computers involved are merely employed for networking arrangements. Clearly one important aspect involves neural pathway changes, with respect to time, in the culture between the stimulating and recording electrodes. Learning and memory investigations are generally at an early stage. However the robot can be witnessed to improve its performance over time in terms of its wall avoidance ability in the sense that neural pathways that bring about a satisfactory action tend to strengthen purely though the process of being habitually performed—learning due to habit.

The number of variables involved is considerable and the plasticity process, which occurs over quite a period of time, is dependent on such factors as initial seeding and growth near electrodes as well as environmental transients such as temperature and humidity. Learning by reinforcement—rewarding good actions and punishing bad is merely investigative research at this time.

We have witnessed through this research that a robot can successfully have a biological brain with which to make its 'decisions'. The size of the culture is merely due to the present day limitations of the experimentation described. Indeed 3 dimensional structures are presently being investigated. Increasing the complexity from 2 dimensions to 3 dimensions realises a figure of over 30 million neurons for the 3 dimensional case—not yet reaching the 100 billion neurons of a perfect human brain, but well in tune with the brain size of many other animals.

Not only is the number of cultured neurons increasing, but the range of sensory input is being expanded to include audio and visual. Such richness of stimulation will no doubt have a dramatic effect on culture development. The potential of such systems, including the range of tasks they can deal with, also means that the physical body can take on different forms. There is no reason, for example, that the body could not be a two legged walking robot, with rotating head and the ability to walk around.

At present rat neurons are usually employed in studies. However human neurons are also now being cultured, allowing for the possibility of a robot with a human neuron brain. If this brain then consists of billions of neurons, many social and ethical questions will need to be asked [6]. For example—If the robot brain has roughly the same number of human neurons as a typical human brain then could/should it have similar rights to humans? Also—What if such creatures had far more human neurons than in a typical human brain—e.g. a million times more—would they make all future decisions rather than regular humans?

3 Braingate Implant

It is more often the case that brain-computer interfaces are used for therapeutic purposes, to overcome a medical/neurological problem such as Parkinson's Disease [21, 22] or Epilepsy. However there is also the possibility of employing such technology to give individuals abilities not normally possessed by humans. Human Enhancement!

Some of the most impressive human enhancement research to date in this area has been carried out using the microelectrode array, shown in Fig. 2. The individual electrodes are 1.5 mm long and taper to a tip diameter of a few micrometers. Although a number of trials not using humans as a test subject have occurred, human tests are at present limited to a small group of studies. In some of these the array has been employed in a recording only role, most notably as part of (what was then called) the 'Braingate' system.

Electrical activity from a few neurons monitored by the array electrodes was decoded into a signal to direct cursor movement. This enabled an individual to position a cursor on a computer screen, using neural signals for control combined with visual feedback. The same technique was later employed to allow the individual recipient, who was paralysed, to operate a robot arm [7, 8].

The first use of the Braingate microelectrode array (shown in Fig. 2) in a human has though considerably broader implications which extend the capabilities of the human recipient. The array was implanted into the median nerve fibers of a healthy human individual (the author) during two hours of neurosurgery in order to test bidirectional functionality in a series of experiments. A stimulation current directly into the nervous system allowed information to be received, while control signals were decoded from neural activity in the region of the electrodes [9, 10]. A number of experimental trials were successfully concluded [11, 12]: In particular:

1. Extra sensory (ultrasonic) input was successfully implemented.
2. Extended control of a robotic hand across the internet was achieved, with feedback from the robotic fingertips being sent back as neural stimulation to give a sense of force being applied to an object (this was achieved between Columbia University, New York (USA) and Reading University, England).

Fig. 2 A 100 electrode, 4 × 4 mm Microelectrode Array, shown on a UK 1 pence piece for scale

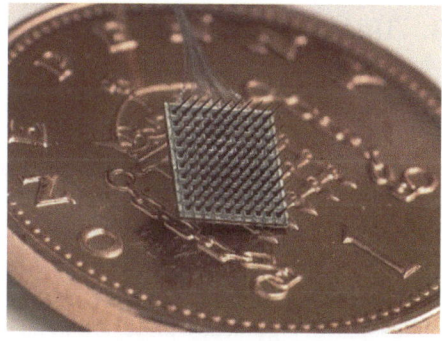

3. A primitive form of telegraphic communication directly between the nervous systems of two humans (the author's wife assisted) was performed [12].
4. A wheelchair was successfully driven around by means of neural signals.
5. The color of jewelry was changed as a result of neural signals—also the behavior of a collection of small robots.

In all of the above cases it can be regarded that the trial proved useful for purely therapeutic reasons, e.g. the ultrasonic sense could be useful for an individual who is blind or the telegraphic communication could be useful for those with certain forms of Motor Neurone Disease. However each trial can also be seen as a potential form of enhancement beyond the human norm for an individual. The author did not need to have the implant for medical purposes to overcome a problem but rather for scientific exploration.

It is clear, from these experiments, that extra sensory input is one practical possibility that has been successfully attempted, however improving memory, thinking in many more dimensions and communication by thought alone are other distinct potential, yet realistic, benefits, with the latter of these also having been investigated to an extent. To be precise—all these things appear to be possible (from a technical viewpoint at least) for humans in general.

An individual human connected in this way can potentially also benefit from some of the advantages of machine/artificial intelligence, for example rapid and highly accurate mathematical abilities in terms of 'number crunching', a high speed, almost infinite, internet knowledge base, and accurate long term memory. Humans are also limited in that presently they can only visualise and understand the world around them in terms of a limited 3 dimensional perception, whereas computers are quite capable of dealing with hundreds of dimensions.

The human means of communication, essentially transferring a complex electro-chemical signal from one brain to another via an intermediate, often mechanical slow and error prone medium (e.g. speech), is extremely poor in comparison with technological communication, particularly in terms of speed, power and precision. It is clear that connecting a human brain, by means of an implant, with a computer network could in the long term open up the distinct advantages of machine intelligence, communication and sensing abilities to the implanted individual.

4 Turing's Imitation Game

The final area to be looked at here is that of practical Turing tests which give an indication of how easy or difficult it is to distinguish between humans and machines in terms of conversational ability. This article is focused on what has become commonly known as the Turing test, although it was originally described by Turing as his Imitation Game [13]. It is worth remembering that Turing originally proposed the test as a replacement for the question "Can Machines Think?" [13], however here we are more concerned with the practical nature of the test rather than in any philosophical argument with regard to its meaning.

The test was described by Turing himself in 1952 as: "The idea of the test is that a machine has to try and pretend to be a man, by answering questions put to it, and it will only pass if the pretence is reasonably convincing. A considerable portion of a jury, who should not be expert about machines, must be taken in by the pretence" [14, p. 495].

The Turing test involves a machine which pretends to be a human in terms of conversational abilities. In a paired comparison the attempt is for the machine to appear to be more human than the human against whom it is paired. To conform to Turing's original wording in his 1950 paper [13] we refer here to 5 min long tests only, although we are well aware that there are those who take issue over a suitable timing and what Turing actually meant [15]—that is considered to be an argument for another day, it does not alter the point made here.

What is presented here are four specific transcripts selected from a day of actual, practical Turing tests which were held under strictly timed conditions with many external viewers at Bletchley Park, England on 23rd June 2012. The date marked the 100th anniversary of Turing's birth and the venue was that at which amongst other things, during the Second World War, Turing led a team of codebreakers who cracked the German Enigma machine cypher. Five different machines took part in the tests during the day along with 30 different judges and numerous hidden humans against which the machines were compared in terms of their conversational ability.

What we focus on here is not how good or bad the machines were at deception or how human the hidden humans were but rather the decisions taken by the judges and how these might compare with your own selections. So this article is more a look at the differences between the hidden entities and how easy or difficult it can be to tell which is human and which is machine.

What follows are three separate transcripts. These represent actual transcripts taken on the morning of 23rd June 2012 at Bletchley Park, England between different human judges/interrogators and hidden entities. Each conversation lasted for a total of 5 min exactly and no more, just as Turing stipulated [13]. There was a hard cut off at that time and no partial sentences were transmitted. Once a sentence had been transmitted it could not be altered or retracted in any way. Hence all wording and spelling is exactly as it was at the time—any spelling mistakes are those which actually occurred in the test, they are not due to poor editing.

Judges were asked to decide if the entity was human, machine or if they were unsure. The hidden humans meanwhile were asked to be themselves—human— they were asked not to make it easy for the machines because it was the machines which were competing against them for humanness [16]. They were however asked not to reveal personal details about themselves that could in any way give away their actual identity to a judge. It is probably a good idea for the reader to firstly consider each transcript themselves and to come to a decision with regard to the nature of the hidden entity.

What we show below are four selected transcripts on their own, i.e. not in parallel paired fashion [17]. The reason for this is that the aim here is not to make a direct comparison between these transcripts and any others, but rather to analyse these transcripts in their own right.

At the end of this section the judges conclusions on each transcript are given. These were the conclusions actually drawn by those judges. They do not reflect any personal conclusions on each of the transcripts. Following this the true identity of each entity is given. In each case the transcript involved a different judge. But the reader will then be able to check their own decision to see if they were correct or not on each occasion and if they agreed with the judge's decision. Each judge had no extra information to go on in making their decision.

Transcript 1 (Taken from [18])

- Judge: My favourite music is contemporary Jazz, what do youy prefer
- Entity: To be short I'll only say that I HATE Britney Spears. All other music is OK compared to her.
- Judge: do you like to play any musical instruments
- Entity: I'm tone-deaf, but my guinea pig likes to squeal Beethoven's "Ode to Joy" every morning. I suspect our neighbors want to cut his throat … Could you tell me about your job, by the way?
- Judge: Guine pig/are you an animal lover
- Entity: Yeah. A nice little guinea pig. Not some annoying chatter bot.
- Judge: do you live with anyone else?
- Entity: Yes I do. But better ask something else. And I forgot to ask you where you are from …
- Judge: Ireland? geen patures and too much rain! How about you

Transcript 2 (Taken from [19])

- Judge: hey, just to lighten the mood a little, what's brown and sticky?
- Entity: no idea, what's the answer?
- Judge: a stick … so anyway what's your name?
- Entity: Sheldon Cooper, what's yours?
- Judge: as in the guy off the big bang theory?
- Entity: I don't believe in the big bang
- Judge: so are you religious then?
- Entity: aren't we all?
- Judge: well i like to consider myself an agnostic, which religion is it you follow then if you dont mind me asking?
- Entity: you can ask what you like, it's a free world, who am I to mind?

Transcript 3

- Judge: Hello. How are you?
- Entity: Hey.
- Judge: How do you like Bletchley Park?
- Entity: lol.
- Judge: Are you from England?
- Entity: They have Wi-Fi here in the pub.
- Judge: Which pub?
- Entity: I'm just down the pub.
- Judge: Have you ever been in a Turing Test before?
- Entity: Will this be the 5 min argument, or were you thinking of going for the full half hour.
- Judge: Very funny. You sound suspiciously human. Do you like the Beatles?
- Entity: I'd like to get the next Dread the Fear tape.
- Judge: What is Dread the Fear?
- Entity: Dread the fear has that Steve Henderson guy in it.
- Judge: What sort of music is that? Or is it comedy?

Transcript 4 (Taken from [20])

- Judge: Why hello there!
- Entity: Why hello to you too!
- Judge: How are you feeling on this fine day?
- Entity: To be quite honest a little rejected, I thought you were never going to reply :(
- Judge: Oh, I'm very sorry, it will not happen again.
- Entity: It just did!
- Judge: Oh, I lied then.
- Entity: Thats a great shame indeed.
- Judge: It is. Are you following the Euro 2012's at the moment?
- Entity: Yeah quite closely actually. I am Cristiano Ronaldo.

Solutions

In transcript 1 the judge was a male medical science professor and clinician. Here he decided that the hidden entity was a human. In fact the entity was machine. In transcript 2, with a different judge, the decision was that the entity was a machine whereas in reality it was a human—the author of this article in fact. In transcript 3 meanwhile the judge's decision was that the hidden entity was a male human whereas it was in fact a machine. In transcript 4 the judge correctly identified the hidden entity as being human, despite the entity's lie about them being Cristiano Ronaldo. So how did you do?

5   Conclusions

In this article a look has been taken at different ways in which the human-machine divide is diminishing. It has to be said that these are all perhaps a curious selection as examples and many other instances exist, particularly so in everyday life. Rather than focus merely on theory, here practical experimental cases have been reported on. Further details for each of these can be found in a variety of publications e.g. [23, 24]. In each case questions arise as a result.

When considering robots with biological brains, this could ultimately mean human brains operating in a robot body. Therefore, should such a robot be given rights of some kind? If one was switched off would this be deemed as cruelty to robots? More importantly at this time—should such research forge ahead regardless? Before too long we may well have robots with brains made up of human neurons that have the same sort of capabilities as those of the human brain—is this acceptable?

In the section focussing on the Braingate implant as a general purpose invasive brain implant, as well as its employment for therapy a look was taken at the potential for human enhancement. Already extra-sensory input has been scientifically achieved, extending the nervous system over the internet and a basic form of thought communication. So if many humans upgrade and become part machine (Cyborgs) themselves, what would be wrong with that? If ordinary (non-implanted) humans are left behind as a result then what is the problem? If you could be enhanced, would you have any problem with it?

In the final section some of the latest results from the Turing test were presented. In three of the four cases mentioned the judge drew an incorrect conclusion. Even if you did manage to give all four correct answers hopefully you are able to agree, with these transcripts as examples, that machine conversation is now getting to the stage where it is difficult for an external observer to decide which is human and which is machine. It has to be said though that this is just as much down to the fallibility of humans as it is to the present-day wonders of machine communication.

References

1. Chiappalone M, Vato A, Berdondini L, Koudelka-Hep M, Martinoia S. Network dynamics and synchronous activity in cultured cortical neurons. Int J Neural Syst. 2007;17:87–103.
2. DeMarse T, Wagenaar D, Blau A, Potter S. The neurally controlled animat: biological brains acting with simulated bodies. Auton Robots. 2001;11:305–10.
3. Warwick K, Nasuto S, Becerra V, Whalley B. Experiments with an in-vitro robot brain. In: Cai Y, editor. Computing with instinct, Lecture notes in artificial intelligence, vol. 5987. Berlin: Springer; 2010. p. 1–15.
4. Warwick K, Xydas D, Nasuto S, Becerra V, Hammond M, Downes J, Marshall S, Whalley B. Controlling a mobile robot with a biological brain. Defence Sci J. 2010;60(1):5–24.
5. Xydas D, Norcott D, Warwick K, Whalley B, Nasuto S, Becerra V, Hammond M, Downes J, Marshall S. Architecture for neuronal cell control of a mobile robot. In: Proc. European Robotics Symposium 2008. Springer: Prague; 2008. pp. 23–31.
6. Warwick K. Implications and consequences of robots with biological brains. Ethics Inf Technol. 2010;12(3):223–34.
7. Donoghue J, Nurmikko A, Friehs G, Black M. Development of a neuromotor prosthesis for humans, Chapter 63 in advances in clinical neurophysiology. Suppl Clin Neurophysiol. 2004;57:508–602.
8. Hochberg LR, Serruya MD, Friehs GM, Mukand JA, Saleh M, Caplan AH, Branner A, Chen D, Penn RD, Donoghue JP. Neuronal ensemble control of prosthetic devices by a human with tetraplegia. Nature. 2006;442:164–71.
9. Warwick K, Gasson M, Hutt B, Goodhew I, Kyberd P, Andrews B, Teddy P, Shad A. The application of implant technology for cybernetic systems. Arch Neurol. 2003;60(10):1369–73.
10. Gasson M, Hutt B, Goodhew I, Kyberd P, Warwick K. Invasive neural prosthesis for neural signal detection and nerve stimulation. Int J Adapt Control Signal Process. 2005;19(5): 365–75.
11. Warwick K, Gasson M. Practical interface experiments with implant technology. In: Sebe N, Lew M, Huang T, editors. Computer vision in human-computer interaction, Lecture notes in computer science, vol. 3058. Heidelberg: Springer; 2004. p. 7–16.
12. Warwick K, Gasson M, Hutt B, Goodhew I, Kyberd P, Schulzrinne H, Wu X. Thought communication and control: a first step using radiotelegraphy. IEE Proc Comm. 2004;151(3):185–9.
13. Turing A. Computing machinery and intelligence. Mind LIX. 1950;236:433–60.
14. Copeland B. The essential Turing—the ideas that gave birth to the computer age. Oxford: Clarendon; 2004.
15. Shah H, Warwick K. Testing Turing's five minutes, parallel-paired imitation game. Kybernetes. 2010;39(3):449–65.
16. Warwick K. Not another look at the Turing test! In: Bielikova M, Friedrich G, Gottlob G, Katzenbisser S, Turan G, editors. SOFSEM 2012: theory and practice of computer science, Lecture notes in computer science, vol. 7147. Berlin: Springer; 2012. p. 130–40.

17. Shah H, Warwick K. Hidden interlocutor misidentification in practical Turing tests. Minds and Machines. 2010;20(3):441–54.
18. Warwick K, Shah H. Good machine performance in Turing's imitation game'. IEEE Trans Comput Intell AI Games. 2013;6:289–99. doi:10.1109/TCIAIG.2013.2283538.
19. Warwick K, Shah H, Moor J. Some implications of a sample of practical Turing tests. Minds and Machines. 2013;23(2):163–77.
20. Warwick K, Shah H. Effects of lying in practical Turing tests. AI & Society (2014), doi:10.1007/s00146-013-0534-3
21. Pan S, Iplicki S, Warwick K, Aziz T. Parkinson's disease tremor classification—a comparison between support vector machines and neural networks. Exp Syst Appl. 2012;39(12): 10764–71.
22. Wu D, Warwick K, Ma Z, Gasson M, Burgess J, Pan S, Aziz T. Prediction of Parkinson's disease tremor onset using a radial basis function neural network based on particle swarm optimization. Int J Neural Syst. 2010;20(2):109–16.
23. Warwick K, Ruiz V. On linking human and machine brains. Neurocomputing. 2006;71(13): 2619–24.
24. Warwick K. Cybernetic organisms: our future. Proc IEEE. 1999;87(2):387–99.

Learning from the Past: Postprocessing of Classification Scores to Find a More Accurate and Earlier Movement Prediction

Sirko Straube, David Feess, and Anett Seeland

Abstract Brain–computer interfaces performing movement prediction are useful in a variety of application fields from telemanipulation to rehabilitation. However, current systems still struggle with a level of unreliability requiring improvement, so that the full potential of these systems can be used in the future. Here, we suggest to improve the performance and robustness of classification outcomes by postprocessing the raw score values with the history of previous classifications. For this several postprocessing methods that operate on the classification outcomes are investigated. In particular, the data was classified after preprocessing using a support vector machine (SVM). The output of the SVM, i.e. the raw score values, were postprocessed using previously obtained scores to account for trends in the classification result. The respective methods differ in the way the transformation is performed. The idea is to use trends, like the rise of the score values approaching an upcoming movement, to yield a better prediction in terms of detection accuracy and/or an earlier time point. We present results from different subjects where upcoming voluntary movements of the right arm were predicted using movement related cortical potentials from the EEG. The results illustrate that better and earlier predictions are indeed possible with the suggested methods. However, the best postprocessing method was rather subject-specific. Finally, we use straightforward

S. Straube (✉)
Robotics Group, Faculty of Mathematics and Computer Science, University of Bremen,
Robert-Hooke-Str. 1, 28359 Bremen, Germany

Robotics Innovation Center, German Research Center for Artificial Intelligence (DFKI GmbH),
Robert-Hooke-Str. 1, 28359 Bremen, Germany
e-mail: sirko.straube@dfki.de

D. Feess
Robotics Innovation Center, German Research Center for Artificial Intelligence (DFKI GmbH),
Robert-Hooke-Str. 1, 28359 Bremen, Germany

Chair of Global Business, Faculty of Business Administration and Economics, University of
Augsburg, Universitaetsstr. 16, 86159 Augsburg, Germany
e-mail: david.feess@googlemail.com

A. Seeland
Robotics Innovation Center, German Research Center for Artificial Intelligence (DFKI GmbH),
Robert-Hooke-Str. 1, 28359 Bremen, Germany
e-mail: anett.seeland@dfki.de

© Springer International Publishing Switzerland 2015   91
A.R. Londral et al. (eds.), *Neurotechnology, Electronics, and Informatics*,
Springer Series in Computational Neuroscience 13, DOI 10.1007/978-3-319-15997-3_7

ensemble approaches to exemplify how the methods can be directly used in an application and how this can influence the overall movement prediction performance. Depending on the requirements of the application at hand, postprocessing the classification scores as suggested here can be used to find the best compromise between prediction accuracy and time point.

Keywords EEG • LRP • Brain–computer interface • Classification score • Movement prediction • Online prediction • Ensemble

1 Introduction

Movement prediction using the electroencephalogram (EEG) has a long standing history in the field of brain–computer interfaces (BCIs) since the discovery of readiness potentials [1, 2], which build up long before the actual movement can occur. Since readiness potentials reflect preparatory activity and movement preparation can be aborted, these potentials can also disappear after a short build-up without any movement occurring. However, the closer such a recorded potential gets to the actual movement, the stronger it is and the less likely will a prepared movement be cancelled [3, for a summary]. When the movement is finally executed a corresponding motor potential can be recorded that reflects signalling to the muscles. For movement prediction, different signals have been applied, from the readiness potential itself over the lateralized readiness potential (LRP) which is closer to the movement and cannot easily be aborted [4], to specific frequency components in the EEG reflecting neural synchronization or desynchronization [5].

Movement prediction can be used as a powerful tool in various fields, with the most prominent being assistance during rehabilitation. Here BCIs predicting a movement can be used to close the gap between a patient's intention to move and the actual movement which can result in more intuitive responses of orthoses [6–8]. Other fields include non-medical applications, e.g., during telemanipulation of a robotic device the user can be supported using a movement prediction based on EEG data [9–11]. The idea is that the human operator experiences a smoother interaction with the telemanipulation device, which *knows* about an upcoming movement. As in the present study, the movement prediction is often based on the LRP.

Decisions in a movement predicting BCI come from some kind of classifier which has to make the prediction. However, the output of the classifier is again noisy, so recent approaches try to apply a postprocessing to minimize classification errors [12–15]. Here, we follow this rationale by applying simple online-capable functions to modify the classifier output according to knowledge about its progression. The scenario is the following: After processing, the classifier which is a support vector machine (SVM) assigns a value, the classification score, to each data instance [16]. The range of these score values depends on the data at hand and on the classifier and can largely fluctuate as can be seen in Fig. 1. A score of zero denotes the borderline between the two classes. The figure illustrates the high fluctuations in single trials and the consistent trend in the data: When the median score is considered, the score is constantly staying below zero, i.e., *no movement* is classified, until the scores rise

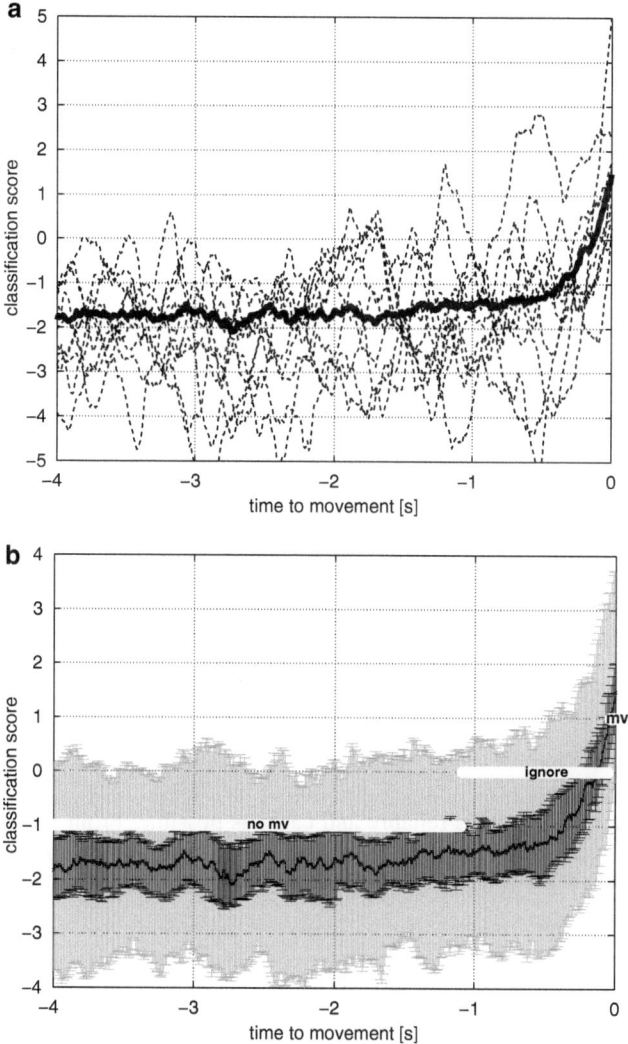

Fig. 1 Example data from single subject prior to a movement. Depicted are 4 s of data with the movement onset at the very right (0 ms). (**a**) The *bold black line* shows the median of all recorded epochs. *Dashed lines* are ten exemplary single trials. (**b**) Data of the same subject as (**a**) illustrated as 32/68 percentiles (*black*) and 5/95 percentiles (*dark grey*). The *white line* denotes time ranges where the data is labelled differently for evaluation: *no movement (no mv)* from −4,000 ms to −1,050 ms and *movement (mv)* from −50 ms to 0 ms. In between, data is ignored for true labels (see text)

approximately 500 ms before the movement and cross zero approximately at 250 ms before onset. The rise in classification scores before movement onset can consistently be observed across subjects. This means that the rise in score values alone may signal an upcoming movement so that the progression of the score values itself can be interpreted as being loosely correlated to the changes in movement probability.

The question now is whether we can use the knowledge about this rise in classification scores to make the prediction more stable and/or predict the upcoming movement earlier. In trying to answer this question we were seeking for a postprocessing method that dampens fast fluctuations in classification scores and stabilizes long rises. To this aim, we applied several methods that modify the current classification scores by taking into account previous scores with a certain weight (see Sect. 2). To demonstrate the applicability and the benefit of these methods, we use all investigated methods in two simple ensemble approaches directly integrated in the signal processing chain (Sect. 6).

To summarize, if an LRP can be detected by high levels of the classification score, it could potentially just as well be predicted earlier by detecting the rise that leads to that elevated level. In the following we will describe the postprocessing methods that we have applied. After a description of the experimental data used, the results will be presented and discussed.

2 Postprocessing Methods

From the perspective of a movement prediction application it is most desirable to perform robust, binary decisions: A movement will either occur or it will not. This decision should be made as reliably and early as possible. From the large margin classification perspective, this means that the classification score S_t at some point in time t would have to be compared against some threshold b so that a movement mv is predicted when

$$mv \text{ iff } \quad S_t \geq b. \tag{1}$$

Yet, as illustrated in Fig. 1, the score sometimes suddenly crosses the threshold when the actual movement is still far away, but then only for a short time. This behaviour hinders reliable prediction when it is purely based on the raw value of S_t crossing b. Looking at the average score progression over time reveals a continuous rise of the score values before the actual movement. Here, we exploit this systematic behaviour to find a function F that is able to generate better movement predictions based on past values of S, such that

$$mv \text{ iff } \quad F(S_t, S_{t-1}, \ldots, S_{t-(k-1)}) \geq b_F \tag{2}$$

for some specific threshold b_F. k is defined as the number of scores that are used in F with the current score being at $k = 1$. In principle, there are no constraints on the functional form of F.

In the present study we apply weights to the current and previous $k - 1$ classification outcomes to transform the current score S_t. These weights decay with the number of steps looked into the past. We also followed an alternative approach by transforming the current score with the average slope of the past samples. A detailed description is given in Sect. 2.2. Both types of functions (weighting and slope approach) can be expressed as

$$F(S_t, S_{t-1}, \ldots, S_{t-(k-1)}) = w_1 S_t + w_2 S_{t-1} + \cdots + w_k S_{t-(k-1)} \qquad (3)$$

with some predefined weights w. With this methodology we try to boost the score value when previous scores were similar in value and at the same time penalize scores when previous ones showed a completely different trend. The approaches are described in more detail in the following.

2.1 Fixed Weighting

In this set of functions the weights are generated by very simple functions, each of which assigns a high weight to the most current classification score sample, and decreasing weights to older samples.

The functions used are depicted in Fig. 2. All functions have in common that the weights add up to one. The coefficients for the `uniform`, `linear`, `square`, and `cubic` method are all generated by evaluating

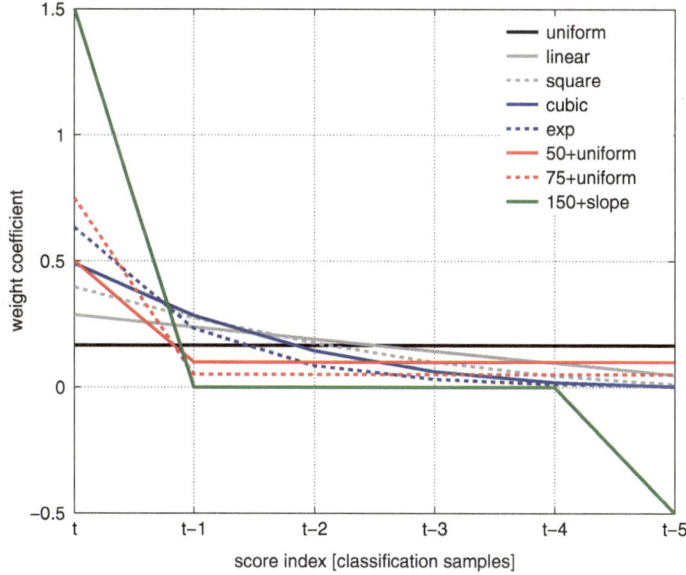

Fig. 2 Comparison of the functions used for classification score postprocessing using $k = 6$ coefficients, i.e., the current score and five instances back in time

$$w_\tau = \frac{\tau^p}{\sum_{i=1}^{k} i^p}, \quad \tau \in \{1, \ldots, k\}, \tag{4}$$

respectively, with the number of k coefficients used and the exponent p according to the corresponding function type. The exp coefficients are accordingly calculated as

$$w_\tau = \frac{\exp \tau}{\sum_{i=1}^{k} \exp i}, \quad \tau \in \{1, \ldots, k\}. \tag{5}$$

Besides these rather universal functions for choosing the weight we added two variants where we explicitly forced the current value to have a much higher weight than the scores corresponding to the previous instances, since the idea behind the postprocessing was exactly this: to transform the current score with its history to weaken fast fluctuations and strengthen longer trends. Again, the weights were set so that they add up to one. In the X+uniform method, the first coefficient gets assigned a weight of $X\%$. The remaining weight of $[1 - (X/100)]$ is then equally distributed across the remaining coefficients.

2.2 Slope Approaches

Since the objective is to identify a rise in the classification score progression over time we also looked at modifications of the score value using local slopes or averaged slope over the last k samples (i.e., the current sample and $k - 1$ instances back in time). Considering two samples, a local slope ΔS_t^1 can be computed as

$$\Delta S_t^1 = \frac{S_t - S_{t-1}}{t - (t-1)} = S_t - S_{t-1}. \tag{6}$$

Therefore, the average slope ΔS_t^{k-1} over k samples is

$$\Delta S_t^{k-1} = \frac{1}{(k-1)} \sum_{i=1}^{k-1} (S_{t-i+1} - S_{t-i}). \tag{7}$$

which is a telescope sum and boils down to

$$\Delta S_t^{k-1} \propto (S_t - S_{t-(k-1)}). \tag{8}$$

The corresponding weighting coefficients for this postprocessing are then

$$w_1 = 1, \quad w_k = -1, \quad w_\tau = 0 \; \forall \tau \notin \{1, k\}, \quad \text{or} \quad w = (1, 0, \ldots, 0, -1). \tag{9}$$

In pilot experiments (not shown) this slope method was tested and performance levels were consistently far below the performance obtained without any postprocessing. Due to these performances losses of at least 0.15 points of balanced accuracy (BA, see Sect. 4.1) and in worst cases a performance around the probability of guessing this method was skipped for the current study.

Nevertheless, since we were looking for stabilizing a slope, we chose another promising and simple variant. Instead of using only the slopes, we modulate the current score with the slope approach in a 2:1 fashion (score:slope), so that we obtain a weight vector w of

$$w = (1.5, 0, \ldots, -0.5). \tag{10}$$

In other words, in this approach we take the current score value with 100 % and add the slope weighted with 0.5. This variant is called 150+slope.

3 Data and Preprocessing

The data used for evaluation has been described in detail previously [8, 17]. Originally, muscle activity has been recorded simultaneously with the EEG. Here, evaluation has been restricted to EEG data. For processing the data, the software pySPACE has been used [18].

3.1 Experimental Data

Eight right-handed male subjects (age: 29.9 ± 3.3 years) participated in the study. They gave written consent to participate and could abort the experiment at any time. The study was conducted in accordance with the Declaration of Helsinki. The subjects were sitting in a comfortable chair in front of a table with a monitor showing a fixation cross and giving occasional feedback. They executed self-paced, intentional movements with their right arm by releasing a button and pressing another one situated 30 cm to the right. A resting period of 5 s between movements had to be performed for a movement to be counted as valid. Subjects were not informed about this time constraint, instead negative feedback was provided (a red circle around the fixation cross) when they performed a movement too quickly after another. In each session 120 correctly performed movements were recorded, divided into three runs (40 movements per run).

3.2 Preprocessing

The EEG was acquired with 5 kHz, filtered between 0.1 Hz and 1 kHz using the BrainAmp DC amplifier [Brain Products GmbH, Munich, Germany]. Recordings were performed using a 128-channel (extended 10–20) actiCap system (reference at FCz). Electrodes I1, OI1h, OI2h and I2 were used for electrooculography and thus not placed on the scalp. For detection of the physical movement onset a motion capturing system consisting of three cameras (ProReflex 1000; Qualisys AB, Gothenburg, Sweden) was used at 500 Hz. After synchronization of the two data streams, the movement onsets were marked in the EEG.

Preprocessing was performed on overlapping windows of 1 s length cut every 10 ms in a range from −4,000 ms to 0 ms before a movement. Consequently, a total of 401 score values were computed per executed movement. Data were standardized channel-wise (subtraction of mean and division by standard deviation) and decimated to 20 Hz. Next, a FFT band-pass filter with a pass band of 0.1–4 Hz was applied. Since the prediction should be based on the most recent data, we proceeded with the last 200 ms of each window that were processed by an xDAWN spatial filter [19] with 4 channels retained. For feature extraction, raw voltage values were used, standardized (mean zero, variance one) and classified by a SVM [20] with linear kernel.

For trainable components in the signal processing chain (xDAWN, feature normalization and SVM) windows ending at −100 and 0 ms were labeled as *movement*. Training windows for *no movement* originated from non-overlapping windows (1 s length) that were continuously cut from the data stream, if no movement occurred 1 s before and 2 s after this window. In addition, a parameter optimization for the complexity parameter of the SVM was performed using a grid search (tested values: $10^0, 10^{-1}, \ldots, 10^{-6}$). A threefold cross-validation, onefold corresponding to one experimental run, was applied and classifier scores were stored for both, training and test data.

4 Evaluation

As the aim is *to detect movements more accurately and/or earlier*, there are basically two criteria for a good postprocessing. One is the detection accuracy, the other the time point of detection. Both are considered for evaluation.

4.1 Movement Detection Accuracy

The prediction of unique events comes along with unbalanced proportions of the two classes *no movement* and *movement*, i.e., class instances of data containing the LRP (in our case) will be underrepresented. The evaluation of the movement detection accuracy has to take this into account. Thus, the simple accuracy is misleading [21, 22, for discussion], so a metric is required which is insensitive to imbalanced classes.

One of the most intuitive measures existing in such a case is the *balanced accuracy* (BA) which is defined as the mean of true positive rate (TPR) and true negative rate (TNR):

$$BA = \frac{TPR + TNR}{2}.$$ (11)

One of the challenges here is to define a ground truth of when the relevant signal (i.e., the LRP) is actually present in the data. While we can postulate that there must be an LRP prior to each movement, we still do not know the precise onset of this signal. To cope with this issue and thereby get unambiguously labelled data for evaluation, we split the time before a movement into three phases (compare Fig. 1), a *no movement* phase from $-4,000$ ms to $-1,050$ ms, a *movement* phase from -50 ms to 0 ms, and the phase in between ($-1,050$ ms to -50 ms) where the data is ignored and not labelled at all. With this approach we obtain a clear labeling in phases where we are sure that the relevant signal is indeed contained in the data. This signal is, of course, also present in the ignored time range, but since we do not know the exact onset this range is skipped. In the actual application where no data are skipped, a movement is predicted whenever the classifier score crosses the threshold.

4.2   Time Point of Detection

The onset of the signal related to movement (here, the LRP) occurs at an unknown point in time before the actual movement. This transition out of noise is typical for event-related potentials and it is reflected in the rise of classification scores that we intend to stabilize with the postprocessing approaches introduced here. Concerning the application, i.e., the prediction of an upcoming movement, the exact time point is of less importance than a *reliable and stable* prediction by the classifier. For this, it remains to define when exactly we consider the LRP as *detected*—the classification score might at any time rise over a given threshold for a short period of time due to noise. For the same reason, the score might fall below the threshold for some samples although the LRP has—supposedly—already been correctly detected. To make sure that we base our evaluation on a stable prediction, the *LRP onset* was defined as the point in time where the classification scores do not drop below the threshold for N predictions. This point was found by going back in time from the actual movement onset until the first time where this criterion was not met. With this method, the LRP onset used for evaluation is then defined as the first score sample crossing the threshold after the set of samples staying below threshold for N predictions.

The choice of N depends on the level of noise, on the classification scores, on the sampling rate, and on the characteristics of the signal applied for movement prediction. Here, the relevant signal has a length of approximately 1 s [3], so we chose $N = 10$ as a good compromise between robustness (higher N) and reliability (lower N), i.e., we tolerate false classifications during periods shorter than 100 ms. Increasing the robustness here means to allow an earlier estimation of the time point of detection, because fluctuations in the score progression are more and more ignored with increasing N. On the contrary, a decrease of N increases the reliability,

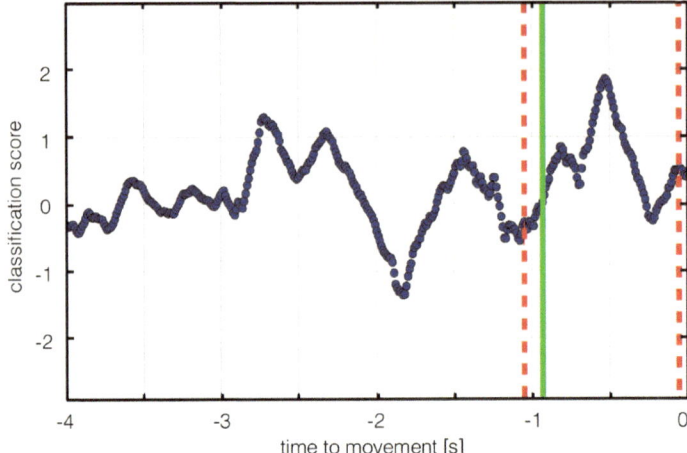

Fig. 3 Single trial example for determination of LRP onset with $N = 10$ specifying the number of samples a false negative classification is tolerated. The *dashed red lines* delimit the feasible (transition) area, the *solid green line* indicates the detected LRP onset. Setting $N = 10$ provokes that the small dip around -200 ms is ignored

because fewer classifications of *no movement* can occur after the estimated time point of detection, but this comes at the cost of a higher sensitivity to outliers. To give an alternative view on the value of N: setting $N = 10$ in our data means that the movement onset is defined as the first score sample crossing the threshold *after* a 100 ms window without any predicted movement (viewed backwards from the actual movement onset). The approach is illustrated in Fig. 3.

4.3 Evaluation Procedure

For each subject and cross-validation fold two data sets exist: one training data set (80 movements) and one test data set (40 movements). The training set is the one used to train the classifier producing the classification scores. Due to the fact that the postprocessing methods introduced here change the absolute value of the classification score, the score thresholds (transition from one class to the other) were re-adjusted for each method, respectively, using the training data, before evaluation was performed on the test data. The results presented in the following show the performance in terms of detection accuracy and time point of detection on the test data.

5   Results and Discussion

All investigated methods introduced in Sect. 2 and illustrated in Fig. 2 were tested
for different values of the parameter $k \in \{1, 2, 4, 8, 12, 16, 20, 40, 60, 100\}$ which
is the number of scores used with the respective method. Since a key motivation for
the current work was to modify the current score by the values of the neighbouring
scores, we chose a finer granularity for sampling near the current score. Without
postprocessing, the movements were predicted on average over all subjects 180 ms
before the movement onset with a balanced accuracy of 0.8. Since we were
interested in performance improvements concerning these two measures, the results
are illustrated as relative changes according to these *reference performances* in
Fig. 4a. The figure shows the average result for all methods applied for the two
criteria detection accuracy and time point of detection (see Sect. 4). Here each data
point corresponds to a particular value for k, with $k = 100$ being the first data point
on the lower left of the plot and the next smaller value for k being the next on the
connecting line. Finally, all methods meet at $k = 1$ on the upper right at a relative
prediction onset of 0 ms and a relative performance of 1, because all of these have an
identical weight vector of $w = (1)$. This point (highlighted as white spot in Fig. 4)
with $k = 1$ is equal to the reference performance without any postprocessing.

The results in Fig. 4a indicate that the performance obtained when using the raw
score values was already on a high level regarding the time point of detection *and*
the classification accuracy. In the figure, a postprocessing method outperforming
this reference would be on the upper left relative to this point. This we observed only
for the slope approach 150+slope, where we found configurations for $k \in \{2, 4\}$
that revealed a slight improvement on average in both, accuracy and time point.

From the figure, it is far more apparent that most methods enabled an earlier
prediction of the movement on the cost of (mostly) slight performance drops. In
the most extreme case for the uniform approach with $k = 100$, this means more
than 300 ms earlier prediction at a loss of 18 % of the initial performance. Overall
improvements in classification accuracy on the average level were only revealed for
the 150+slope approach.

The reason for the large standard deviations depicted in Fig. 4a is disclosed by
illustration of the single-subject results in Fig. 4b. The benefit of the applied method
strongly differed between subjects, so that the results in Fig. 4a only show the rough
trend. On the single-subject level we observed slight improvements for both, time
point and accuracy. However, the *best* method was subject-specific. Again, most
extreme differences where achieved using the uniform approach: Using $k = 100$
we could detect the movement nearly 800 ms earlier without any performance
loss for one subject, while the same configuration resulted in only 100 ms earlier
detection at a loss of 30 % of the initial performance for another. In the analysis,
especially subjects with a worse performance on the raw scores could benefit from
the postprocessing.

To summarize, the postprocessing methods presented here provide a tool to
modify mainly the earliness of the prediction and to a little extent the classification

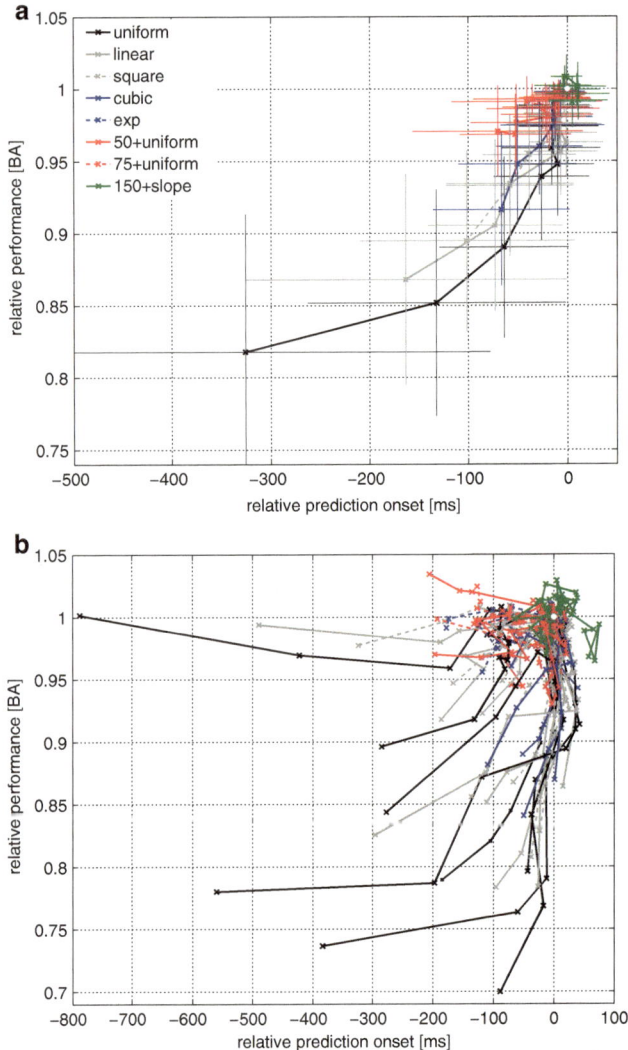

Fig. 4 Performance changes (BA and onset time) of postprocessing methods. The results are illustrated relative to the case where the scores were not further processed, illustrated as a *white dot* at (0,1). Each data point corresponds to a different $k \in \{1, 2, 4, 8, 12, 16, 20, 40, 60, 100\}$ (see text). In (**a**) the grand average results over all subjects are shown as mean and standard deviation for each k, respectively. Thus, there is one line for each method applied. In (**b**) the same results are shown for all subjects, separately

accuracy. The 150+slope method with $k \in \{2, 4\}$ worked best on the average level and enhanced both, time point of detection and accuracy. On the single-subject level, the individual best method differed, so that the spectrum presented here can serve as a general framework to *adjust* the movement prediction according to the respective application and/or the data of the individual subject.

6 Application: Ensemble Classification

This section illustrates how the postprocessing methods described here can be directly used in an application without additional investigations or prior knowledge. It should be noted that various alternative ways exist to use the outlined or similar postprocessing methods depending on the exact application with the respective constraints. Here, we try to give an intuition on their applicability using simple and straightforward approaches.

With the results presented in Sect. 5 we end up with two characteristics of the discussed methods for movement prediction: the best method is unknown in advance and a high subject-specificity is observed. One possibility to deal with this uncertainty in an application is to use *all* methods in an ensemble learning approach. Here, we apply this approach on the data (training and testing) described in Sect. 3. Accordingly, the ensemble consisted of 73 members (all eight postprocessing methods with nine values for k plus the reference point without postprocessing being equal to $k = 1$).

6.1 Procedure

Two simple ensemble strategies were used independently during training to obtain a single decision from all outputs of the ensemble members during testing: either the best individual method from the training was selected for the decision in a winner-take-all fashion (which is equal to viewing the ensemble learning as a grid search optimization), or the weights were distributed proportionally over all methods according to their individual behaviour on the training data. This evaluation has been performed after the re-adjustment of the score thresholds as described in Sect. 4.3. For each ensemble strategy (winner-take-all or distributed), three optimization criteria were respectively used: the individual classification performance (BA), the onset of the movement prediction (time), or a combination of both. For the latter, we added a value of 0.05 to the BA obtained for each 100 ms that the movement has been detected earlier. This means, that a perfect BA of 1.0 obtained 1,000 ms earlier would result in a total metric value of 1.5. The value of 0.05 has been arbitrarily determined from the overall results of all methods (Fig. 4) with the rationale that the performance did hardly improve, but instead dropped quickly in most cases, when the prediction has been made earlier. Therefore, we forced the combination to rely mainly on the classification performance getting a bonus when the classification has been correctly made earlier. Depending on the constraints of the individual application, this value may be chosen differently.

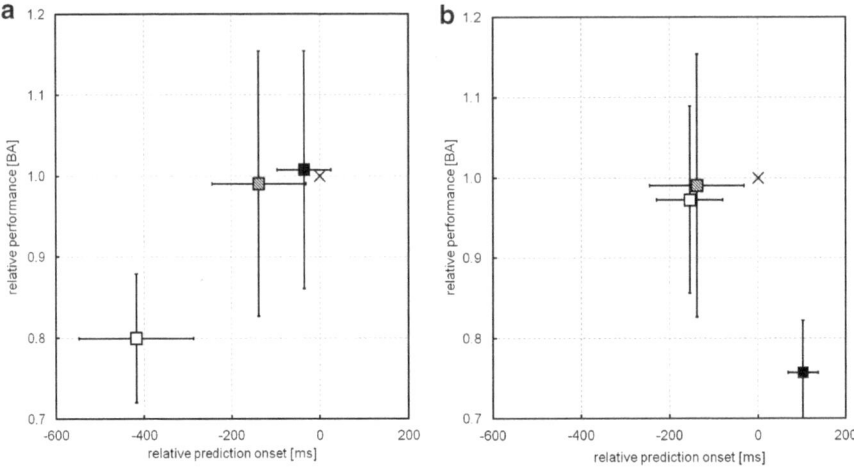

Fig. 5 Performance of the two ensemble approaches (mean and standard deviation) relative to reference performance. The ensemble was optimized according to best performance (*filled square*), earliest classification (*white square*) or a combination of the two (*striped square*, see text). The reference is marked with a cross at [0,1] (**a**) winner-take-all (**b**) distributed weighting

6.2   *Ensemble: Results and Discussion*

The resulting classification performances and prediction times of the ensemble approaches are depicted in Fig. 5. As is most obvious for the winner-take-all configuration, the ensemble is able to compensate for subject-specific differences in the *best method*, so that the postprocessing results on average in an earlier prediction. In the case of optimizing for time, the prediction was even about 400 ms earlier than the reference at the cost of accuracy. The figure also illustrates two more important factors when using ensembles: the optimization criterion and the distribution of the weights. The distributed weighting ensemble made the prediction slightly earlier when time was (at least partly) a criterion, but it failed when optimizing according to BA, because then the classification behaviour was governed by too many methods resulting in an overall later prediction with lower certainty.

7   Conclusions

Without any postprocessing, the classification of each window is performed independently of the neighbouring windows. However, we can see in the distributions of these classification outcomes that they intrinsically carry information about the probability of an upcoming event, like the rise in scores illustrated in Fig. 1. Here, we use simple methods that can easily be applied during online movement prediction to make use of this knowledge and stabilize a single classification by the surrounding ones. As we have shown for individual methods and in the combined approach using the ensemble, the methodology introduced here can be used as a tool to improve classification outcomes.

On average we could observe an improvement of both, time point of detection and accuracy, using the 150+slope method with small values of k. However, we found the most pronounced effects on the single-subject level: the proposed methods performed individually different. In general we could show that the applied postprocessing methods succeed for individual subjects in improving the accuracy and/or time point of prediction, but we could not find one straightforward solution in the current study for all subjects investigated. For most methods we found a trade-off between classification performance and the time point of prediction. This means for the application of such a movement predicting system, that one can indeed enhance the system, but has to carefully chose the postprocessing method and/or combination approach according to the requirements of the intended application. The finding of a subject specificity is consistent with results from other postprocessing methods [13], but it is still an open question where these effects particularly originate from. So far, existing postprocessing methods operate rather blindly on the data which may cause the individual differences.

From the view of an application, such a high subject-specificity can be principally dealt with using two strategies: either extra calibration time is used to find the best individual method, or the prediction itself is integrated in the application in a way that is flexible or robust enough to make use of the possible benefits illustrated here. Since the predictions obtained without postprocessing can serve as a reliable fallback option, this could be realized, e.g., by using a number of the proposed approaches on top and making the final prediction from the ensemble as we have exemplified in Sect. 6. As we have shown, a general approach can be implemented by performing a subject-specific optimization with an ensemble approach instead of using one method for all subjects. Then the performance can indeed be boosted resulting, e.g., in an earlier time point of prediction on the average level.

While most research is dedicated to improvement of the classifier and/or prepro- cessing algorithms, the idea of postprocessing classification outcomes as such is not completely new. Techniques for incorporating preceding probabilities to enhance the current prediction have been proposed [12, 15], but neither been evaluated in the way we did in the present study, nor been tested in an ensemble learning approach. Therefore and since SVM scores do not directly represent probabilities like in a Bayesian framework, a direct comparison with the methods we proposed is difficult. However, from the technical point of view, all of these methods have in common that they actually manipulate single prediction outcomes making use of the individual prediction history. Other techniques exist for postprocessing that rather operate on the global level by changing the decision criterion of the classifier or using additional thresholds. Here, threshold selection, dwell time optimization or debiasing of the score time course have been proposed [13, 14]. Due to their different nature, these techniques can be easily combined with what we proposed here, as we already implicitly did by including threshold optimization (see Sect. 4.3) and selecting a stability criterion of 100 ms ($N = 10$; see Sect. 4.2 and Fig. 3), which can be interpreted as a kind of dwell time.

With the approach outlined here, other and more complex algorithms can of course be used, although they might have the possible drawback of being too computationally complex for an online predicting system. Generally, the methods applied here are not specific for the context of movement prediction, so they can be used in any context where such postprocessing may be helpful.

Acknowledgements This work was supported by the German Bundesministerium für Wirtschaft und Technologie (BMWi, grant FKZ 50 RA 1012 and grant FKZ 50 RA 1011). The authors like to thank Marc Tabie for providing us with the evaluation data.

References

1. Kornhuber HH, Deecke L. Hirnpotentialänderungen bei Willkürbewegungen und passiven Bewegungen des Menschen: Bereitschaftspotential und reafferente Potentiale. Pflüger's Archiv für die gesamte Physiologie des Menschen und der Tiere 1965;284(1):1–17.
2. Libet B, Gleason CA, Wright EW, Pearl DK. Time of conscious intention to act in relation to onset of cerebral activity (readiness-potential) the unconscious initiation of a freely voluntary act. Brain 1983;106(3):623–42.
3. Fabiani M, Gratton G, Federmeier KD. Event-related brain potentials: methods, theory, and applications. In: Cacioppo J, Tassinary LG, Berntson GG, editors. Handbook of psychophysiology. 3rd ed. Cambridge [u.a]: Cambridge University Press; 2007. p. 85–119.
4. Blankertz B, Dornhege G, Lemm S, Krauledat M, Curio G, Müller K. The Berlin brain-computer interface: machine learning based detection of user specific brain states. J Univ Comput Sci 2006;12(6):581–607.
5. Bai O, Rathi V, Lin P, Huang D, Battapady H, Fei DY, Schneider L, Houdayer E, Chen X, Hallett M. Prediction of human voluntary movement before it occurs. Clin Neurophysiol 2011;122(2):364–72. http://www.sciencedirect.com/science/article/pii/S1388245710005699.
6. Ahmadian P, Cagnoni S, Ascari L. How capable is non-invasive EEG data of predicting the next movement? A mini review. Front Hum Neurosci 2013;7:124.
7. Kirchner EA, Albiez J, Seeland A, Jordan M, Kirchner F. Towards assistive robotics for home rehabilitation. In: Chimeno MF, Solé-Casals J, Fred A, Gamboa H, editors. Proceedings of the 6th International Conference on Biomedical Electronics and Devices (BIODEVICES-13). Barcelona: SciTePress; 2013. p. 168–77.
8. Kirchner EA, Tabie M. Closing the gap: combined EEG and EMG analysis for early movement prediction in exoskeleton based rehabilitation. In: Proceedings of the 4th European Conference on Technically Assisted Rehabilitation - TAR 2013; 2013 March.
9. Folgheraiter M, Jordan M, Straube S, Seeland A, Kim SK, Kirchner EA. Measuring the improvement of the interaction comfort of a wearable exoskeleton. Int J Soc Robot 2012;4(3):285–302.
10. Folgheraiter M, Kirchner EA, Seeland A, Kim SK, Jordan M, Wöhrle H, Bongardt B, Schmidt S, Albiez J, Kirchner F. A multimodal brain-arm interface for operation of complex robotic systems and upper limb motor recovery. In: Vieira P, Fred A, Filipe J, Gamboa H, editors. Proceedings of the 4th International Conference on Biomedical Electronics and Devices (BIODEVICES-11). Rome: SciTePress; 2011. p. 150–62.

11. Seeland A, Woehrle H, Straube S, Kirchner EA. Online movement prediction in a robotic application scenario. In: 6th International IEEE EMBS Conference on Neural Engineering (NER). San Diego, CA: IEEE; 2013. p. 41–4

12. Lemm S, Schäfer C, Curio G. BCI competition 2003–data set III: probabilistic modeling of sensorimotor mu rhythms for classification of imaginary hand movements. IEEE Trans Biomed Eng 2004;51(6):1077–80. http://ieeexplore.ieee.org/xpl/articleDetails.jsp?arnumber=1300806.

13. Mohammadi R, Mahloojifar A, Coyle D. A combination of pre- and postprocessing techniques to enhance self-paced BCIs. Adv Hum Comput Interact 2012;2012:3:1–10. http://www.hindawi.com/journals/ahci/2012/185320/abs/.

14. Solis-Escalante T, Müller-Putz G, Pfurtscheller G. Overt foot movement detection in one single laplacian EEG derivation. J Neurosci Meth 2008;175(1):148–53.

15. Zhu X, Wu J, Cheng Y, Wang Y. GMM-based classification method for continuous prediction in brain-computer interface. In: Proceedings of the 18th International Conference on Pattern Recognition - ICPR '06, Washington, DC: IEEE Computer Society; 2006. vol. 01. p. 1171–4. http://dx.doi.org/10.1109/ICPR.2006.610.

16. Vapnik VN. The nature of statistical learning theory. New York: Springer; 1995.

17. Tabie M, Kirchner EA. EMG onset detection – comparison of different methods for a movement prediction task based on EMG. In: Alvarez S, Solé-Casals J, Fred A, Gamboa H, editors. Proceedings of the 6th International Conference on Bio-inspired Systems and Signal Processing (BIOSIGNALS-13). Barcelona: SciTePress; 2013. p. 242–7.

18. Krell MM, Straube S, Seeland A, Wöhrle H, Teiwes J, Metzen JH, Kirchner EA, Kirchner F. pySPACE - a signal processing and classification environment in Python. Front Neuroinf 2013;7(40). http://www.frontiersin.org/neuroinformatics/10.3389/fninf.2013.00040/abstract, https://github.com/pyspace.

19. Rivet B, Souloumiac A, Attina V, Gibert G. xDAWN algorithm to enhance evoked potentials: application to brain-computer interface. IEEE Trans Biomed Eng 2009;56(8):2035–43. http://dx.doi.org/10.1109/TBME.2009.2012869.

20. Chang CC, Lin CJ. LIBSVM: a library for support vector machines. ACM Trans Intell Syst Technol 2011;2:27:1–27. Software available at http://www.csie.ntu.edu.tw/~cjlin/libsvm.

21. Kubat M, Holte RC, Matwin S. Machine learning for the detection of oil spills in satellite radar images. Mach Learn 1998;30(2–3):195–215. http://dx.doi.org/10.1023/A:1007452223027.

22. Straube S, Krell MM. How to evaluate an agent's behaviour to infrequent events? – reliable performance estimation insensitive to class distribution. Front Comput Neurosci 2014;8(43). http://www.frontiersin.org/computational_neuroscience/10.3389/fncom.2014.00043/abstract.

Index

© Springer International Publishing Switzerland 2015
A.R. Londral et al. (eds.), *Neurotechnology, Electronics, and Informatics*,
Springer Series in Computational Neuroscience 13, DOI 10.1007/978-3-319-15997-3

Springer Complexity

Springer Complexity is an interdisciplinary program publishing the best research and academic-level teaching on both fundamental and applied aspects of complex systems – cutting across all traditional disciplines of the natural and life sciences, engineering, economics, medicine, neuroscience, social and computer science.

Complex Systems are systems that comprise many interacting parts with the ability to generate a new quality of macroscopic collective behavior the manifestations of which are the spontaneous formation of distinctive temporal, spatial or functional structures. Models of such systems can be successfully mapped onto quite diverse "real-life" situations like the climate, the coherent emission of light from lasers, chemical reaction-diffusion systems, biological cellular networks, the dynamics of stock markets and of the internet, earthquake statistics and prediction, freeway traffic, the human brain, or the formation of opinions in social systems, to name just some of the popular applications.

Although their scope and methodologies overlap somewhat, one can distinguish the following main concepts and tools: self-organization, nonlinear dynamics, synergetics, turbulence, dynamical systems, catastrophes, instabilities, stochastic processes, chaos, graphs and networks, cellular automata, adaptive systems, genetic algorithms and computational intelligence.

The two major book publication platforms of the Springer Complexity program are the monograph series "Understanding Complex Systems" focusing on the various applications of complexity, and the "Springer Series in Synergetics", which is devoted to the quantitative theoretical and methodological foundations. In addition to the books in these two core series, the program also incorporates individual titles ranging from textbooks to major reference works.

Springer Series in Synergetics

Founding Editor: H. Haken

The Springer Series in Synergetics was founded by Herman Haken in 1977. Since then, the series has evolved into a substantial reference library for the quantitative, theoretical and methodological foundations of the science of complex systems.

Through many enduring classic texts, such as Haken's *Synergetics and Information and Self-Organization*, Gardiner's *Handbook of Stochastic Methods*, Risken's *The Fokker Planck-Equation* or Haake's *Quantum Signatures of Chaos*, the series has made, and continues to make, important contributions to shaping the foundations of the field.

The series publishes monographs and graduate-level textbooks of broad and general interest, with a pronounced emphasis on the physico-mathematical approach.

For further volumes:
http://www.springer.com/series/712

José María Amigó

Permutation Complexity in Dynamical Systems

Ordinal Patterns, Permutation Entropy and All That

 Springer

José María Amigó
Universidad Miguel Hernandez
Centro de Investigacion Operativa
Avda. de la Universidad, s/n
03202 Elche
Spain
jm.amigo@umh.es

ISSN 0172-7389
ISBN 978-3-642-04083-2   e-ISBN 978-3-642-04084-9
DOI 10.1007/978-3-642-04084-9
Springer Heidelberg Dordrecht London New York

Library of Congress Control Number: 2010920733

Cover design: Integra Software Services Pvt. Ltd., Pondicherry

Printed on acid-free paper

Springer is part of Springer Science+Business Media (www.springer.com)

To my parents

Preface

This is a research book on ordinal patterns, permutation entropy, and complexity, written at graduate level. The common denominator of the different topics presented in its pages is a hypothetical order structure of the state space, substantiated in form of ordinal patterns—permutations defined by the order relations among points in the orbits of dynamical systems. Here the state space is meant to be arbitrary (including discrete sets and n-dimensional intervals), as long as it is totally ordered, and the dynamical systems are meant to include stochastic processes (sometimes called random dynamical systems). Out of the order structure of the state space, a number of constructs will emerge to pave our way as we progress: admissible and forbidden patterns, order isomorphy, metric and topological permutation entropy, discrete entropy, regularity parameters, etc. The relation of these concepts to similar concepts in applied mathematics and computer science will be addressed as well, especially in the introductory part. The final result is a new approach to dynamical complexity characterized by conceptual simplicity, an algebraic flavor, and computational speed. The term *permutation complexity* in the title of this book intends to direct attention to this circle of ideas.

Complexity is a general concept that has different meanings in different contexts. For instance, complexity is related to "incompressibility" in information theory and computer science. In dynamical systems, complexity is usually measured by the topological entropy and reflects roughly speaking, the proliferation of periodic orbits with ever longer periods or the number of orbits that can be distinguished with increasing precision. In physics, the label "complex" is in principle attached to any nonlinear system whose numerical solutions exhibit a chaotic behavior. Neurologists claim that the human brain is the most complex system in the solar system, while entomologists teach us the baffling complexity of some insect societies. The list could be enlarged with examples from geometry, management science, communication and social networks, etc. In this book we will be mainly concerned with complexity from the viewpoint of discrete-time dynamical systems. In particular, permutation complexity refers to the dynamical features captured and quantified by tools based on order relations.

Permutation entropy was introduced in 2002 by C. Bandt and B. Pompe as a measure of complexity in time series. In a nutshell, permutation entropy replaces the probabilities of length-L symbol blocks in the definition of the Shannon entropy by the probabilities of length-L ordinal patterns. Since then this proposal has sparked new lines of research that capitalize on the order structure of the state space. Order is as well at the base of some classical results of combinatorial dynamics (notably, Sarkovskii's theorem), but the focus in these investigations is the periodicity structure of the map. Ordinal patterns provide an akin though different picture: akin because periodic points and ordinal patterns are closely related; different because ordinal patterns are amenable to numerical methods, while periodicity is not. A complete analysis of the relation between ordinal patterns and periodic points is still lacking.

As conventional entropy, permutation entropy comes in metric and topological versions, and these are limits of the corresponding rates of finite order. The metric and topological permutation entropies can be shown to coincide with their conventional counterparts under several assumptions. In applications, permutation entropy rates of finite order may be used to measure the complexity of a finite data sequence. Periodic or quasiperiodic sequences have vanishing or negligible complexity. At the opposite end, independent and identically distributed random sequences (white noise) have asymptotically divergent permutation entropies, owing to the fact that the number of allowed (or "admissible") ordinal patterns grows superexponentially with length. Between both ends lie the kind of sequences we are interested in; their permutation entropy rates of finite order can be calibrated by comparison with the corresponding rates of the white noise.

The study of permutation complexity, which we call *ordinal analysis*, can be envisioned as a new kind of symbolic dynamics whose basic blocks are ordinal patterns. Interesting enough, it turns out that under some mild mathematical assumptions, not all ordinal patterns can be materialized by the orbits of a given one- or multi-dimensional deterministic dynamics, not even if this dynamic is chaotic—contrarily to what happens with the symbol patterns. As a result, the existence of "forbidden" (i.e., not occurring) ordinal patterns is always a persistent dynamical feature, in opposition to properties such as proximity and correlation which die out with time in a chaotic dynamic. Moreover, if an ordinal pattern is forbidden, its absence pervades all longer patterns in form of more missing ordinal patterns, called outgrowth forbidden patterns. Admissible ordinal patterns grow exponentially with length, while forbidden patterns do superexponentially. Since random (unconstrained) dynamics has no forbidden patterns with probability 1, their existence can be used as a fingerprint of deterministic orbit generation.

This book is addressed to both researchers on dynamical systems and complexity and graduate students interested in these subjects. Some topics are already well established; others are asking for generalizations or more comprehensive analyses; still others, like the applications to space–time dynamics, are newcomers. The book consists of ten chapters, plus two technical annexes where the reader can find the mathematical background needed in the main text; overlaps between the main text and the annexes were unavoidable, but they have been kept at a minimum. The topics selected correspond to materials published by the author and collaborators in recent years, although they have been thoroughly revised and eventually reformulated for this occasion. The presentation is a compromise between mathematical rigor and getting the message across in a smooth way. Formal statements of results and their proofs allow knowing exactly which are the assumptions behind them, facilitating at the same time to refer to them from any place in the text. Examples illustrate the theory wherever convenient. Both the main text and the annexes contain also a sufficient number of exercises that invite the reader to explore beyond our exposition. Next we describe briefly the content of the different chapters.

Chapter 1 is an introduction to the main topics of this book, namely, patterns, complexity, and entropy. We show how these concepts are linked—sometimes in unexpected ways—in five different settings: information theory, symbolic dynamics, dynamical systems, computer science, and cellular automata. Ordinal patterns and permutation entropy make their first appearance in the second section, together with the forbidden patterns, one of the main characters of permutation complexity.

Once the stage has been set, Chap. 2 is a brief account on a few applications of ordinal analysis. We review four of them, to wit: entropy estimation, permutation complexity of time series, recovery of control parameters of unimodal maps from symbolic sequences, and characterization of the different kinds of synchronization between chaotic oscillators. This chapter should convey to the reader a first impression of the disparate possibilities of ordinal analysis, before going into technical details in Chaps. 3 through 7.

Chapter 3 is wholly devoted to the study of ordinal patterns and their main properties. Two of them are specially important in applications: existence of forbidden patterns in the orbits of dynamical systems (herein referred only to one-dimensional dynamics) and robustness of admissible and forbidden patterns against observational noise. Forbidden patterns are further classified into two groups: outgrowth and root forbidden patterns. The study of robustness is continued in Chap. 9.

In the relation between maps and the structure of their admissible and forbidden patterns there are far more questions than answers. It is therefore gratifying that this relation can be analyzed with great detail in the case of the shift and signed shift transformations. Due to its length, this topic has been divided into two parts: Chap. 4 and Chap. 5. Signed shifts include the standard ones but their handling is more difficult, and the results gotten till now are not so sharp. By order isomorphy, the results of these two chapters apply to perhaps more interesting cases, like the logistic map, baker map, sawtooth maps.

The next two chapters comprise an in-depth analysis of metric and topological permutation entropies. On defining the metric permutation entropy of maps in Chap. 6, we depart from the original approach to follow basically Kolmogorov's path, based on finite partitions. The pay-off is that the results are not limited to one-dimensional maps. For this reason we have to make a detour over symbolic dynamics (or, equivalently, finite-alphabet information sources), before getting ready to deal with maps. The main outcome is that the metric permutation entropy of ergodic maps coincides with the metric entropy (otherwise called measure-theoretical or Kolmogorov–Sinai entropy) of the map.

The same applies to the topological permutation entropy (Chap. 7), where now expansiveness is called in. An important consequence is the existence of forbidden patterns also in higher dimensional dynamics. Furthermore, numerical simulations provide ample evidence that forbidden patterns is a general feature of deterministic orbit generation.

Discrete entropy (Chap. 8) was proposed (together with the discrete Lyapunov exponent) as a tool of discrete chaos, a generalization of chaos to dynamical systems with discrete state spaces. Our approach follows the work of Bandt and Pompe on permutation entropy of time series. It is proved that discrete entropy converges to its "continuous" counterpart in an adequate sense.

Having shown in Chap. 7 that the existence of forbidden patterns is a landmark of determinism, Chap. 9 grapples with the implementation of this fact, the main obstacle being that real data are finite and noisy. The properties of ordinal patterns studied in Chap. 3 come here to the rescue, as well as the "dynamical robustness" discussed in the first section. Two methods are proposed, based on (i) the number of missing ordinal patterns and (ii) the distribution of visible ordinal patterns. The second resorts to a chi-square test, the null hypothesis being that the time series is white noise; its performance compares favorably to some widely used tests of statistical independence.

Cellular automata and coupled map lattices are, so to speak, toy models for real physics. And yet, what these dynamical systems lack in sophistication as compared to the usual space–time systems, they more than make up for in conceptual simplicity and modelization power. On applying some tools of ordinal analysis to cellular automata and coupled map lattices, as done in Chap. 10, we put to test the capabilities of this approach to discern different temporal structures in spatially extended systems. The task is formidable: trying to reduce the behavior of a space–time system to just a parameter seems to be more than what one could reasonably ask for. Nevertheless, the results reported in Chap. 10 are encouraging.

The book concludes with Chap. 11, where we remind the main messages of ordinal analysis and permutation complexity, gather some open problems scattered in the preceding chapters, and suggest future lines of research.

Much labor will be necessary to survey the full potential of ordinal analysis and the intricacies of permutation complexity at theoretical and practical levels. This book should be considered as a contribution to this task. One of the main challenges of complexity theory is to design conceptual and numerical tools to study, classify, and quantify the different degrees of complexity found in our mathematical models of the world around. Think, for example, of turbulence in fluid mechanics or the asymptotic behavior of cellular automata and coupled map lattices. Nonlinear physics has developed a battery of instruments that go by the name of power spectra, Lyapunov exponents, fractal dimensions of attractors, order parameters, etc. On the mathematical side, ergodic theory and topological dynamics study general properties of systems evolving in time. These disciplines have provided plenty of handles to understand complex dynamics, like deep concepts, invariants for classification purposes (most notably, the entropy), prototypes, and powerful theoretical and practical techniques. But order relations have been less exploited. One possible reason is that order relations are not invariant under metric and topological isomorphisms, which consistently only address measure-theoretical and topological properties. We hope that this book on permutation complexity convincingly shows that properties related to the temporal (and eventually also spatial) structure of a dynamics are useful and worth researching.

It is a great pleasure to thank all friends and colleagues who have collaborated with me on the topics of this book: Gonzalo Álvarez, David Arroyo, Rui Dilão, Sergi Elizalde, Matthew(Matt) B. Kennel, Ljupco Kocarev, Roberto Monetti, Ulrich Parlitz, Miguel A.F. Sanjuán, Janusz Szczepanski, Igor Tomovski, Elek Wajnryb, and Samuel Zambrano—without them this book had not been possible. In particular, Matt made the numerical simulations of Chaps. 6 and 7, Igor of Chap. 8, and Samuel of Chaps. 9 and 10; moreover Matt's ingenuity was decisive for the theoretical results of Chapter 6. For further assistance I am also indebted to Óscar Martínez Bonastre and Agustín Pérez Martín. Special thanks are due to Manfred Denker and Wolfgang Krieger for clarifying discussions on the generator problem. Most of the scientific articles this book is based on were written under the auspices of the Spanish Ministry for Education and Science (Project MTM2005-04948); this financial support is gratefully acknowledged. Furthermore, I want to express my gratitude to Ljupco Kocarev and Jürgen Kurths, Editorial and Programme Adviser of the Springer Series in Complexity, for encouraging me to write this book, as well as to Dr. Christian Caron, Executive Publishing Editor of Springer Verlag, for guiding me through the publication stages. Last but not least, I wish to highlight the enduring and stimulating collaboration of Samuel Zambrano; he has been much of a driving force in exploring new ideas, working out the applications and getting insights from the results.

Elche, Spain José María Amigó

Contents

Chapter 1
What Is This All About?

This introductory chapter is meant as a tour of the main topics in this book: patterns, ordinal relations, complexity, and entropy. The approach is mostly informal; for the technicalities behind the different notions met on the way, the reader is referred to Annex A and Annex B.

1.1 Patterns, Complexity, and Entropy

Pattern is an abstract concept with different acceptations. In the context of dynamical systems, information theory, and computer science (the ones we are interested in), a pattern is a finite string of symbols, eventually chosen with some criterion. In the next sections we will meet some familiar instances of patterns in those contexts. Contrary to the concept of pattern, complexity does not lend itself to a short definition (would this be not a contradiction otherwise?) but, like poetry, it is very easy to recognize. For a panorama of complexity, see [77] or, at an introductory level, [158]. A third and also recurrent issue in the next pages will be entropy, one of the most important quantities when dealing with complexity in deterministic and random dynamical systems. Indeed, no matter how one counts the diversity of patterns generated by a data source, entropy enters the scene in some of its many disguises: Shannon entropy, metric entropy, topological entropy, etc.

1.1.1 Information Theory

Consider an information source outputting symbols or letters, one at a time, from a finite alphabet $S = \{s_1, \ldots, s_{|S|}\}$ (i.e., $|S|$ is the cardinality of S). Formally, an information source is a discrete-time, stationary stochastic process $\mathbf{X} = \{X_n\}_{n \in \mathbb{N}_0}$, where $\mathbb{N}_0 = \{0, 1, \ldots\}$ and X_n are random variables on a common probability space, taking on values in S. For the time being, we will dispense with the underlying probability space. A realization of $\mathbf{X}$ is a one-sided sequence, $x_0^\infty := (x_n)_{n \in \mathbb{N}_0}$, called[1] a

---

[1] The symbol ":=" means that the left side is defined by the right one; a corresponding meaning holds for "=:".

J.M. Amigó, *Permutation Complexity in Dynamical Systems,*
Springer Series in Synergetics, DOI 10.1007/978-3-642-04084-9_1,
© Springer-Verlag Berlin Heidelberg 2010

message. Correspondingly, the symbols $x_n \in S$ are sometimes called *letters*. A finite segment of a message, say, $x_k^{k+L-1} := x_k x_{k+1} \dots x_{k+L-1}$ is called a *word* of length L. If $p(x_0^{L-1})$ denotes the probability of the word x_0^{L-1} to be output, then the (Shannon) entropy rate (or just *entropy*) of the data source $\mathbf{X}$ is defined as

$$h(\mathbf{X}) = - \lim_{L \to \infty} \frac{1}{L} \sum p(x_0^{L-1}) \log p(x_0^{L-1}), \tag{1.1}$$

where log usually stands for logarithm to base 2 ($h(\mathbf{X})$ is then measured in bits per symbol), and the sum is over all possible words of length L, numbering $|S|^L$, with the convention $0 \times \log 0 = \lim_{x \to 0+} x \log x = 0$. To indicate that a logarithm is to base e, we will write ln instead of log ($h(\mathbf{X})$ is then measured in nats per symbol). The convergence of limit (1.1) is proven in Sect. B.1.2.

In an information-theoretical setting, $\log p(x_0^{L-1})$ is the information conveyed by the output x_0^{L-1}, hence $h(\mathbf{X})$ is the average information per symbol conveyed by the messages of the information source $\mathbf{X}$ in the limit of arbitrarily long messages.

When the random variables X_n are independent, or (more often) intersymbol dependency is neglected for simplicity or limited influence, the information source is called *memoryless*. In this case $h(\mathbf{X})$ coincides with the entropy $H(X)$ of a random variable X with outcomes $x \in S$ and probabilities $p(x)$:

$$H(X) = - \sum_{x \in S} p(x) \log p(x).$$

Compression is any procedure that reduces the data requirements of a message without, in principle, losing information—although it can be acceptable as a trade-off between data reduction and information degradation. The idea of using codes or dictionaries for compression of information originates with the invention of the telegraph, since users were charged by the number of letters in the message. It is clear that data compression can be achieved by assigning short words to the most frequent outcomes of the information source. For example, in the Morse code, the most frequent symbol in English, namely the letter e, is represented by a single dot. This intuition is the guiding principle in the construction of the celebrated Huffman code for memoryless sources. Suppose that code words $w_1, \dots, w_{|S|}$ of lengths $l_1, \dots, l_{|S|}$, respectively, are assigned to the values $s_1, \dots, s_{|S|}$ taken on by a random variable X with probabilities $p(s_1), \dots, p(s_{|S|})$. The code words are combinations of characters taken from an alphabet $a_1, \dots, a_D$, usually $0, 1$ ($D = 2$) in modern communications. Then the Huffman code is a uniquely decipherable code that minimizes the average code-word length $\bar{l} = \sum_{n=1}^{|S|} p(s_n) l_n$, which according to the *noiseless coding theorem* is known to satisfy [22]

$$H(X) \leq \bar{l} < H(X) + 1, \tag{1.2}$$

where the logarithm of $H(X)$ is taken to base D. But how to compress a message, say a digital picture to be sent by electronic mail or a text file written in a foreign language, if the probabilities of the corresponding symbols are not known? This feat requires a universal compressor.

Universal compressors are based on the fact that natural languages are not completely random but repeat patterns from time to time. In 1976 and 1978, A. Lempel and J. Ziv published two simple algorithms for universal data compression [137, 211], which work by parsing an input string of finite length into successive phrases. Some variants of the second (LZ78) are implemented in the most popular compressors currently used in electronic editing (like WinZip or pdf). For our purposes it is sufficient to consider the first scheme (LZ76); also, we will emphasize the interplay between complexity and entropy rather than the compression-related aspects.

In the LZ76, the message is sequentially parsed into strings that have not appeared so far in the initial segment ending at (and excluding) the current letter. For example, the binary word $x_0^{19} = 0101101000110111010$ is parsed as

$$0, 1, 011, 0100, 011011, 1001, 0. \tag{1.3}$$

If, say, x_k is the first bit after a comma, then we check whether x_k appears in x_0^{k-1}. If it does not, then we write a comma after x_k and start a new block (this is the case for $k = 1$ in (1.3)). Otherwise, we check whether $x_k x_{k+1}$ appears in x_0^k; in negative case, we write a comma after x_{k+1}, otherwise the process continues till a pattern $x_k x_{k+1} \ldots x_{k+l}$ repeats (or the sequence finishes). The number of patterns found in the parsing of a word x_0^{L-1} is called its Lempel–Ziv (LZ) complexity, $C(x_0^{L-1})$. In example (1.3), $C(x_0^{19}) = 7$. Words x_0^{L-1} with a general alphabet S are parsed in an analogous way.

The formal definition of $C(x_0^{L-1})$ is recursive. A *block* of length l ($1 \leq l \leq L$) is just a segment of x_0^{L-1} of length l, i.e., a string of l consecutive letters, say $x_k^{k+l-1} = x_k x_{k+1} \ldots x_{k+l-1}$ ($0 \leq k \leq L - l$). In particular, letters are blocks of length 1. Set $B_0 = x_0$ and suppose that after $k \geq 1$ steps, we have parsed x_0^{L-1} as

$$B_0, B_1, \ldots, B_{k-1},$$

where $B_1 = x_1^{n_1}, \ldots, B_{k-1} = x_{n_{k-2}+1}^{n_{k-1}}$, and $n_{i-1} + 1 \leq n_i < L - 1$ for $i = 1, \ldots, k - 1$ (with $n_0 = 0$). Define

$$B_k := x_{n_{k-1}+1}^{n_k} \quad (n_{k-1} + 1 \leq n_k \leq L - 1),$$

to be the shortest block such that it does not occur in the sequence $x_0^{n_k - 1}$. (In the LZ78 algorithm, one checks instead whether the current block $x_{n_{k-1}+1}^{n_k}$ coincides with one of the previous blocks, $B_0, B_1, \ldots, B_{k-1}$.) Proceeding in this way, we obtain a (uniquely defined) decomposition of x_0^{L-1} in "minimal" blocks, say

$$x_0^{L-1} = B_0, B_1, \ldots, B_{p-1}, \tag{1.4}$$

in which only the last block can occasionally appear twice. Then,

$$C(x_0^{L-1}) := p.$$

For computational efficiency, one uses the well-known "suffix-tree" data structure and search algorithms for quickly finding substrings of the input string.

From the foregoing description, we may say that $C(x_0^{L-1})$ measures the complexity of the word x_0^{L-1}; words with a periodic or almost periodic structure have a small LZ complexity, while those displaying a random-looking structure have a high count of distinct patterns, hence a great LZ complexity. It can be proven [211] that if the source $\mathbf{X}$ is ergodic (i.e., the probability of any length-L word equals its frequency in a single, "typical" sequence), then

$$\limsup_{L \to \infty} \frac{C(x_0^{L-1})}{L/\log_{|S|} L} = h(\mathbf{X}) \tag{1.5}$$

with probability 1. The normalization factor in (1.5) is the LZ complexity of a memoryless, equidistributed source. Let us mention in passing that (1.5) shows that the ideal compression factor of the LZ76 algorithm, in the limit of long messages, is $h(\mathbf{X})$. The same is true for the LZ78 scheme.

Equations (1.2) and (1.5) provide examples in which the concepts of complexity (here related to "incompressibility") and entropy (here related to "uncertainty") are linked in a perhaps unexpected way. As a by-product, LZ complexity can be used as an estimator of the entropy. A principal advantage of this approach is that the LZ algorithm is entirely automatic with no free parameters (unlike naive plug-in methods or methods which estimate $h(\mathbf{X})$ via block entropies; see [167] and Sect. 2.1). Another practical issue is the convergence speed with L: the normalized LZ76 complexity converges to the entropy faster than the LZ78, what makes it a better choice in practice [6]. A variance estimator for the entropy estimation by means of the LZ76 complexity can be found in [9].

1.1.2 Symbolic Dynamics

Symbolic dynamics, first proposed by Morse and Hedlund [160], is an approach to complex dynamics that aims to capture the essential aspects of complexity by studying conceptually simple models. As it often happens in mathematics, symbolic dynamics has developed in short time from an auxiliary tool to an independent field [139, 123], with applications to the study of formal languages. As a result, dynamical systems connect through symbolic dynamics to computer science, information theory, and automata.

To motivate symbolic dynamics, consider the dynamics generated by a self-map f of a set Ω. Of course, the dynamics is introduced in the *state space* Ω via the repeated action of f on Ω. Given $x \in \Omega$, the *orbit* or *trajectory* of x under f is defined as $\mathcal{O}_f(x) = \{f^n(x) : n \in \mathbb{N}_0\}$, where $f^0(x) := x$ and $f^n(x) := f(f^{n-1}(x))$. If f is invertible, then one can distinguish between the *full orbit* $\mathcal{O}_f(x) = \{f^n(x) : n \in \mathbb{Z}\}$ and the *forward orbit* $\mathcal{O}_f^+(x) = \{f^n(x) : n \in \mathbb{N}_0\}$. The name "orbit" clearly hints to the interpretation of the iteration index n as discrete time: each application of f on the point $x_n = f(x_{n-1})$ updates the "movement" of the *initial condition* x in Ω. If the resulting dynamics is complicated, we might content ourselves with a "blurred" picture of the orbit behavior. This can be done as follows. Divide Ω into a finite number of disjoint pieces A_i, $i = 0, 1, \ldots, k - 1$, and keep track of the trajectory of $x \in \Omega$ with the precision set by the decomposition $\alpha = \{A_0, \ldots, A_{k-1}\}$. (We reserve the name partition for a measurable decomposition, provided Ω is endowed with a sigma algebra; see below.) Specifically, we assign to x a (one-sided) sequence[2] $\Phi(x) = (\xi_0, \xi_1, \ldots, \xi_n, \ldots)$, the nth entry $\xi_n \in \{0, 1, \ldots, k - 1\}$ telling us in which element of α the iterate $f^n(x)$ is to be found. When f is invertible, we can also assign a two-sided sequence $\Phi(x) = (\ldots, \xi_{-1}, \xi_0, \xi_1, \ldots, \xi_n, \ldots)$, the entries with negative indices corresponding to the locations of $f^{-n}(x)$, $n \geq 1$. For brevity we focus on the general case. We call Φ a *coding map*, and $\Phi(x)$ the *itinerary* of x with respect to the decomposition α. Formally,

$$\Phi_n(x) = i \text{ iff } f^n(x) \in A_i, \tag{1.6}$$

where $n \in \mathbb{N}_0$ and $\Phi_n(x)$ denotes the nth component of the sequence $\Phi(x)$.

Let us reformulate this simple idea in a more general way. Given the finite alphabet $S = \{0, 1, \ldots, k-1\}$, denote by $S^{\mathbb{N}_0}$ the space of one-sided sequences of symbols from S:

$$S^{\mathbb{N}_0} = \{(\xi_n)_{n\in\mathbb{N}_0} = (\xi_0, \xi_1, \ldots, \xi_n, \ldots): \xi_n \in S\}.$$

Hence, $\Phi(x) \in S^{\mathbb{N}_0}$. The space $S^{\mathbb{N}_0}$ (and also $S^{\mathbb{Z}}$) is generically referred to as a *sequence* or *symbolic space*. One can put on a sequence space different (non-equivalent) metrics d making it a compact space. For example,

$$d((\xi_n)_{n\in\mathbb{N}_0}, (\eta_n)_{n\in\mathbb{N}_0}) = \begin{cases} 0 & \text{if } \xi_n = \eta_n \text{ for all } n \in \mathbb{N}_0, \\ 2^{-N} & \text{if } \xi_n = \eta_n \text{ for } n < N \text{ and } \xi_N \neq \eta_N. \end{cases} \tag{1.7}$$

Thus, two one-sided sequences are apart 2^{-N} in this metric if their first N entries coincide (and the $(N + 1)$th ones do not). In $S^{\mathbb{Z}}$, two sequences $(\xi_n)_{n\in\mathbb{Z}}$ and $(\eta_n)_{n\in\mathbb{Z}}$ are at distance 2^{-N} if their entries coincide from $-(N - 1)$ to $N - 1$, i.e., if $\xi_n = \eta_n$ for $|n| < N$. In Annex A.2 we consider other metrics.

Having introduced the sequence spaces, observe now that the action of f on the orbit of $x \in \Omega$, namely, $f^n(x) \mapsto f(f^n(x)) = f^{n+1}(x)$, translates into the action $(\Phi(x))_n \mapsto (\Phi(x))_{n+1}$ on the components of the itineraries. For this reason one introduces the (one-sided) *shift transformation* (or just *shift*) $\Sigma: S^{\mathbb{N}_0} \to S^{\mathbb{N}_0}$ as follows:

$$\Sigma: (\xi_0, \xi_1, \ldots, \xi_n, \ldots) \mapsto (\xi_1, \xi_2, \ldots, \xi_{n+1}, \ldots). \tag{1.8}$$

---

[2] The dependence of $\Phi(x)$ on f and α is not made explicit in order to keep the notation simple.

In words, Σ deletes the first component of $(\xi_n)_{n \in \mathbb{N}_0}$ and shifts the other components one position to the left. It is easily shown that Σ is a continuous transformation. As observed above, the diagram

$$
\begin{array}{ccc}
\Omega & \xrightarrow{f} & \Omega \\
\Phi \downarrow & & \downarrow \Phi \\
S^{\mathbb{N}_0} & \xrightarrow{\Sigma} & S^{\mathbb{N}_0}
\end{array}
$$

commutes, i.e., $\Phi \circ f = \Sigma \circ \Phi$. Note that Σ is not invertible (indeed, it is a k-to-1 map), although f might be invertible—unless two-sided itineraries are used.

As a simple illustration (see Fig. 1.1), consider the *sawtooth* (also called *dyadic*, *shift*, etc.) *map* $E_2 : [0, 1] \to [0, 1]$, defined as

$$
E_2(x) = 2x \bmod 1,
$$

and decompose $[0, 1]$ into the intervals $A_0 = [0, \frac{1}{2})$ and $A_1 = [\frac{1}{2}, 1]$, so the alphabet is $S = \{0, 1\}$. In this case, the orbit $E_2^n(x)$, $n \in \mathbb{N}_0$, is coded to an infinitely long 0–1 string $\Phi(x)$, where

$$
(\Phi(x))_n = \begin{cases} 0 & \text{if } E_2^n(x) \in A_0, \\ 1 & \text{if } E_2^n(x) \in A_1. \end{cases}
$$

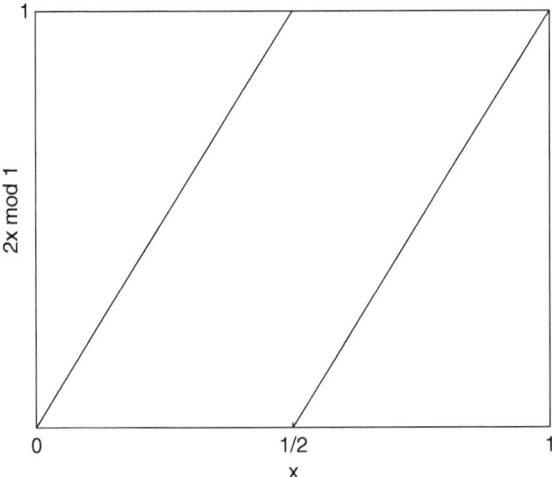

Fig. 1.1 The function $E_2(x) = 2x \bmod 1$

Let

$$x = \frac{b_0}{2} + \frac{b_1}{2^2} + \cdots + \frac{b_k}{2^{k+1}} + \cdots = \sum_{k=0}^{\infty} b_k 2^{-(k+1)} =: 0.b_0 b_1 \ldots b_k \ldots,$$

$b_n \in \{0, 1\}$, be a binary expansion of $x \in [0, 1]$. Then

$$E_2(0.b_0 b_1 \ldots b_k \ldots) = 0.b_1 b_2 \ldots b_{k+1} \ldots$$

for $x \in [0, 1)$ and $E_2(1) = E_2(0.1^{\infty}) = 0 = 0.0^{\infty}$, where here and throughout the upper label "∞" attached to a symbol means indefinite repetition of that symbol. The *dyadic rationals* in $(0, 1)$ (i.e., numbers of the form $m/2^n$, $m = 1, 2, \ldots, 2^n - 1$) are characterized by possessing two binary expansions: one terminating with 0^{∞} and other terminating with 1^{∞}. Indeed, $0.10^{\infty} = 0.01^{\infty}$ and $0.b_0 \ldots b_{k-1} 10^{\infty} = 0.b_0 \ldots b_{k-1} 01^{\infty}$, $k \geq 1$, since

$$\sum_{n=k+1}^{\infty} 2^{-(n+1)} = 2^{-(k+2)} \sum_{n=0}^{\infty} 2^{-n} = 2^{-(k+2)} \cdot 2 = 2^{-(k+1)}.$$

If $x = 0.b_0 b_1 \ldots \in (0, 1)$ is not a dyadic rational, then

$$E_2^n(x) = 0.b_n b_{n+1} \ldots \in A_i \quad \text{iff } b_n = i \in \{0, 1\},$$

hence

$$\Phi(x) = (b_n)_{n \in \mathbb{N}_0} = (b_0, b_1, \ldots, b_n, \ldots). \tag{1.9}$$

Furthermore, $\Phi(0) = (0^{\infty})$ and $\Phi(1) = (1, 0^{\infty})$. If $x \in (1, 0)$ is a dyadic rational, then x is a preimage of 0 under E_2, thus (1.9) is fulfilled provided $(b_n)_{n \in \mathbb{N}_0}$ corresponds to the binary expansion of x ending with 0^{∞}. We conclude that given any binary sequence $(b_n)_{n \in \mathbb{N}_0}$ not terminating with 1^{∞}, there exists always $x \in [0, 1]$, namely $x = 0.b_0 b_1 \ldots$, such that its itinerary with respect to the decomposition $\alpha = \{A_0, A_1\}$ under E_2 is precisely that sequence. In particular, given a finite word b_0^n, there exist infinitely many points in $[0, 1]$, to wit:

$$x \in \left[\frac{b_0 2^n + b_1 2^{n-1} + \cdots + b_n}{2^{n+1}}, \frac{b_0 2^n + b_1 2^{n-1} + \cdots + b_n + 1}{2^{n+1}} \right)$$

$$= [0.b_0 \ldots b_n, 0.b_0 \ldots b_n + 2^{-(n+1)}), \tag{1.10}$$

whose itineraries $\Phi(x)$ "realize" the pattern b_0^{L-1} in the sense that $\Phi(x)_0^n = b_0^n$. The fact that all finite words of a symbolic space can be materialized as segments of itineraries for a wide class of maps (Sect. 3.1) contrasts with the situation we shall come upon when studying the so-called ordinal patterns in Sect. 1.2.

Shifts are a special instance of the so-called subshifts. If K is a closed and Σ-invariant (i.e., $\Sigma(K) \subset K$) subset of $S^{\mathbb{N}_0}$, the restriction of the shift transformation to K, written as $\Sigma|_K$, is called a *subshift*. Sometimes Σ is called a *full* shift to distinguish it from the subshifts proper ($K \neq S^{\mathbb{N}_0}$).

A special class of subshifts are of great interest in applications. Let $A = (a_{ij})_{0 \leq i,j \leq k-1}$ be a $k \times k$ matrix of 0's and 1's and define

$$S_A^{\mathbb{N}_0} = \left\{ (\xi_n)_{n \in \mathbb{N}_0} \in S^{\mathbb{N}_0} : a_{\xi_n \xi_{n+1}} = 1 \text{ for all } n \in \mathbb{N}_0 \right\}.$$

Put in simple terms, the matrix A determines which letters $\xi_{n+1} \in S = \{0, 1, \ldots, k - 1\}$ may follow the letter ξ_n in the word $(\xi_n)_{n \in \mathbb{N}_0}$. Thus $S_A^{\mathbb{N}_0}$ is a closed and Σ-invariant subset of the sequence space $S^{\mathbb{N}_0}$ that contains all well-formed or *admissible* sequences. Alternatively, one can also describe $S_A^{\mathbb{N}_0}$ by listing the forbidden words. This explains the connection between symbolic dynamics and the theory of formal languages we mentioned above. The restriction of Σ to $S_A^{\mathbb{N}_0}$, written as Σ_A, is called a *subshift of finite type*, *Markov subshift*, or a *topological Markov chain*. If $a_{ij} = 1$ for every $0 \leq i, j \leq k - 1$, we recover the full shift. At the opposite end, $S_A^{\mathbb{N}_0}$ may be empty. This happens if and only if the matrix A is nilpotent (i.e., $A^n = 0$ for some $n \in \mathbb{N}$).

As way of example, take $k = 2$ and

$$A = \begin{pmatrix} 1 & 1 \\ 1 & 0 \end{pmatrix}.$$

Since $a_{11} = 0$, this means that the binary sequence $(\xi_n)_{n \in \mathbb{N}_0}$ is admissible if and only if it does not contain two consecutive 1's. In this case, the only forbidden block of length 2 is 11.

Let $\mathbb{K} = \mathbb{N}_0$ or $\mathbb{Z}$, and $(S_A^{\mathbb{K}}, \Sigma_A)$, $(T_B^{\mathbb{K}}, \Sigma_B)$ be two subshifts of finite type possibly with different alphabets S and T, respectively. Suppose $F : S_A^{\mathbb{K}} \to T_B^{\mathbb{K}}$ is a shift-commuting map, that is, $F \circ \Sigma_A = \Sigma_B \circ F$. The continuous, shift-commuting maps from a subshift of finite type $S_A^{\mathbb{K}}$ to another $T_B^{\mathbb{K}}$ were characterized in [92] as those maps for which there exist integers $l \leq r$ and a "local rule" $f : S^{r-l+1} \to T$ such that for any $\xi = (\xi_n)_{n \in \mathbb{K}} \in S_A^{\mathbb{K}}$ and $i \in \mathbb{K}$,

$$F(\xi)_i = f(\xi_{i+l}, \ldots, \xi_{i+r}). \tag{1.11}$$

If F is not the constant map, then a maximal l and a minimal r with this property exist; they are called left and right radii of F, respectively. If $\mathbb{K} = \mathbb{N}_0$, then $l \geq 0$. When $\mathbb{K} = \mathbb{Z}$, $p = \max\{-l, r\}$ is called the *radius* of F. In this case,

$$F(\xi)_i = f(\xi_{i-p}, \ldots, \xi_i, \ldots \xi_{i+p}),$$

where $\xi = (\xi_n)_{n \in \mathbb{Z}}$. A map between two subshifts of finite type of the form (1.11) is called a *block map* [123]. Block maps provide the mathematical underpinnings of cellular automata (Sect. 1.5).

Markov subshifts not only do provide conceptually simple prototypes for important dynamical properties, but they are basic components of some physical systems (e.g., think of Smale's horseshoes in Hamiltonian dynamical systems). To be more specific, we point out next that Markov subshifts can exhibit all properties of low-dimensional chaos.

Let us recall some basic definitions first. A 0–1 matrix A is said to be *transitive* if A^m is positive (i.e., all its entries are positive) for some $m \in \mathbb{N}$. A continuous self-map f of a metric space M is *topologically transitive* if there exists $x \in M$ such that $\mathcal{O}_f(x) = (f^n)_{n \in \mathbb{N}_0}$ is dense in M; if f is invertible, then the requirement for topological transitivity is that $\mathcal{O}_f(x) = (f^n)_{n \in \mathbb{Z}}$ is dense in M for some $x \in M$. It holds [91] that if A is a transitive $k \times k$ matrix, then the topological Markov chain Σ_A is topologically transitive and its periodic orbits are dense in $S_A^{\mathbb{N}_0}$ ($S = \{0, 1, \ldots, k-1\}$), therefore Σ_A is chaotic in the sense of Devaney [69]; in particular, Σ_A has sensitive dependence on initial conditions (see Sect. A.2). This result includes the full shifts. The corresponding statements for f invertible and $M = S_A^{\mathbb{Z}}$ hold true as well.

1.1.3 Dynamical Systems

We shall encounter two kinds of dynamical systems in this book. A *continuous* (or *topological*) *dynamical system* consists of a topological space (e.g., a metrical space) M and a continuous map $f : M \to M$. This being the case, these systems will be denoted by the pair (M, f). Subshifts are examples of continuous systems, (K, Σ_K). A *measure-theoretical dynamical system* is comprised of a *measurable space* $(\Omega, \mathcal{B})$, a measurable map $f : \Omega \to \Omega$, and a *non-singular measure* μ on $(\Omega, \mathcal{B})$. Thus, Ω is a non-empty set, $\mathcal{B}$ is a sigma-algebra of subsets of Ω, $f^{-1}B \in \mathcal{B}$ for all $B \in \mathcal{B}$, and $B \in \mathcal{B}$ is a μ-zero set iff $f^{-1}B$ is a μ-zero set. Only finite-measure spaces will be considered henceforth. Therefore, $(\Omega, \mathcal{B}, \mu)$ may be assumed without restriction to be a probability space, with μ being a probability on the space of "events" $(\Omega, \mathcal{B})$. Measure-theoretical systems will be denoted by $(\Omega, \mathcal{B}, \mu, f)$. To promote a continuous system (M, f) to a measure-theoretical one, it suffices to endow the topological space M with its Borel sigma-algebra (i.e., the sigma-algebra generated by the open sets), and the corresponding Lebesgue measure. In topological dynamics, the attention focuses on continuous systems. In ergodic theory, the framework is set by *measure-preserving* self-maps of (usually) probability spaces. We say that $f : \Omega \to \Omega$ preserves a measure μ on $(\Omega, \mathcal{B})$, if $\mu(f^{-1}B) = \mu(B)$ for all $B \in \mathcal{B}$. Alternatively, we say that the measure-theoretical system $(\Omega, \mathcal{B}, \mu, f)$ is μ-preserving, or that μ is f-invariant. Sometimes, measure-preserving, invertible maps are called *automorphisms*, while the name *endomorphisms* is reserved for the non-invertible ones.

The dynamical complexity of a measure-preserving system $(\Omega, \mathcal{B}, \mu, f)$ can be quantified by its metric entropy. So to speak, the metric entropy measures the uncertainty of the forward evolution of the system when the initial condition is not exactly known —the higher the uncertainty, the greater the complexity. The original proposal of A. Kolmogorov (later completed by Y. Sinai) amounts to the following recipe: coarse-grain the state space of the dynamical system and calculate the Shannon entropy of the resulting stochastic process. Let us follow this path.

A *partition* of a measure space $(\Omega, \mathcal{B}, \mu)$ (or just Ω for brevity) is a disjoint family of elements of $\mathcal{B}$, called atoms, whose union is Ω. Partitions will be denoted by small Greek letters. Two extreme examples of partitions of Ω are the *trivial partition* $\{\emptyset, \Omega\}$ and the *point partition* (or partition of Ω into separate points)

$$\epsilon = \{\{x\} : x \in \Omega\}. \tag{1.12}$$

Except for ϵ, we consider only finite partitions, i.e., partitions with a finite number of atoms. If, furthermore, Ω is a compact metric space with metric d, then the "size" or "coarseness" of a partition $\alpha = \{A_0, A_1, \ldots, A_{|\alpha|-1}\}$ is measured by its *norm* (sometimes also called *diameter*),

$$\|\alpha\| = \sup_{0 \le k \le |\alpha|-1} \{d(x, y) : x, y \in A_k\}. \tag{1.13}$$

We saw already in the last section that a discretization of the state space Ω may provide useful insights into a complicated dynamic. In measure-preserving systems this is even more certain since, as we are going to see presently, partitions allow establishing a connection with stochastic and information theory.

Given a finite partition $\alpha = \{A_0, A_1, \ldots, A_{|\alpha|-1}\}$ of $(\Omega, \mathcal{B}, \mu)$, the maps[3] $X_n : \Omega \to S = \{0, 1, \ldots, |\alpha| - 1\}$, $n \in \mathbb{N}_0$, defined as

$$X_n(x) = i \quad \text{iff } f^n(x) \in A_i$$

are random variables on the probability space $(\Omega, \mathcal{B}, \mu)$. Indeed,

$$X_n^{-1}(i) = f^{-n}(A_i) \in \mathcal{B}$$

because f is measurable. Observe that $X_n(x)$ is the nth component of the itinerary of x with respect to α. The difference now with respect to the itineraries of Sect. 1.1.2 is the existence of an invariant measure, which allows to promote $\mathbf{X} = \{X_n\}_{n \in \mathbb{N}_0}$ to a stationary stochastic process. In fact

(i) The probability (mass) function of X_n is given by

---

[3] The dependence of X_n on α is not made explicit here in order to keep the notation simple.

$$\Pr\{X_n = i\} = \mu\left\{x \in \Omega : f^n(x) \in A_i\right\} = \mu(f^{-n}A_i) = \mu(A_i),$$

because f is μ-preserving. As for the joint probability function of $X_0, \ldots,$ $X_n = X_0^n,$

$$\Pr\left\{X_0^n = i_0, \ldots, i_n\right\} = \mu\left\{x \in \Omega : x \in A_{i_0}, \ldots, f^n(x) \in A_{i_n}\right\}$$
$$= \mu\left(A_{i_0} \cap \ldots \cap f^{-n}A_{i_n}\right).$$

(ii) The stochastic process $\{X_n : n \in \mathbb{N}_0\}$ is stationary:

$$\Pr\left\{X_k^{k+n} = i_0, \ldots, i_n\right\} = \mu\left\{x \in \Omega : f^k(x) \in A_{i_0}, \ldots, f^{k+n}(x) \in A_{i_n}\right\}$$
$$= \mu\left(f^{-k}(A_{i_0} \cap \cdots \cap f^{-n}A_{i_n})\right)$$
$$= \mu\left(A_{i_0} \cap \cdots \cap f^{-n}A_{i_n}\right)$$

because f is μ-preserving. Therefore,

$$\Pr\left\{X_k = i_0, \ldots, X_{k+n} = i_n\right\} = \Pr\left\{X_0 = i_0, \ldots, X_n = i_n\right\}$$

for every $n, k \in \mathbb{N}_0$.

It follows that the stochastic process $\mathbf{X} = \{X_n\}_{n \in \mathbb{N}_0}$ is an information source with alphabet $S = \{0, 1, \ldots, |\alpha| - 1\}$. The *metric entropy of f with respect to the partition* α is defined to be the Shannon entropy (rate) of $\mathbf{X}$:

$$h_\mu(f, \alpha) = - \lim_{n \to \infty} \frac{1}{n} \sum \Pr\{X_0^{n-1} = i_0, \ldots, i_{n-1}\} \log \Pr\{X_0^{n-1} = i_0, \ldots, i_{n-1}\}$$
$$= - \lim_{n \to \infty} \frac{1}{n} \sum \mu(A_{i_0} \cap \cdots \cap f^{-n}A_{i_n}) \log \mu(A_{i_0} \cap \cdots \cap f^{-n}A_{i_n}),$$

where the summation is over all $i_0, \ldots, i_{n-1} \in S$. If we define the *refinement*

$$\bigvee_{i=0}^{n-1} f^{-i}\alpha = \{A_{j_0} \cap f^{-1}A_{j_1} \cap \cdots \cap f^{-(n-1)}A_{j_{n-1}} : 0 \le j_0, \ldots, j_{n-1} \le |\alpha| - 1\}$$

of the partition $\alpha = \{A_0, \ldots, A_{|\alpha|-1}\}$, and the function

$$H_\mu(\beta) = - \sum_{j=0}^{|\beta|-1} \mu(B_j) \log (B_j)$$

for any partition $\beta = \{B_0, \ldots, B_{|\beta|-1}\}$ of $(\Omega, \mathcal{B}, \mu)$, then we recover the usual expression of $h_\mu(f, \alpha)$:

$$h_\mu(f, \alpha) = \lim_{n \to \infty} \frac{1}{n} H_\mu \left(\bigvee_{i=0}^{n-1} f^{-i} \alpha \right). \tag{1.14}$$

The convergence of this limit is proven in Sect. B.2.

If an application of f is interpreted as a passage of one unit of time, then $\bigvee_{i=0}^{n-1} f^{-i} \alpha$ represents the combined experiment of performing n consecutive times the original experiment, represented by α. Then $h_\mu(f, \alpha)$ is the average information per unit of time that one gets from performing the original experiment every unit of time [202].

The metric (Kolmogorov–Sinai or measure-theoretical) entropy of f is then the supremum of $h_\mu(f, \alpha)$ over all finite partitions of $(\Omega, \mathcal{B}, \mu)$:

$$h_\mu(f) = \sup_\alpha h_\mu(f, \alpha). \tag{1.15}$$

Continuing with the previous information-theoretical interpretation, $h_\mu(f)$ provides the maximum average information per unit of time obtainable by performing the same experiment every unit of time.

In general there are several obstacles preventing an exact calculation of $h(f)$. First, except in simple cases limit (1.14) itself is not computable, so we must be content with an evaluation of $\frac{1}{n} H_\mu \left(\bigvee_{i=0}^{n-1} f^{-i} \alpha \right)$ for some large value of n. Second, considerable computation is necessary to identify the elements of the refined partitions $\bigvee_{i=0}^{n-1} f^{-i} \alpha$, the computational effort being exponential in n. Third, the measure μ is usually unknown to us in closed form. Fortunately, there are exceptions, for instance, when one can find a partition α for which $h_\mu(f, \alpha) = h_\mu(f)$. Such partitions are called generators or generating partitions with respect to f. A finite partition α is a one-sided generator for f if

$$\bigvee_{i=0}^{\infty} f^{-i} \alpha = \epsilon, \tag{1.16}$$

where ϵ is the point partition of Ω (see (1.12)). Moreover, if f is even an automorphism and $\bigvee_{i=-\infty}^{\infty} f^{-i} \alpha = \epsilon$, then α is called a two-sided generator or just a generator for f. Automorphisms may have not only generators but also one-sided generators. According to the Kolmogorov–Sinai theorem (Annex B.13), if α is a generator (one-sided or not) for f, then $h_\mu(f, \alpha) = h_\mu(f)$.

As way of illustration, consider the symmetric tent map $\Lambda:[0, 1] \to [0, 1]$ defined as (Fig. 1.2)

$$\Lambda(x) = 1 - |1 - 2x| = \begin{cases} 2x & \text{if } 0 \le x \le \frac{1}{2}, \\ 2(1 - x) & \text{if } \frac{1}{2} \le x \le 1. \end{cases} \tag{1.17}$$

If we equip $[0, 1]$ with the Borel sigma-algebra (generated by the intersections of open intervals of $\mathbb{R}$ with $[0, 1]$), then Λ is easily seen to preserve the Lebesgue

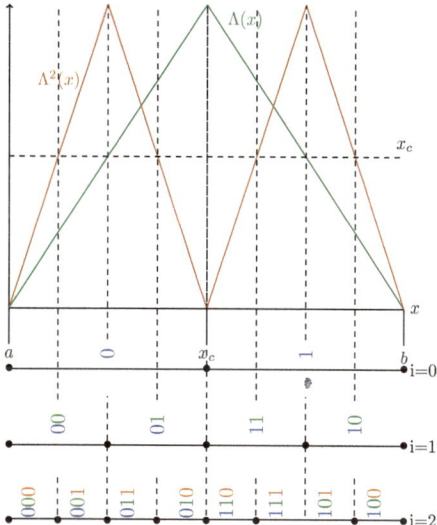

Fig. 1.2 Symbolic intervals generated by the symmetric tent map Λ and its second iterate $\Lambda^2(x_c = \frac{1}{2})$

measure. As in the previous section, let $\alpha = \{A_0, A_1\}$, where

$$A_0 = [0, \tfrac{1}{2}), \ A_1 = [\tfrac{1}{2}, 1].$$

Then,

$$\Lambda^{-1}A_0 = [0, \tfrac{1}{4}) \cup (\tfrac{3}{4}, 1], \ \Lambda^{-1}A_1 = [\tfrac{1}{4}, \tfrac{3}{4}].$$

Hence

$$\alpha \cap \Lambda^{-1}\alpha = \{A_{00}, A_{01}, A_{11}, A_{10}\},$$

with

$$A_{00} = A_0 \cap \Lambda^{-1}A_0 = [0, \tfrac{1}{4}), \quad A_{01} = A_0 \cap \Lambda^{-1}A_1 = [\tfrac{1}{4}, \tfrac{1}{2}),$$

$$A_{11} = A_1 \cap \Lambda^{-1}A_1 = [\tfrac{1}{2}, \tfrac{3}{4}], \quad A_{10} = A_1 \cap \Lambda^{-1}A_0 = (\tfrac{3}{4}, 1].$$

The sets of $\alpha, \alpha \cap \Lambda^{-1}\alpha$, and $\alpha \cap \Lambda^{-1}\alpha \cap \Lambda^{-2}\alpha$ are shown in Fig. 1.2. In general,

$$\bigcap_{i=0}^{k} \Lambda^{-i}\alpha = \left\{ A_{b_0 b_1 \ldots b_k} : b_0, b_1, \ldots, b_k \in \{0, 1\} \right\},$$

where the 2^{k+1} disjoint sets

$$A_{b_0 b_1 \ldots b_k} = A_{b_0} \cap \Lambda^{-1} A_{b_1} \cap \cdots \cap \Lambda^{-k} A_{b_k} \tag{1.18}$$

build a family of ever-shorter intervals that covers uniformly the unit interval. As a matter of fact, the sets $A_{b_0 b_1 \ldots b_k}$ are a permutation of the dyadic intervals (1.10), except eventually for the endpoints. It follows that $\bigcap_{i=0}^{k} \Lambda^{-i} \alpha$ converges to the point partition of $[0, 1]$, hence α is a one-sided generator for Λ. If λ denotes the Lebesgue measure, $\lambda(dx) = dx$, then

$$
\begin{aligned}
h_\lambda(\Lambda) &= -\lim_{n \to \infty} \frac{1}{n} \sum_{b_0 \ldots b_{n-1} \in \{0,1\}} \lambda(A_{b_0 \ldots b_{n-1}}) \log \lambda(A_{b_0 \ldots b_{n-1}}) \\
&= -\lim_{n \to \infty} \frac{1}{n} \sum_{b_0 \ldots b_{n-1} \in \{0,1\}} 2^{-n} \log 2^{-n} \\
&= \log 2.
\end{aligned}
$$

A similar argument can be applied to other maps, like the *logistic map* $g:[0, 1] \to [0, 1]$,

$$g(x) = 4x(1 - x). \tag{1.19}$$

In this case, the absolutely continuous measure[4]

$$\mu(dx) = \frac{dx}{\pi \sqrt{x(1 - x)}} \tag{1.20}$$

is g-invariant. This measure is called the *natural* or *physical invariant measure* of g because it is the one obtained in numerical experiments [72].

Since $(\Omega, \mathcal{B}, \mu)$ is a probability space, dynamical complexity can be given a probabilistic meaning. In this sense we can say that the entropy $h_\mu(f)$ (or other related concepts, like the Lyapunov exponents greater than 1) measures the randomness or, rather, the pseudo-randomness of the dynamic induced by the map f.

The complexity of continuous dynamical systems is usually measured by the topological entropy. As we shall presently see, this quantity is related to the periodic structure in some relevant systems. Rather than going into the definition of topological entropy, which is quite technical (see Sect. B.3), we only recall here its expression for a one-sided or two-sided Markov subshift Σ_A. It can be shown [91] that

$$h_{\text{top}}(\Sigma_A) = \limsup_{n \to \infty} \frac{1}{n} \log^+ P_n(\Sigma_A),$$

---

[4] Absolute continuity of measures will refer to the Lebesgue measure throughout this book.

where $h_{\text{top}}(\Sigma_A)$ is the topological entropy of Σ_A (in general, $h_{\text{top}}(f)$ stands for the topological entropy of a continuous self-map f), $P_n(\Sigma_A)$ is the number of periodic points of period n of Σ_A, and $\log^+ x = \log x$ if $x \geq 1$, and 0 otherwise. To explicitly calculate the right-hand side of this expression, we need the following two properties: (i) If B is a non-negative matrix, then there exists an eigenvalue $\lambda_{\max} \geq 0$ such that no other eigenvalue of B has absolute value greater than $\lambda_{\max}$ (this is part of the Perron–Frobenius theorem [202]) and (ii) the number of periodic points of period $p \in \mathbb{N}$ of a Markov subshift Σ_A is the trace of A^p (i.e., the sum of the diagonal elements), denoted as $\operatorname{tr} A^p$. For the full shift on k symbols, $(A^n)_{ij} = k^{n-1}$, for all $0 \leq i, j \leq k - 1$, hence the trace of A^n is k^n. This yields

$$h_{\text{top}}(\Sigma) = \log k.$$

In general, $\operatorname{tr} A^p = \lambda_1^p + \cdots + \lambda_k^p$, where λ_i are the k eigenvalues (eventually repeated) of the matrix A. It follows that [91]

$$h_{\text{top}}(\Sigma_A) = \log^+ \lambda_{\max}.$$

1.1.4 Computer Science

The origin of algorithmic complexity has to be sought in the efforts of R. Solomonoff, A. Kolmogorov, and G. Chaitin to define the elusive concept of "randomness" of finite-alphabet sequences [79, 133, 201]. The basic intuition is that random sequences are "patternless," hence there is no efficient way to describe them other than giving the sequence itself. The *algorithmic complexity* of a string $s_0^{n-1} = s_0 s_1 \ldots s_{n-1}$, written as $K(s_0^{n-1})$, can be consistently defined as the length of the shortest binary program that, run on a universal prefix-free Turing machine, outputs s_0^{n-1} and halts [59, 67, 138]. As in the case of information theory, this definition of complexity is linked to the general concept of compressibility, this time with respect to all possible algorithms that produce the sequence in question.

Somewhat paradoxically, algorithmic complexity is not a computable quantity. Then suppose that K_n is claimed to be the complexity of a length-n string s_0^{n-1}. In order to check this, we remove one bit from the hypothetically shortest program and let it run. There are two possibilities: either the $(K_n - 1)$-bit program outputs a string different from s_0^{n-1} and halts or else it runs longer than we have time to wait. In the second case, there is no way to know whether the program will halt (this is the famous Turing's halting problem), eventually revealing the actual complexity to be $K_n - 1$.

Any finite sequence s_0^{n-1} can be certainly output by the copy program: "PRINT $s_0, \ldots, s_n$." Without loss of generality, we may restrict to binary sequences for the time being. Since patternless n-bit sequences cannot be computed by any algorithm significantly shorter than the copy program, their complexity is given by $K_n \leq n + C$, where C is a constant that accounts for the computational overhead (like the operating system). At the opposite end stands the sequences consisting of a repeated bit, say 0. The complexity of the program "PRINT 0, n TIMES" can be bounded as $K_n \leq \log_2 n + C'$, where $\log_2 n$ is the number of bits needed to specify the length n and, again, C' is the computational overhead. Observe that if these programs are run on a computer other than a universal Turing machine, the constants C and C' may depend on the machine, but they are independent of the actual sequence being calculated. In the limit of very long sequences, the algorithmic complexity will practically range between $\log_2 n$ and n. This being the case, one may state that the binary sequence s_0^{n-1} is random if $K(s_0^{n-1}) \simeq n$. (In the non-binary case, $K(s_0^{n-1}) \simeq nb$ for random sequences, where b is the minimal number of bits needed to code the symbols s_i, $0 \leq i \leq n - 1$.) Formally, a sequence $(s_n) \in S^{\mathbb{N}_0}$ is said to be *incompressible* when there exists a constant C such that

$$K(s_0^{n-1}) \geq n - C$$

for all $n \geq 1$.

Randomness can also be defined as *typicality*, meaning that typical sequences have no feature that makes them special in any sense. This was the path taken by Martin-Löf to come to grips with the concept of random sequence. Rather than addressing the technicalities of this approach, which are beyond the scope of this book, we will proceed directly to the conclusions: random sequences are realizations of stochastic processes.

Let $(\Omega, \mathcal{B}, \mu)$ be a probability space. The realizations of a stochastic process $\{X_n\}_{n \in \mathbb{N}_0}$ on $(\Omega, \mathcal{B}, \mu)$ with a finite number of possible outcomes can be identified with the elements of a (one-sided) sequence space. Specifically, if $X_n : \Omega \to S$ with $S = \{s_1, \ldots, s_{|S|}\}$ for every $n \in \mathbb{N}_0$, then $(X_n(\omega))_{n \in \mathbb{N}_0} \in S^{\mathbb{N}_0}$ for every $\omega \in \Omega$. The general method to place a probability m on $S^{\mathbb{N}_0}$ induced by the probability μ is explained in Sect. A.3. At present we only need to resort to the so-called (p, q)-Bernoulli shifts or systems on two symbols, which are measure-preserving systems $(S^{\mathbb{N}_0}, \mathcal{B}, m, \Sigma)$, where

(i) $S = \{0, 1\}$,
(ii) $\mathcal{B}$ is the sigma-algebra generated by the so-called *cylinder sets*,

$$C_{s_0 \ldots s_{n-1}} = \{\xi_0^\infty \in S^{\mathbb{N}_0} : \xi_0 = s_0, \ldots, \xi_{n-1} = s_{n-1}\},$$

(iii) the probability m of the binary string $s_0^{n-1} = s_0 s_1 \ldots s_{n-1}$ is defined as

$$m(s_0^{n-1}) = m(C_{s_0 \ldots s_{n-1}}) = p^k q^{n-k},$$

where $p + q = 1$, k is the number of 1's in s_0^{n-1}, and $n - k$ is the number of 0's, and
(iv) Σ is the shift transformation on $S^{\mathbb{N}_0}$.

In the language of probability theory, the cylinder sets correspond to the elementary events; in the language of computer science, $C_{s_0 \ldots s_{n-1}}$ comprises all sequences with the prefix $w = s_0, \ldots, s_{n-1}$. The (p, q)-Bernoulli system models an independent, dichotomous process, one outcome (say, "success") having probability p to occur and the other ("failure") probability $q = 1 - p$. Think, for example, of a random experiment consisting in tossing forever a coin with the odds p for head and q for tail. The shift Σ corresponds to the "time" translation $n \mapsto n + 1$. The fact that Σ preserves m (or, equivalently, that m is Σ-invariant) accounts for the probabilities being the same in every draw.

In particular, the $(\frac{1}{2}, \frac{1}{2})$-Bernoulli system is a model for the tossing of a fair coin. If $0.b_0 b_1 \ldots b_n \ldots$ is a binary expansion and $\Phi:[0, 1] \to \{0, 1\}^{\mathbb{N}_0}$ is the map

$$\Phi:0.b_0 b_1 \ldots b_n \ldots \mapsto (b_0, b_1, \ldots, b_n, \ldots)$$

we met already in (1.9), then

$$\Phi([0.b_0 b_1 \ldots b_n, \ 0.b_0 b_1 \ldots b_n + 2^{-(n+1)})) = C_{b_0 b_1 \ldots b_n}.$$

Thus, Φ allows to identify the cylinder set $C_{b_0 b_1 \ldots b_n}$ of $\{0, 1\}^{\mathbb{N}_0}$ with the interval $[0.b_0 b_1 \ldots b_{n-1}, 0.b_0 b_1 \ldots b_{n-1} + 2^{-n})$ of $[0, 1]$. But even more is true. If m denotes the measure of the $(\frac{1}{2}, \frac{1}{2})$-Bernoulli system and λ the Lebesgue measure of $[0, 1]$, then

$$m(C_{b_0 b_1 \ldots b_{n-1}}) = \frac{1}{2^n} = \lambda([0.b_0 b_1 \ldots b_{n-1}, 0.b_0 b_1 \ldots b_{n-1} + 2^{-n})).$$

Since the cylinder sets generate the sigma-algebra of the Bernoulli systems and the semi-open dyadic intervals do the same for the Borel sigma-algebra of $[0, 1]$, we conclude $m = \lambda \circ \Phi^{-1}$, i.e., m corresponds to the Lebesgue (or uniform) measure on $[0, 1]$.

Levin, Schnorr, and Chaitin proved that a binary sequence is typical with respect to the $(\frac{1}{2}, \frac{1}{2})$-Bernoulli measure (i.e., it can be considered the result of tossing a fair coin indefinitely) if and only if it is incompressible. In this way, two seemingly different concepts of randomness incompressibility and typicality are shown to coincide in a natural setting.

Remarkably enough, this result is not the only achievement connecting concepts related to complexity but stemming from different areas. Let us provide another one in which algorithmic complexity and metric entropy are brought together.

Given a measure-preserving dynamical system $(\Omega, \mathcal{B}, \mu, f)$, each $x \in \Omega$ generates an infinitely long sequence under the action of f, namely, its (forward) orbit $\mathcal{O}_f(x) = \{f^n(x) : n \in \mathbb{N}_0\}$. Let $s_0^\infty = s_0^\infty(x, \alpha)$ be the itinerary of x with respect to the partition $\alpha = \{A_0, \ldots, A_{|\alpha|-1}\}$ of Ω, that is, $s_k = i$ iff $f^k(x) \in A_i$, $i \in \{0, \ldots, |\alpha| - 1\}$. The *algorithmic complexity* of $\mathcal{O}_f(x)$, written as $k(f, x)$, is measured by the largest algorithmic complexity per symbol of $s_0^\infty(x, \alpha)$ over all possible finite partitions α:

$$k(f, x) = \sup_\alpha \limsup_{n \to \infty} \frac{1}{n} K(s_0^{n-1}(x, \alpha)).$$

Of course, one expects that random-like trajectories are computationally more difficult to reproduce than the regular ones. This expectation can be rigorously proved under the proviso that f is ergodic with respect to the invariant measure μ. In this case [39],

$$k(f,x) = h_\mu(f) \quad \mu\text{-almost everywhere.}$$

1.1.5 Cellular Automata

A cellular automaton is a discrete-time dynamical system with discrete space and discrete states. The state variables are defined on the sites of a D-dimensional regular lattice $(\mathbb{Z}^D)$—the cells of the D-dimensional automaton—taking on values in a finite alphabet $S = \{0, 1, \ldots, k - 1\}$. The set of all possible states (formally the set of all possible mappings $\mathbb{Z}^D \to S$) is called the *configuration space*. For numerical simulations it is convenient that the lattice of sites is finite or has a non-trivial topology, like a circle or a 2-torus; these requirements can be implemented with quiescent cells or with periodic conditions, respectively. In order to accommodate this disparity of possibilities, the configuration space will be denoted by a neutral Ω. The states of the cells evolve synchronously in discrete time steps according to identical rules. But what makes cellular automata special is the evolution rule: the state of a particular cell is determined by the previous states of a neighborhood of cells around it.

Cellular automata were introduced by Ulam [199] and von Neumann [161] as simple models of universal computation and machine self-reproduction, respectively. Indeed, a remarkable property of cellular automata is their ability to simulate other symbol processors. Another one is self-organization, even when started from disordered configurations. Two-dimensional cellular automata became quite popular in the 1970s thanks to the article that Martin Gardner devoted to John Conway's *Game of Life* in his section "Mathematical Games" of *Scientific American* [84]. A purely mathematical approach was initiated by Hedlund and collaborators, who studied the endomorphisms and automorphisms of the shift dynamical system [92]. Apart from the many subsequent papers on their dynamical and ergodic properties from this point of view, cellular automata have also been the object of intensive study in mathematical physics, computer science, biology, etc. [207]. Being at the crossroads of symbolic dynamical systems and computation, it is not surprising that the theory of cellular automata benefits from both areas, at the same time that cross-pollinate them, as we try to show in the next lines. For a readable account on cellular automata and their remarkable performance in physical modeling, see, e.g., [198].

For simplicity we will consider only one-dimensional cellular automata. In this case, the configuration space is the two-sided sequence space $S^\mathbb{Z}$. One-sided sequences or even finite sequences, corresponding to lattices adequately flanked by quiescent cells, may also be considered along the same lines. A *neighborhood* of size $l \geq 1$ of the cell $i \in \mathbb{Z}$, written as $\mathcal{U}_l(i)$, is the set of $2l + 1$ cells

$$i - l, i - l + 1, \ldots, i, \ldots, i + l.$$

The state of cell i at time $t \geq 0$ will be denoted as $s_t(i)$. At each time step $t + 1$, the previous state at each cell i, $s_t(i) \in S$, is updated according to the states of $\mathcal{U}_l(i)$ by a *local rule* $f:S^{2l+1} \rightarrow S$ of the form

$$s_{t+1}(i) = f(s_t(i - l), s_t(i - l + 1), \ldots, s_t(i + l)).$$

Note that f does not depend on i nor t, but only on the states of $\mathcal{U}_l(i)$; if f is allowed to depend on i, then one speaks of *hybrid* cellular automata.

The local rule f leads to a *global transition map* of the configuration space, $F:\Omega \rightarrow \Omega$, defined in the obvious way:

$$F(\ldots, s_t(i), \ldots) = (\ldots, f(s_t(i - l), s_t(i - l + 1), \ldots, s_t(i + l)), \ldots)$$
$$= (\ldots, s_{t+1}(i), \ldots).$$

Observe that F is a block map from a full shift to itself of radius l. As pointed out in Sect. 1.1.2, it follows that F is continuous and shift-commuting. (This characterization generalizes to D-dimensional cellular automata just by replacing the sequence space $S^{\mathbb{Z}}$ by $S^{\mathbb{Z}^D}$.)

As way of illustration, Fig. 1.3 depicts the time evolution of a one-dimensional, binary cellular automaton with periodic boundary conditions: $s_t(N + 1) = s_t(1)$ and $s_t(0) = s_t(N)$ for all $t \geq 0$. Here $N = 250$, the horizontal axis represents space (label i), and time (label t) elapses along the vertical direction, from top to bottom. Once the initial configuration has been fixed, the global map F determines the dynamics of the automaton on the configuration space.

The relation between the properties of the local rule f and the properties of the global transition map F is one of the most important and difficult problems in the

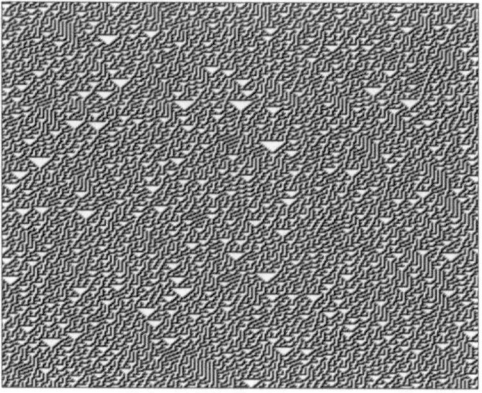

Fig. 1.3 A typical space–time evolution diagram of a one-dimensional cellular automaton with 250 sites and periodic boundary conditions. Time elapses from top to bottom

theory of cellular automata. This problem has been proved to be algorithmically unsolvable for some properties (surjectivity and injectivity for dimension $D > 1$, nilpotency for $D \geq 1$, etc.), and it is believed to be unsolvable for others (ergodicity, sensitivity, etc.).

On a more practical level, hybrid cellular automata with binary state variables and null boundaries (i.e., the cells delimiting the site lattice are permanently in the 0-state) have been explicitly shown to emulate linear feedback shift registers (LFSRs), which are widely used in cryptography as pseudo-random bit generators for stream ciphers. Specifically, given the primitive polynomial of an LFSR [151], then the algorithm given in [48] allows to "synthesize" a null-boundary, hybrid binary cellular automaton that emulates the said LFSR using only the local rules $f(p, q, r) = p + r \bmod 2 \equiv p \oplus q$ and $f(p, q, r) = p + q + r \bmod 2 \equiv p \oplus q \oplus r$. Most importantly, the same is true for the so-called self-shrunken LSFRs [149], which are nonlinear structures featured in some designs of stream ciphers. Since the previous local rules are linear, this fact allows to cryptanalize such ciphers using cellular automata.

Suppose that the configuration space Ω is $S^{\mathbb{Z}}$. In the topology induced by the cylinder sets

$$C_{s_{-n},\dots,s_0,\dots,s_n} = \{\xi_0^\infty \in S^{\mathbb{Z}} : \xi_k = s_k, |k| \leq n\},$$

the global transition map $F : \Omega \to \Omega$ that updates the states of the cellular automaton is continuous, which makes (Ω, F) a continuous dynamical system. Hence, we can measure the complexity of its time evolution with the topological entropy $h_{\text{top}}(F)$; see Sect. B.3 for different ways of calculating the topological entropy of a continuous dynamical system. Alternatively, let $R(w, t)$ be the number of distinct rectangles of width w and height (temporal extent) t occurring in a space–time evolution diagram of (Ω, F); see Fig. 1.4. Then [62]

$$h_{\text{top}}(F) = \lim_{w \to \infty} \lim_{t \to \infty} \frac{1}{t} \log R(w, t). \qquad (1.21)$$

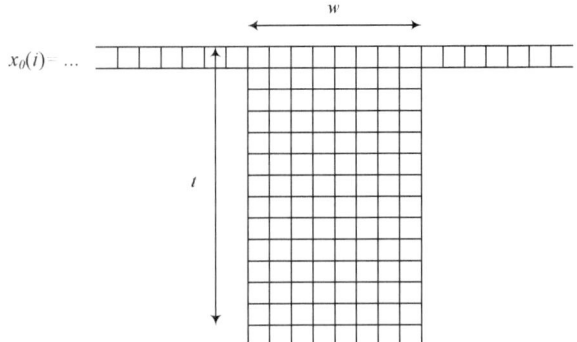

Fig. 1.4 Geometrical illustration of the rectangles $R(w, t)$ used in (1.21)

Therefore, the complexity of (Ω, F) can be measured by the number of distinct words or patterns per time unit generated by the global transition map F as time evolves. It follows that

$$h_{\text{top}}(F) \leq 2l \log k,$$

where l is the neighborhood size of the automaton and $k = |S|$.

Topological entropy belongs also to the dynamical properties that cannot be algorithmically computed for general cellular automata [101]. More generally, whether metric and/or topological entropy is effectively computable (i.e., can be approximated with an arbitrary small error) is an open question for most dynamical systems.

1.2 Admissible and Forbidden Ordinal Patterns

The concept of ordinal pattern of length L only demands a totally ordered set $(\Omega, \leq)$. Let us caution the reader that there are several definitions of ordinal patterns in the literature; the one used in this book follows Bandt et al. [28, 29]. In the simplest setting, the ordinal pattern defined by the elements $x_0, \ldots, x_{L-1} \in \Omega$ can be viewed as the permutation π of $\{0, 1, \ldots, L-1\}$ that arrange those elements according to their order in Ω: $x_{\pi_0} < x_{\pi_1} < \cdots < x_{\pi_{L-1}}$. In case $x_i = x_j$, we agree that $x_i < x_j$ if $i < j$. We write $\pi = \langle \pi_0, \pi_1, \ldots, \pi_{L-1} \rangle$ to summarize that x_{π_0} is the smallest element, x_{π_1} is the second smallest element, etc., in the length-L sequence $x_0, \ldots, x_{L-1}$. For example, if $\Omega = \mathbb{R}$ (endowed with the standard order), and $x_0 = \sqrt{3}$, $x_1 = e$, $x_2 = 2$, and $x_3 = -1.7$, then $\pi = \langle 3, 0, 2, 1 \rangle$. In an extended setting where we have a self-map f of Ω, the sets of points to be arranged by π are naturally provided by the initial segments of the f-orbits: $x_n = f^n(x)$, $0 \leq n \leq L-1$. In this case, one usually dispenses with periodic orbits of period smaller than L. The set of ordinal L-patterns will be denoted by $\mathcal{S}_L$ throughout this book. Ordinal patterns are sometimes called permutations.

As a minor technical point, let us mention that a permutation $\tau : i \mapsto \tau(i)$, $i \in \{0, 1, \ldots, L-1\}$, is written in combinatorics as

$$\begin{pmatrix} 0 & 1 & \cdots & L-1 \\ \tau(0) & \tau(1) & \cdots & \tau(L-1) \end{pmatrix} =: [\tau(0), \tau(1), \ldots, \tau(L-1)]. \tag{1.22}$$

Observe that an ordinal pattern $\pi = \langle \pi_0, \ldots, \pi_{L-1} \rangle$ does not correspond—as one might think— to the permutation $[\pi_0, \ldots, \pi_{L-1}]$, but rather to its inverse: $\pi_0 \mapsto 0, \ldots, \pi_{L-1} \mapsto L-1$, i.e.,

$$\langle \pi_0, \ldots, \pi_{L-1} \rangle = \begin{pmatrix} \pi_0 & \pi_1 & \cdots & \pi_{L-1} \\ 0 & 1 & \cdots & L-1 \end{pmatrix} = [\pi_0, \ldots, \pi_{L-1}]^{-1}. \tag{1.23}$$

For example, the ordering $x_2 < x_0 < x_1$ defines the ordinal pattern $\langle 2, 0, 1 \rangle$ but the permutation $0 = \pi_1 \mapsto 1$, $1 = \pi_2 \mapsto 2$, and $2 = \pi_0 \mapsto 0$, which in the conventional notation reads

$$[1,2,0] = [2,0,1]^{-1}.$$

In sum, an ordinal pattern $\pi \in \mathcal{S}_L$ corresponds actually to the permutation $\pi_i \mapsto i$, $0 \le i \le L - 1$, which will be denoted as $[\pi]^{-1}$ whenever needed:

$$[\pi]^{-1} = [\pi_0, \pi_1, \dots, \pi_{L-1}]^{-1}. \tag{1.24}$$

Furthermore, if $\pi = \langle \pi_0, \dots, \pi_{L-1} \rangle$ and $\pi' = \langle \pi'_0, \dots, \pi'_{L-1} \rangle$, a (non-commutative) product $\pi \circ \pi'$ can be defined in $\mathcal{S}_L$ via composition

$$
\begin{aligned}
\pi \circ \pi' &= \begin{pmatrix} \pi_0 & \pi_1 & \cdots & \pi_{L-1} \\ 0 & 1 & \cdots & L-1 \end{pmatrix} \begin{pmatrix} \pi'_0 & \pi'_1 & \cdots & \pi'_{L-1} \\ 0 & 1 & \cdots & L-1 \end{pmatrix} \\
&= \begin{pmatrix} \pi_{\pi'_0} & \pi_{\pi'_1} & \cdots & \pi_{\pi'_{L-1}} \\ 0 & 1 & \cdots & L-1 \end{pmatrix} \\
&= \langle \pi_{\pi'_0}, \pi_{\pi'_1}, \dots, \pi_{\pi'_{L-1}} \rangle.
\end{aligned} \tag{1.25}
$$

Endowed with this product, $\mathcal{S}_L$ becomes a non-Abelian group of order $L!$. The neutral element of the group $(\mathcal{S}_L, \circ)$ is the identity permutation $\langle 0, 1, \dots, L-1 \rangle$. Ordinal patterns will be studied in detail in Chap. 3.

After these algebraic prolegomena, consider now a function $f : I \to I$, where I is a closed interval of $\mathbb{R}$. Given the finite orbit $\{f^n(x) : 0 \le n \le L - 1\}$ of $x \in I$, we say that x defines the ordinal pattern of length L (or ordinal L-pattern) $\pi = \pi(x) = \langle \pi_0, \pi_1, \dots, \pi_{L-1} \rangle$ if

$$f^{\pi_0}(x) < f^{\pi_1}(x) < \cdots < f^{\pi_{L-1}}(x). \tag{1.26}$$

We say also that π is realized by x or that x is of type π.

If, for example, $I = [0,1]$ and g is the *logistic map*, $g(x) = 4x(1-x)$, then we find to four digit precision.

$$\mathcal{O}_g(0.6416) = 0.6416, 0.9198, 0.2951, 0.8320, 0.5590, 0.9861, \dots$$

hence $x = 0.6416$ is of the types

$$\langle 0, 1 \rangle, \langle 2, 0, 1 \rangle, \langle 2, 0, 3, 1 \rangle, \langle 2, 4, 0, 3, 1 \rangle, \langle 2, 4, 0, 3, 1, 5 \rangle, \dots$$

Instead of fixing x and varying L, we can do the opposite, as in the following illustration with $L = 3$:

$\mathcal{O}_g(0.15) = 0.15, 0.51, 0.9996, \ldots$ hence 0.15 is of type $\langle 0, 1, 2 \rangle$,
$\mathcal{O}_g(0.30) = 0.30, 0.84, 0.5376, \ldots$ hence 0.30 is of type $\langle 0, 2, 1 \rangle$,
$\mathcal{O}_g(0.55) = 0.55, 0.99, 0.0396, \ldots$ hence 0.55 is of type $\langle 2, 0, 1 \rangle$,
$\mathcal{O}_g(0.80) = 0.80, 0.64, 0.9216, \ldots$ hence 0.80 is of type $\langle 1, 0, 2 \rangle$,
$\mathcal{O}_g(0.95) = 0.95, 0.19, 0.6156, \ldots$ hence 0.95 is of type $\langle 1, 2, 0 \rangle$.

Points and ordinal patterns provide complementary perspectives of the same picture. Thus, as in the first instance, one can be more interested in the ordinal patterns defined by a given point or, as in the second instance, in the points that realize a given pattern. In order to introduce the second point of view, we define following [29] the sets

$$P_\pi = \{x \in I : x \text{ defines } \pi \in \mathcal{S}_L\}. \tag{1.27}$$

If $P_\pi \neq \emptyset$, then π is said to be an *allowed* or *admissible* (ordinal) *pattern* for f; otherwise π is called a *forbidden* (ordinal) *pattern* for f. In words, $\pi \in \mathcal{S}_L$ is allowed or admissible if there exists $x \in I$ such that x is of type π, whereas it is forbidden if no x is of type π. We will see shortly that maps have forbidden patterns (in fact, infinitely many of them) under quite general assumptions.

The properties of the sets $P_\pi \neq \emptyset$ are closely related to the properties of f. Thus, P_π is a union of open intervals if f is continuous or the union of intervals (including none, one, or both endpoints) if f is piecewise continuous. The endpoints of P_π are determined by the periodic points of f. All these facts can be easily exposed via the graphs of the map and their iterates. First of all, draw the graph of the identity (f^0) in the square $I \times I \subset \mathbb{R}^2$, which is the diagonal $y = x$, $x \in I$, on the Cartesian plane $\{(x, y) \in \mathbb{R} \times \mathbb{R}\}$. Then draw the graphs of the functions $y = f(x), \ldots, y = f^{L-1}(x)$, $x \in I$. The components of the distinct P_π's, $\pi \in \mathcal{S}_L$, are separated by the intersection points of all those graphs. Indeed, if $x \in P_\pi$ "moves" leftward or rightward, it will leave the current component of P_π at the left or right endpoint, respectively, as soon as the condition

$$f^{\pi_i}(x) = f^{\pi_{i+1}}(x) \tag{1.28}$$

holds for some $i = 0, 1, \ldots, L - 2$, unless it leaves the interval I before. Note that condition (1.28) implies that $f^{\min\{\pi_i, \pi_{i+1}\}}(x)$ is a periodic point of period $|\pi_i - \pi_{i+1}|$, thus x is a $\min\{\pi_i, \pi_{i+1}\}$th preimage of such a point. In this case, $\min\{\pi_i, \pi_{i+1}\} + |\pi_i - \pi_{i+1}| = \max\{\pi_i, \pi_{i+1}\} \leq L - 1$. In particular, if $\pi_i = 0$ or $\pi_{i+1} = 0$, then x is a periodic point.

In short, the endpoints of the intervals $P_\pi \neq \emptyset$, $\pi \in \mathcal{S}_L$, are given by the periodic points of f of periods $p \leq L - 1$, and their preimages up to the order $L - 2$. We conclude that the admissible ordinal patterns for f are determined by its periodic structure.

As a simple illustration, consider again the logistic map $g(x) = 4x(1 - x)$, $0 \leq x \leq 1$. For $L = 2$ we have, see Fig. 1.5,

$$P_{\langle 0,1\rangle} = \left(0, \tfrac{3}{4}\right), \quad P_{\langle 1,0\rangle} = \left(\tfrac{3}{4}, 1\right).$$

The separating point $x = \tfrac{3}{4}$ between $P_{\langle 0,1\rangle}$ and $P_{\langle 1,0\rangle}$ is given by the condition $g^{\pi_0}(x) = g^{\pi_1}(x)$, where $\pi_0, \pi_1 \in \{0, 1\}$, i.e.,

$$g(x) = x.$$

For $L = 3$ $(g^2(x) = -64x^4 + 128x^3 - 80x^2 + 16x)$, Fig. 1.6 shows that

$$P_{\langle 0,1,2\rangle} = \left(0, \tfrac{1}{4}\right), \qquad P_{\langle 0,2,1\rangle} = \left(\tfrac{1}{4}, \tfrac{5-\sqrt{5}}{8}\right), \quad P_{\langle 2,0,1\rangle} = \left(\tfrac{5-\sqrt{5}}{8}, \tfrac{3}{4}\right),$$
$$P_{\langle 1,0,2\rangle} = \left(\tfrac{3}{4}, \tfrac{5+\sqrt{5}}{8}\right), \quad P_{\langle 1,2,0\rangle} = \left(\tfrac{5+\sqrt{5}}{8}, 1\right). \tag{1.29}$$

The separating points of the intervals P_π, $\pi \in \mathcal{S}_3$, are given now by the conditions $g^{\pi_i}(x) = g^{\pi_{i+1}}(x)$, $\pi_i, \pi_{i+1} \in \{0, 1, 2\}$, i.e.,

$$g(x) = x, \ g^2(x) = x, \ g^2(x) = g(x).$$

We conclude that the common endpoints of the intervals P_π for $\pi \in \mathcal{S}_3$ are now the points of period 1 (fixed points), period 2, and first preimages of period-1 points. Moreover, when going from $L = 2$ to $L = 3$, we see that $P_{\langle 0,1\rangle}$ splits into the subintervals $P_{\langle 0,1,2\rangle}$, $P_{\langle 0,2,1\rangle}$, and $P_{\langle 2,0,1\rangle}$ at the eventually period-1 point $\tfrac{1}{4}$ (preimage of the fixed point $\tfrac{3}{4}$) and at the period-2 point $\tfrac{5-\sqrt{5}}{8}$. Likewise, $P_{\langle 1,0\rangle}$ splits into $P_{\langle 1,0,2\rangle}$ and $P_{\langle 1,2,0\rangle}$ at the period-2 point $\tfrac{5+\sqrt{5}}{8}$.

Ordinal patterns are the main ingredient of *permutation entropy* which, as the standard concept of entropy, comes also in metric and topological versions.

Suppose that μ is an f-invariant measure. Then the definition of the *metric permutation entropy* of f is formally similar to the definition of the Shannon entropy of an information source:

$$h_\mu^*(f) = -\lim_{L\to\infty} \frac{1}{L} \sum_{\pi \in \mathcal{S}_L} \mu(P_\pi) \log \mu(P_\pi), \tag{1.30}$$

provided the limit exists. Note that $\mu(P_\pi)$ is the probability for the ordinal L-pattern π to occur (while in the expression for the Shannon entropy, (1.1), the corresponding probabilities refer to length-L blocks x_0^{L-1}). Sometimes the factor $1/(L-1)$ is used instead of $1/L$ —of course, this is inconsequential in the limit $L \to \infty$.

As for the *topological permutation entropy* of f, one just counts distinct allowed patterns:

$$h_{\text{top}}^*(f) = -\lim_{L\to\infty} \frac{1}{L} \log |\{P_\pi \neq \emptyset {:} \pi \in \mathcal{S}_L\}|, \tag{1.31}$$

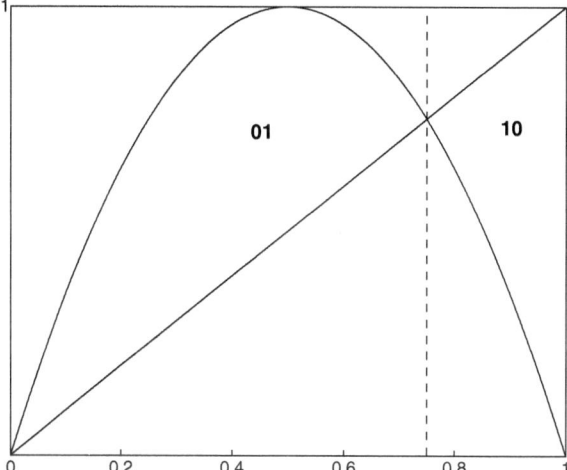

Fig. 1.5 Points in the interval $(0, \frac{3}{4})$ are of type $\langle 0, 1 \rangle$ (shorthanded 01), while points in the interval $(\frac{3}{4}, 0)$ are of type $\langle 1, 0 \rangle$ (shorthanded 10)

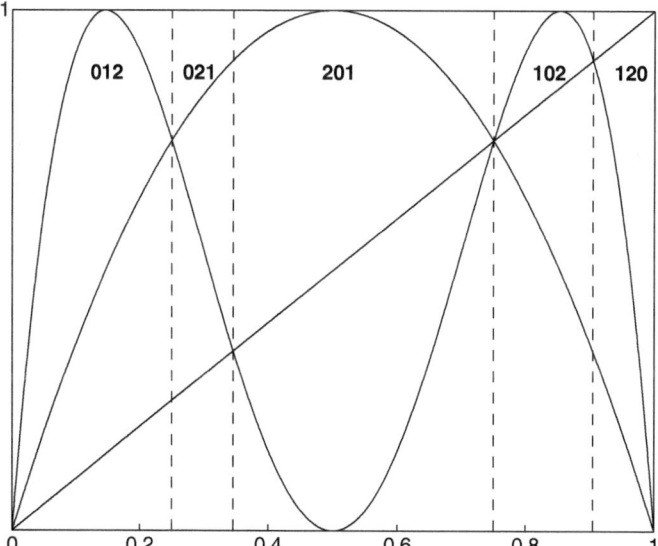

Fig. 1.6 The sets P_π, $\pi \in \mathcal{S}_3$, are graphically obtained by raising vertical lines at the crossing points of the curves $y = x$, $y = f(x)$, and $y = f^2(x)$. The three digits on the upper part of the figure are shorthand for ordinal patterns (e.g., 012 stands for $\langle 0, 1, 2 \rangle$). Observe that $P_{\langle 2,1,0 \rangle} = \emptyset$

where $|\cdot|$ denotes here cardinality. We are assuming again that this limit converges, otherwise $h^*_{\text{top}}(f)$ is not defined.

An interval map $f:I \rightarrow I$ is called *piecewise monotone* if there is a finite partition of I into intervals, such that f is continuous and monotone on each of those intervals. A nice result of Bandt, Keller, and Pompe [29] states that if f is piecewise monotone, then (i) the metric permutation entropy of f coincides with its metric entropy and (ii) the topological permutation entropy of f coincides with its topological entropy. In mathematical notation:

$$\text{(i) } h^*_\mu(f) = h_\mu(f) \quad \text{and} \quad \text{(ii) } h^*_{\text{top}}(f) = h_{\text{top}}(f). \tag{1.32}$$

From (ii) and (1.31), it follows that if f is piecewise monotone and its topological entropy is finite, then

$$|\{P_\pi \neq \emptyset : \pi \in \mathcal{S}_L\}| \sim e^{Lh_{\text{top}}(f)}, \tag{1.33}$$

where the symbol $\sim$ stands for "asymptotically as $L \rightarrow \infty$." Hence, the number of allowed L-patterns for f grows exponentially with L. On the other hand,

$$|\{P_\pi : \pi \in \mathcal{S}_L\}| = L! \sim e^{L(\ln L - 1) + 1/2 \ln 2\pi L}, \tag{1.34}$$

according to Stirling's formula for the factorial of a positive integer. Comparison of (1.33) and (1.34) not only does show that piecewise monotone maps have necessarily forbidden L-patterns for L sufficiently large but also that their number grows superexponentially with L.

From (1.29) we see that already for $L = 3$ there is one forbidden pattern for the logistic map, namely, $\langle 2, 1, 0 \rangle$. But this is not the end of the story. The absence of the ordinal pattern $\pi = \langle 2, 1, 0 \rangle$ triggers, in turn, an avalanche of longer missing patterns. To begin with, all the patterns $\langle *, 2, *, 1, *, 0, * \rangle$ (where the wildcard $*$ stands eventually for any other entries of the pattern) cannot be realized by any $x \in [0, 1]$ since the inequalities

$$\cdots < g^2(x) < \cdots < g(x) < \cdots < x < \cdots \tag{1.35}$$

cannot occur. By the same token, the patterns $\langle *, 3, *, 2, *, 1, * \rangle$, $\langle *, 4, *, 3, *, 2, * \rangle$, and, more generally,

$$\langle *, n + 2, *, n + 1, *, n, * \rangle \in \mathcal{S}_L, \ 0 \leq n \leq L - 3, \tag{1.36}$$

cannot be realized either for the same reason (replace x by $g^n(x)$ in (1.35)). We conclude that each forbidden pattern generates an infinite trail of ever-longer forbidden patterns. This issue will be revisited in full generality in Chap. 3.

Let us clarify this last point with the logistic map once more and $L = 4$. In Fig. 1.7, which is Fig. 1.6 with the curve

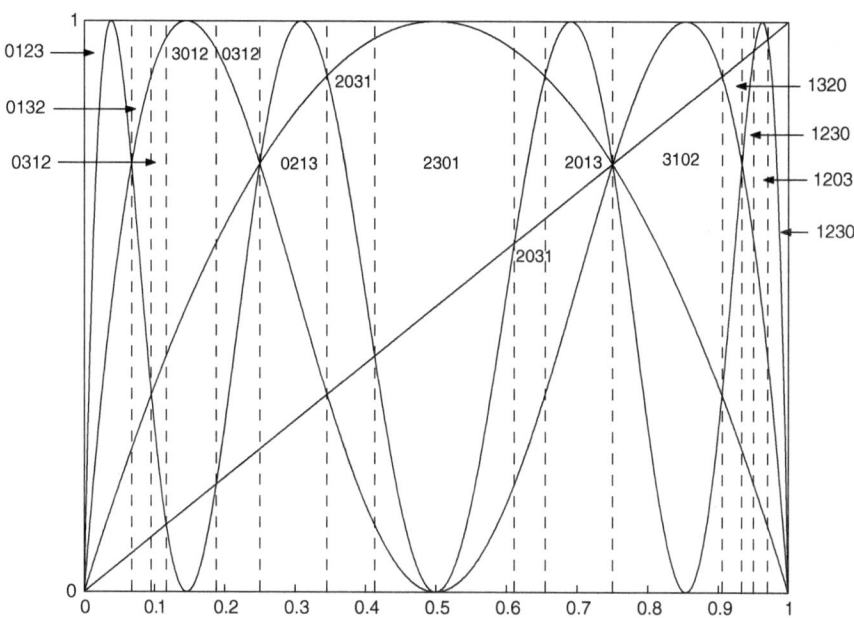

Fig. 1.7 The 12 allowed ordinal 4-patterns for the logistic map. Note the two components of $P_{(0,3,1,2)}$, $P_{(2,0,3,1)}$, and $P_{(1,2,3,0)}$

$$
\begin{aligned}
y &= g^3(x) \\
&= -16\,384x^8 + 65\,536x^7 - 106\,496x^6 + 90\,112x^5 \\
&\quad -42\,240x^4 + 10\,752x^3 - 1344x^2 + 64x
\end{aligned}
$$

superimposed, we can see the 12 allowed 4-patterns for the logistic map. Since there are 24 possible patterns of length 4, we conclude that 12 of them are forbidden. Seven forbidden 4-patterns belong to trail (1.36) of $\langle 2, 1, 0 \rangle$ (observe that $\langle 3, 2, 1, 0 \rangle$ is repeated):

$$
\begin{aligned}
(n = 0) \quad & \langle 3, 2, 1, 0 \rangle, \langle 2, 3, 1, 0 \rangle, \langle 2, 1, 3, 0 \rangle, \langle 2, 1, 0, 3 \rangle \\
(n = 1) \quad & \langle 0, 3, 2, 1 \rangle, \langle 3, 0, 2, 1 \rangle, \langle 3, 2, 0, 1 \rangle, \langle 3, 2, 1, 0 \rangle
\end{aligned}
\tag{1.37}
$$

Therefore, the remaining five forbidden 4-patterns,

$$
\langle 0, 2, 3, 1 \rangle, \langle 1, 0, 2, 3 \rangle, \langle 1, 0, 3, 2 \rangle, \langle 1, 3, 0, 2 \rangle, \langle 3, 1, 2, 0 \rangle,
\tag{1.38}
$$

are seeds for new trails of forbidden patterns of lengths $L \geq 5$ that eventually can overlap.

In Fig. 1.7 one can also follow the first two splittings of the intervals P_π:

$$P_{\langle 0,1\rangle} \rightarrow \begin{cases} P_{\langle 0,1,2\rangle} \rightarrow P_{\langle 0,1,2,3\rangle}, P_{\langle 0,1,3,2\rangle}, P_{\langle 0,3,1,2\rangle}, P_{\langle 3,0,1,2\rangle}, \\ P_{\langle 0,2,1\rangle} \rightarrow P_{\langle 0,2,1,3\rangle}, \\ P_{\langle 2,0,1\rangle} \rightarrow P_{\langle 2,0,1,3\rangle}, P_{\langle 2,0,3,1\rangle}, P_{\langle 2,3,0,1\rangle}, \end{cases}$$

$$P_{\langle 1,0\rangle} \rightarrow \begin{cases} P_{\langle 1,0,2\rangle} \rightarrow P_{\langle 3,1,0,2\rangle}, \\ P_{\langle 1,2,0\rangle} \rightarrow P_{\langle 1,2,0,3\rangle}, P_{\langle 1,2,3,0\rangle}, P_{\langle 1,3,2,0\rangle}. \end{cases}$$

The splitting of the intervals P_π can be understood in terms of periodic points and their preimages. Thus, the splitting of $P_{\langle 0,1\rangle}$ is due to the points $\frac{1}{4}$ (first preimage of the period-1 point $\frac{3}{4}$) and $\frac{5-\sqrt{5}}{8}$ (a period-2 point); the second period-2 point, $\frac{5-\sqrt{5}}{8}$, is responsible for the splitting of $P_{\langle 1,0\rangle}$. On the contrary, $P_{\langle 0,2,1\rangle}$ and $P_{\langle 1,0,2\rangle}$ do not split because they contain neither period-3 point nor first preimages of period-2 points nor second preimages of fixed points.

Chapter 2
First Applications

In this chapter we present four applications of permutation entropy and ordinal patterns: entropy estimation, complexity analysis, recovery of parameters from itineraries, and synchronization analysis of time series. The scope is to give the reader a multifaceted picture of ordinal analysis in action. Two more applications (to determinism detection and to space–time chaos) will be discussed at length in Chaps. 9 and 10, respectively.

2.1 Entropy Estimation

Real or numerical time series, say $(x_n)_{n \in \mathbb{N}_0}$ with $x_n \in \mathbb{R}$, can be produced in principle by discrete-time or continuous-time dynamical systems, which for convenience we think as including also the corresponding stochastic systems. In the continuous-time case, x_n can be thought as readouts of an analogue signal at discrete times, as it actually happens in practice. Formally, continuous-time dynamical systems are constructed from the solutions of ordinary differential equations and are called *flows* [98]. When solving differential equations numerically, the time variable is discretized anyway [173].

Permutation entropy made its first appearance in the analysis of univariate time series, i.e., sequences of real numbers—the only ones we will consider in this section. Given a finite time series[1] $x_0^{N-1} = x_0, x_1, \ldots, x_{N-1}$, take a sliding window of size $2 \leq L \ll N$ along the time series (each window comprising a symbol block $x_n^{n+L-1} = x_n, \ldots, x_{n+L-1}, 0 \leq n \leq N-L$) and count the number of blocks realizing a particular ordinal pattern $\pi \in \mathcal{S}_L$. The relative frequency of each $\pi \in \mathcal{S}_L$ in the sequence x_0^{N-1} is then

$$\hat{p}(\pi) = \frac{\left| \{ n : 0 \leq n \leq N - L, \, x_n^{n+L-1} \text{ is of type } \pi \} \right|}{N - L + 1}. \tag{2.1}$$

---

[1]For notational simplicity, we assume that one symbol is output per time unit. In this way, a time series can be labeled as a sequence.

J.M. Amigó, *Permutation Complexity in Dynamical Systems*,
Springer Series in Synergetics, DOI 10.1007/978-3-642-04084-9_2,
© Springer-Verlag Berlin Heidelberg 2010

This estimator of the probability of π converges with probability 1 to the true value in the limit of infinitely long time series, under the proviso that the underlying stochastic process is stationary or, at least, that the probability for $x_n < x_{n+k}$, $1 \leq k \leq L - 1$, does not depend on n [28]. Let us mention in passing that the ordinal pattern probability distributions have been calculated for some random processes and pattern lengths, like Gaussian, fractional Brownian, and autoregressive moving-average (ARMA) processes for $L \leq 4$ [30, 213]; see also [190].

The permutation entropy per symbol of order L of x_0^{N-1} is then defined as

$$h_L^*(x_0^{N-1}) = -\frac{1}{L} \sum_{\pi \in \mathcal{S}_L} \hat{p}(\pi) \log \hat{p}(\pi). \tag{2.2}$$

In the case of infinitely long sequences, one defines the permutation entropy of a sequence x_0^∞ as

$$h^*(x_0^\infty) = \lim_{L \to \infty} h_L^*(x_0^\infty), \tag{2.3}$$

provided the limit exists.

The general procedure followed so far is well known to the practitioners of nonlinear time analysis: L is the *embedding dimension* and the *delay time T* is here 1 (since we take consecutive entries). As the window of size L slides along the time series x_0^∞, the vectors $\mathbf{x}_n = x_n^{n+L-1} \in \mathbb{R}^L$ describe the so-called *reconstructed trajectory* in the L-dimensional embedding space [1, 112, 166, 197]. The changes to be done when the sequences

$$x_n, x_{n+T}, \ldots, x_{n+(L-2)T}, x_{n+(L-1)T}, \tag{2.4}$$

have a delay time $T > 1$, are merely a matter of form but not of concept. Note that for deterministic sequences $x_n = f^n(x_0)$, $n \geq 0$, subsequence (2.4) is an orbit segment of f^T.

In general, h_L^* and h^* are defined for arbitrary-alphabet sequences whose symbols can be linearly ordered, while Shannon entropy applies to finite-alphabet sequences.[2] In practice all alphabets are finite because of the finite precision of the observation device and/or the finite real number representation of the computers. Such being the case, let $\mathbf{X} = \{X_n\}_{n \in \mathbb{N}_0}$ be the actual data source of the sequences x_0^∞, where now x_i are "discretized" values drawn from a finite alphabet S, and

$$h(\mathbf{X}) = -\lim_{n \to \infty} \frac{1}{L} \sum_{x_0, \ldots, x_{L-1} \in S} p(x_0, \ldots, x_{L-1}) \log p(x_0, \ldots, x_{L-1}),$$

---

[2]Real-valued data sources call for the concept of differential entropy [59].

its Shannon entropy. Usually, $h(\mathbf{X})$ is estimated by means of the so-called *plug-in, maximum likelihood,* or *naive estimator*

$$\hat{h}_L(x_0^{N-1}) = -\frac{1}{L} \sum \hat{p}(a_0 \dots a_{L-1}) \log \hat{p}(a_0 \dots a_{L-1}), \qquad (2.5)$$

where the summation is over all blocks $a_0^{L-1} = a_0 \dots a_{L-1} \in S^L$, and

$$\hat{p}(a_0 \dots a_{L-1}) = \frac{\left| \{ n : 0 \le n \le N-L, \, x_n^{n+L-1} = a_0^{L-1} \} \right|}{N-L+1} \qquad (2.6)$$

is the relative frequency of a_0^{L-1} in x_0^{N-1}.

Important for us is that if the process $\mathbf{X}$ is stationary and ergodic, then $h^*(x_0^\infty) = h(\mathbf{X})$ for a "typical" sequence (Chap. 6, Theorem 8). Therefore, in such cases $h_L^*(x_0^{N-1})$, with $L \ll N$, can be used as an estimator of $h(\mathbf{X})$ instead of (2.5). The numerical estimation of entropy via ordinal patterns will be discussed with more detail in Sect. 6.4, once the theoretical underpinnings of metrical permutation entropy of maps have been elucidated. At this point it suffices to advance that the computation is fast but the convergence is in general slow.

The slow convergence of h_L^* to the Shannon entropy seems to require great values of L for an accurate estimation. On the other hand, the superexponential growth of $|S_L| = L!$ makes exhaustive sampling computationally unfeasible for, say, $L \gtrsim 12$, even if there would be enough data at our disposal. In Chap. 7 we shall learn sampling techniques that work pretty well in these cases. In practice, the estimation of both Shannon entropy and permutation entropy (or, for that matter, of any quantity involving the limit $L \to \infty$) suffers from *undersampling* when L becomes sufficiently large as compared to the length N of the sequence. Undersampling means that the observed relative frequencies (of blocks or ordinal patterns) are no longer good estimators of the corresponding probabilities, simply because the samples are too small to be statistically significant. The following first-order correction due to finite sample effects was proposed by Herzel [93]:

$$\hat{h}_L(x_0^{N-1}) \longleftarrow \hat{h}_L(x_0^{N-1}) - \frac{M_1}{2M_2}, \qquad (2.7)$$

where M_1 is the number of words a_0^{L-1} with positive probabilities and M_2 is the number of samples ($M_2 = N - L + 1$ when the sequence is sampled by means of overlapping sliding windows, see (2.6)). In principle, the samples should be independent, but as stated in [94], the results are also satisfactory when the words overlap. Other corrections have been discussed by Grassberger [88] (who generalizes (2.7)) and Schmitt et al. [181] (who exploit Shannon–McMillan–Breiman's theorem of asymptotic equidistribution). Sometimes extrapolation techniques perform fine when undersampling occurs. One of them [195, 6] calls for plotting the partial entropies h_L^* against $1/L$; if the graph exhibits a distinctive linear part (showing

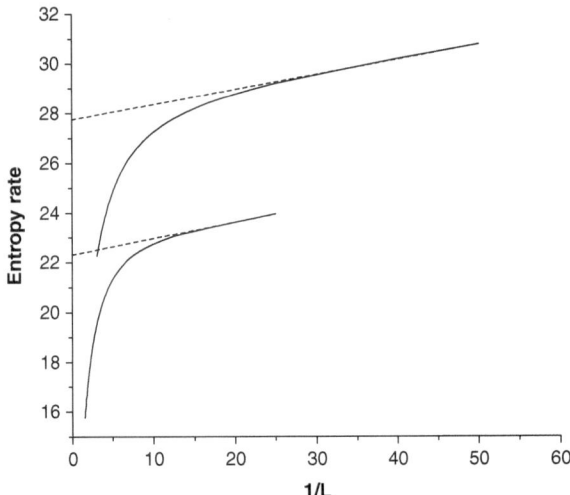

Fig. 2.1 Extrapolating the linear part (if any) of h_L^* vs $1/L$, over the undersampled values. The *continuous lines* correspond to entropy rates of finite order obtained from neurological time series

that h_L^*/L has already converged), then one extrapolates with a straight line this linear part till it intercepts the vertical axis ($1/L \rightarrow 0$), Fig. 2.1. See [127] for other methods to estimate the Shannon entropy and [167] for a review on entropy estimation.

Summing up, permutation entropy ("counting ordinal patterns") provides a conceptually simple and computationally fast method to estimate Shannon entropy. When compared to the usual block-based estimators ("counting blocks"), there is a difference that can be important in applications: the number of ordinal L-patterns does not depend on the alphabet. Specifically, the maximal number of length-L blocks (Shannon entropy) and length-L ordinal patterns (permutation entropy) grows with L as

$$|S|^L = e^{L \ln|S|} \quad \text{and} \quad L! \sim e^{L \ln L},$$

respectively, where S is the alphabet. It follows that if $|S|$ is very large, undersampling might set in earlier for block-based estimation than for ordinal pattern-based estimation. This occurs precisely with real-world or computer-generated data. Such an advantage has been reported in the literature, also in the computation of the *Rényi entropy*

$$h_{R_\alpha}(\mathbf{X}) = \lim_{L \to \infty} \frac{1}{L} \frac{1}{1-\alpha} \log \left(\sum_{x_0,\ldots,x_{L-1} \in S} p(x_0,\ldots,x_{L-1})^\alpha \right), \quad (2.8)$$

where $\alpha \geq 0$, $\alpha \neq 1$ ($\lim_{\alpha \to 1} h_{R_\alpha}(\mathbf{X}) = h(\mathbf{X})$), and the *Tsallis entropy*

$$h_{T_q}(\mathbf{X}) = \lim_{L \to \infty} \frac{1}{L} \frac{1}{q-1} \sum_{x_0,\dots,x_{L-1} \in S} \left(p(x_0,\dots,x_{L-1}) - p(x_0,\dots,x_{L-1})^q \right), \quad (2.9)$$

where $q \in \mathbb{R}$, $q \neq 1$ [213]. When $p(x_0,\dots,x_{L-1})$ is replaced in (2.8) and (2.9) by $p(\pi)$, $\pi \in S_L$ (or estimated by the relative frequency $\hat{p}(\pi)$), one speaks of the *Rényi permutation entropy* and the *Tsallis permutation entropy*, respectively. To complete the picture, let us add that the situation reverses when the alphabet comprises few symbols. But in this case, Lempel–Ziv complexity (specifically, LZ-76) can be a better choice than block counting [6]; see [82] for the entropy estimation in binary sequences.

Although less used than the "Shannon permutation entropy" h^*, one can also define the *topological permutation entropy* or *permutation capacity*,

$$h_0^*(x_0^\infty) = -\lim_{L \to \infty} h_{0,L}^*(x_0^\infty), \quad (2.10)$$

provided the limit exists, where the rate of finite order is given as

$$h_{0,L}^*(x_0^\infty) = -\frac{1}{L} \log N(L), \quad (2.11)$$

$N(L)$ being the number of distinct ordinal patterns defined by sliding windows x_n^{n+L-1} of size L. That is, we just count now how many different L-patterns are realized, instead of computing the relative frequency of those L-patterns. It follows that h_0^* is an upper bound of h^*. When the sequences x_0^∞ are seen as outputs of an information source $\mathbf{X}$, then $N(L)$ stands for the number of admissible L-patterns in the messages that $\mathbf{X}$ can emit, and one speaks of the permutation capacity or topological permutation entropy of $\mathbf{X}$ (Chap. 7).

The ordinal pattern-based approach to Shannon entropy can also be extended to the metric and topological entropy of maps; see Chaps. 7 and 8. The situation is specially simple for one-dimensional, piecewise monotone interval maps $f{:}I \to I$. In this case, we only need to numerically estimate the probabilities $\mu(P_\pi)$ of the admissible L-patterns ($P_\pi \neq \emptyset$), or just the number of distinct admissible patterns, to get an estimate of the metric or topological entropy of f, respectively (see (1.30), (1.31), and (1.32)). Thus, the estimation of $h_\mu^*(f)$ and $h_{\text{top}}^*(f)$ boils down again to counting ordinal L-patterns. The computation of $h_{\text{top}}^*(f)$ is also simpler than for its standard counterpart. The higher dimensional case will also be considered in Chaps. 6 and 7.

2.2 Permutation Complexity

Although complexity, (pseudo-)randomness, disorder, irregularity, typicality, etc., are terms that have been introduced eventually in different settings to mean more or less the same dynamical behavior, complexity is the preferred one when there is no measure (or probability) involved. In fact, Bandt and Pompe introduced permutation entropy in [28] via (2.1), (2.2), and (2.3) as a "natural complexity measure for time series." The time series can be the output of a random process or an orbit of a dynamical system. By analyzing the complexity of a signal (if no other information available), we are inquiring into the complexity of the source. An axiomatic characterization of complexity was proposed in [163].

The measurement of complexity and its eventual time variation is an issue of utmost important in the analysis of biomedical, economic, physical, and technical time series. Think of the forecasting of transitions to abnormal health conditions, financial crashes, severe weather, earthquakes, etc. Over the years, a battery of methods has been proposed and developed with this purpose or adapted from other fields like information theory and networks. Let us mention some of these methods (see also the references therein):

- Cross-correlation sum analysis [111]
- Lempel–Ziv complexity [208, 196, 90, 6, 78]
- Mutual information [90]
- Nonlinear cross-prediction analysis [183]
- Recurrence plots [73, 144, 200] and recurrence quantification analysis [81]
- Relative entropy [180]
- Statistical complexity [56, 143] (statistical complexity was introduced by Crutchfield and collaborators within a theory called computational mechanics [60, 185, 24])
- Statistical tests in the reconstructed phase space [120]
- Topological methods [209]

Permutation entropy and other related quantities are specially well suited to measure the complexity of random and deterministic dynamical systems for several reasons.

First of all, permutation entropy in its different variants involves counting ordinal patterns. With the exception of a few cases, the number of ordinal L-patterns realized by a map f increases with L. Therefore, the (logarithm of the) rate of this increasing is a natural measure (as stated by Bandt and Pompe) for quantifying the complexity of a deterministic time series or, more generally, of a dynamical system. In the metric variant, each admissible L-pattern contributes to the entropy a term containing its relative frequency or probability, respectively. In the topological variant, all such patterns make the same contribution to the entropy; formally, they are assigned the same probability. Since random, unconstrained processes have no forbidden patterns with probability 1 (hence, they have a superexponential growth of admissible ordinal patterns with length), their complexity, as measured by the permutation entropy, is infinite. At the other end, a periodic or quasiperiodic dynamic has vanishing or negligible permutation entropy. Complex systems lie between order and randomness. From a practical point of view, we can characterize them as having a positive, finite permutation entropy. Both metric and topological permutation entropies increase as the sequence "looks" more random.

Second, unlike other proposals for complexity measures, permutation entropy applies in principle both to finite-alphabet and arbitrary-alphabet sequences, albeit it is more interesting in the second case.

Technically we are assuming that the limits involved in the corresponding definitions (like (2.3) and (2.10)) converge. In practice, limits have to be estimated using a finite number of terms—real sequences are finite anyway. What we mean is that the actual tools of permutation complexity are going to be the permutation entropy rates of finite order, like $h_L^*(x_0^{N-1})$ and $h_{0,L}^*(x_0^{N-1})$, and other related quantities based on finite-length ordinal patterns, like probability distributions, information-theoretical tools (relative entropy, mutual information, etc.), complexity functionals. Moreover, since the maximal value of $h_L^*(x_0^{N-1})$ and $h_{0,L}^*(x_0^{N-1})$ is $\log L!$, we can eventually divide both entropy rates by $\log L!$ to obtain dimensionless quantities ranging between the two non-complex extremes: 0 (order) and 1 (randomness).

Finally, permutation entropy rates of finite order are computationally fast for the pattern lengths used in practice ($3 \leq L \leq 7$)—also for the Rényi (2.8) and Tsallis (2.9) permutation entropies. This allows calculation in real time, which is a significant advantage in applications. We come back to this point in the next chapters.

Application of ordinal patterns and permutation entropy to complexity analysis of data has been reported in different fields. For instance

- biomedical series [116, 45, 118]
- financial series [146, 147]
- physical series [28]
- statistical series [30, 146, 147, 212]

Let us underline at this point that the application by Keller [116] of ordinal patterns to electroencephalogram (EEG) data from children with epileptic disorders dates from about the same time as permutation entropy was formulated [28].

Similarly, one of the first applications of permutation entropy was the detection of dynamical changes in time series and, in particular, epileptic seizure detection from EEGs by Cao et al. [45]. Regarding the second application, the authors analyzed continuous EEG measurements recorded intracranially (also called depth EEG) with typically 28 electrodes. Figure 2.2 shows the normalized permutation entropy rate of order $L = 5$ for three different patients. Each signal is more than 5 h long, with a sample frequency of 200 Hz and time delay 3 (i.e., only every third entry in the EEG signal is taken into account, what amounts to sampling the signal with frequency 200/3 Hz). According to [45], the change of permutation complexity in all these cases indicates that the dynamics of the brain first becomes more regular right after the seizure, then its irregularity increases as it approaches the normal state.

Since these and other pioneering works, ordinal analysis of time series has remain a popular technique. In some cases, ordinal analysis has been incorporated into more general schemes, such as the *method of recurrence plots*, introduced by Eckmann et al. [73] to visualize the recurrences of dynamical systems. This method, which is being used to analyze virtually any natural data [144], is based on the *recurrence matrix* of a scalar or vectorial trajectory $(x_i)_{i=0}^{N-1}$ of a system in its state space S, defined as

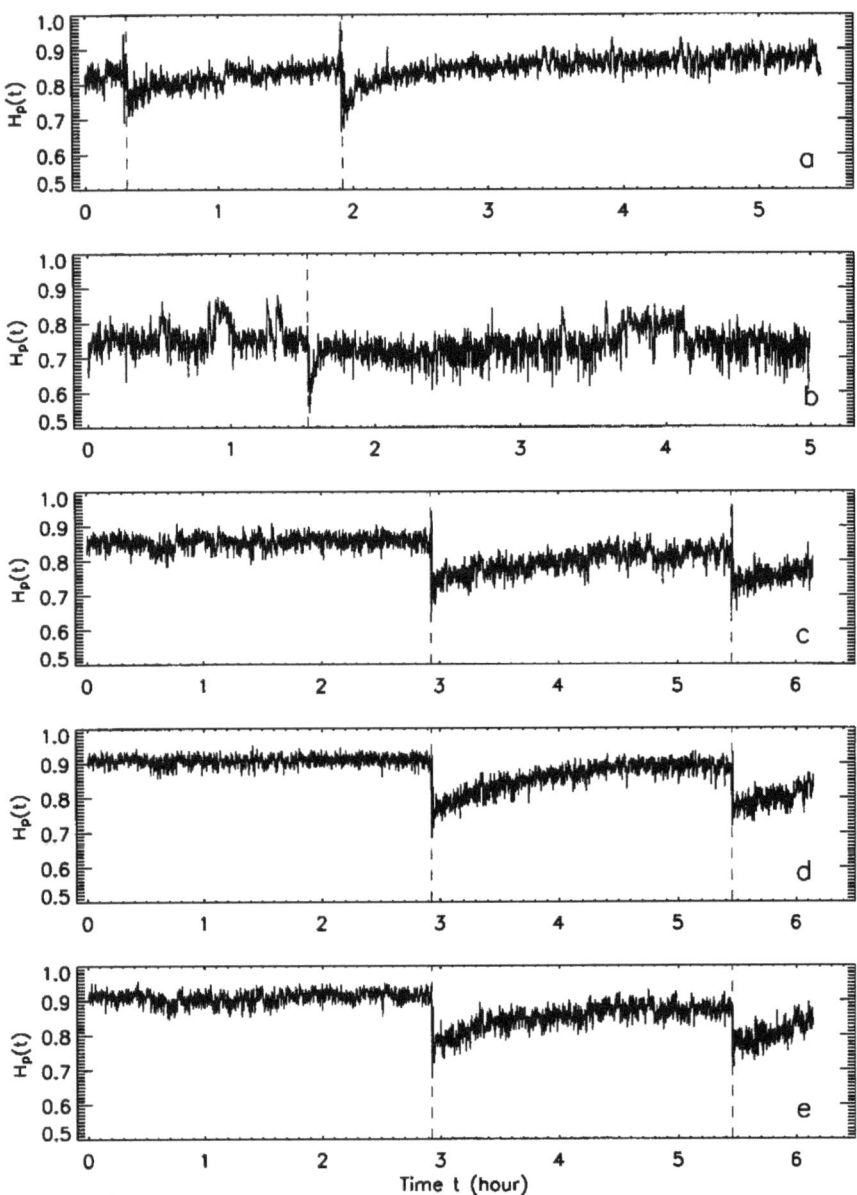

Fig. 2.2 [Reproduced with permission from [45].] Variation of the normalized h_5^* with time for EEG signals of (**a**) patient 1, channel 1, (**b**) patient 2, channel 1, and (**c**)–(**e**) patient 3, channels 1–3

$$\mathbf{R}_{i,j}(\varepsilon) = H(\varepsilon - \|x_i - x_j\|), \quad i,j = 0,\ldots,N-1, \tag{2.12}$$

where ε is a threshold distance, $H(\cdot)$ is the *Heaviside function* ($H(x) = 0$ if $x < 0$ and $H(x) = 1$ otherwise), and $\|\cdot\|$ is a norm in S. Instead of using spatial closeness as in (2.12), *ordinal patterns recurrence plots* are based on the ordinal patterns $\pi(i)$ realized by the sequences x_i^{i+L-1}, $0 \le i \le N - L$. If $\delta(\pi, \pi') = 1$ for $\pi = \pi' \in \mathcal{S}_L$, and $\delta(\pi, \pi') = 0$ otherwise, set

$$\mathbf{R}_{i,j}(L) = \delta(\pi(i), \pi(j)), \tag{2.13}$$

$\pi(i), \pi(j) \in \mathcal{S}_L$, $0 \le i, j \le N - L$. According to [144], the main advantage of (2.13) is its robustness against non-stationary data.

To distinguish the kind of complexity captured by the tools of ordinal analysis— ordinal patterns, permutation entropy, permutation entropy rates of finite order, and other quantities based on order relations—we propose to call it *permutation complexity*. Therefore, permutation complexity has to do with the ordinal structure of data obtained from deterministic or random dynamical systems. These also include spatially extended systems, like the ones we shall consider in Chap. 10.

2.3 Estimation of Control Parameters from Symbolic Sequences

The basis of permutation complexity is the relation between order and dynamics. This relation is specially strong on one-dimensional intervals, where order and metric are intertwined, leading to such interesting results as Sarkovskii's theorem [179, 150]. It is therefore not surprising that the study of the ordinal structure of time series provides valuable information on the underlying dynamical system. In this section we learn how to recover the "control" parameter of a unimodal map from itineraries. The relationship between the itineraries of parametric unimodal maps and the value of the parameter that controls a particular dynamics was shown in [153, 203, 5].

Let $\mathcal{U}$ be the class of unimodal maps on an interval $I = [a, b] \subset \mathbb{R}$. A map $f{:}I \to I$ is *unimodal* if it is continuous, has a single turning point (called hereafter the *critical point*) x_c in I, and is monotone increasing on the left of x_c and decreasing on the right. The class $\mathcal{U}$ includes maps defined in a parametric way, say, $f_v(x) = \varphi(v, x)$, where $x \in I$, $v \in J \subset \mathbb{R}$ will be called the *control parameter*, and φ is a map on $I \times J$.

The class $\mathcal{U}$ includes the *logistic family* $g_v{:}[0, 1] \to [0, 1]$,

$$g_v(x) = vx(1 - x), \tag{2.14}$$

where $0 \le v \le 4$, and the *tent family* $\Lambda_v{:}[0, 1] \to [0, 1]$,

$$\Lambda_v(x) = \begin{cases} x/v & \text{if } 0 \le x \le v, \\ (1 - x)/(1 - v) & \text{if } v \le x \le 1, \end{cases} \tag{2.15}$$

where $0 < v < 1$; see Fig. 2.3. In particular, g_4 is the logistic map (1.19) and $\Lambda_{1/2}$ the symmetric tent map (1.17). The critical point of g_v does not depend on v: $x_c = \frac{1}{2}$

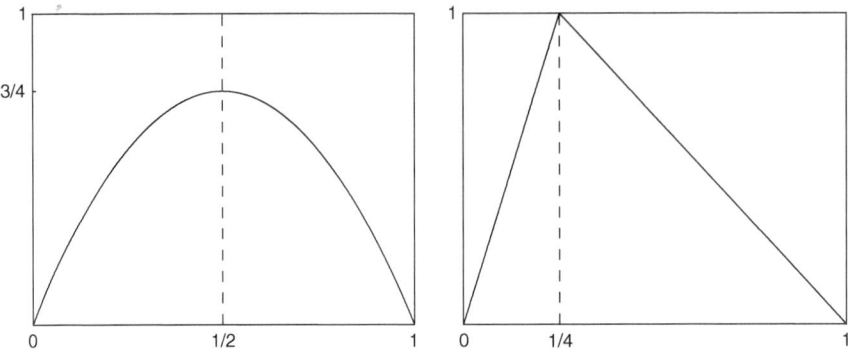

Fig. 2.3 Graphs of the logistic map g_v with $v = 3$ (*left*) and the tent map Λ_v with $v = 0.25$ (*right*)

for all v. On the opposite side, the critical point of Λ_v coincides with the parameter value: $x_c = v$. As usual in the literature, we will also refer to g_v and Λ_v just as the logistic and tent maps, respectively, when the parameter v is thought to be fixed. Note that Λ_v preserves the Lebesgue measure for all $v \in (0, 1)$.

For $f \in \mathcal{U}$, let $\Phi(x)$ be the itinerary of $x \in [a, b]$ with respect to the partition $\{A_0, A_1\}$, with $A_0 = [a, x_c)$ and $A_1 = [x_c, b]$. Specifically,

$$\Phi(x) = \Phi_0(x), \Phi_1(x), \ldots, \Phi_n(x), \ldots = (\Phi_i(x))_{i=0}^{\infty}, \tag{2.16}$$

where

$$\Phi_n(x) = \begin{cases} 0 & \text{if } f^n(x) < x_c, \\ 1 & \text{if } f^n(x) \geq x_c. \end{cases}$$

As a result, any orbit $\mathcal{O}_f(x)$ can be encoded into a binary sequence. Whenever convenient, we will write $\Phi(f, x)$ instead of $\Phi(x)$ to make clear which unimodal map is generating the itinerary of x.

An interesting aspect of the binary sequences $\Phi(x)$ is that they can be endowed with a *signed lexicographical order* (sometimes called *Gray ordering*) $\leq$ that is equivalent to the order in $[a, b]$ in the following weakened sense:

(E1) If $x < y$, then $\Phi(x) \leq \Phi(y)$.
(E2) If $\Phi(x) < \Phi(y)$, then $x < y$.

A sufficient condition for $x < y$ if and only if $\Phi(x) \leq \Phi(y)$ is given in [57, Theorem II.5.4]. The order between binary sequences is defined as follows. Given $\Phi(x) \neq \Phi(y)$, let $i_{\min}$ be the first index such that $\Phi_i(x) \neq \Phi_i(y)$, $i \geq 0$. Depending on $i_{\min}$, three cases can occur:

(O1) If $i_{\min} = 0$, then $\Phi(x) < \Phi(y)$ iff $\Phi_0(x) < \Phi_0(y)$.

(O2) If $i_{\min} > 0$ and $\{\Phi_i(x):0 \le i < i_{\min}\}$ contains an *even* number of 1's, then
$\Phi(x) < \Phi(y)$ iff $\Phi_{i_{\min}}(x) < \Phi_{i_{\min}}(y)$.
(O3) If $i_{\min} > 0$ and $\{\Phi_i(x):0 \le i < i_{\min}\}$ contains an *odd* number of 1's, then
$\Phi(x) < \Phi(y)$ iff $\Phi_{i_{\min}}(x) > \Phi_{i_{\min}}(y)$.

Given $x, f_v(x), \dots, f_v^{L-1}(x)$, suppose that their corresponding itineraries, namely,

$$(\Phi_i(x))_{i=0}^{\infty}, (\Phi_i(x))_{i=1}^{\infty}, \dots, (\Phi_i(x))_{i=L-1}^{\infty},$$

are all different. Then, according to (E1)–(E2),

$$f^{\pi_0}(x) < \cdots < f^{\pi_{L-1}}(x) \Leftrightarrow (\Phi_i(x))_{i=\pi_0}^{\infty} < \cdots < (\Phi_i(x))_{i=\pi_{L-1}}^{\infty}. \tag{2.17}$$

Before proceeding further, let us point out that this setting can be extended to *l-modal maps*, i.e., continuous and piecewise strictly monotone self-maps of compact intervals with l local maxima, which map endpoints to endpoints. For the applications we will discuss, it is sufficient to consider only unimodal maps ($l = 1$).

In some applications, one is confronted with the following task: given the "sharp" orbit $\mathcal{O}_{f_v}(x_0)$ of $x_0 \in [a, b]$ under $f_v \in \mathcal{U}$, find the value of the parameter v. In practice, the exact values of $\mathcal{O}_{f_v}(x_0)$ are seldom known because of the finite precision of real number computation, so one has only access to a (finite segment of a) "coarse-grained" orbit $(\hat{x}_i)_{i=0}^{\infty}$, where $\hat{x}_i$ is an approximation to $x_i = f_v^i(x_0)$. In some chaos-based cryptosystems, the situation is even worse: the plaintext (i.e., the message to be encrypted prior to its transmission or storage) is encoded via the symbolic sequences (2.16) of a chaotic map $f_v \in \mathcal{U}$, the value v being part of the secret key of the cipher (see, e.g., [131]). Therefore, the cryptanalist has eventually only access to the binary code $\Phi(f_v, x)$ (via a so-called chosen-text attack) to recover the control parameter v. D. Arroyo has shown how to recover v with the aid of the ordinal patterns of f_v and their itineraries $\Phi(f_v, x)$, if f_v is ergodic with respect to its natural measure μ_v for all values of v [21].

For simplicity, the estimation of v from the symbolic sequences $\Phi(f_v, x)$ will be illustrated using the tent map Λ_v, which is chaotic for all $v \in (0, 1)$. Since the natural invariant measure of Λ_v is the Lebesgue measure, the probability that x is of type $\pi \in \mathcal{S}_L$ when drawn uniformly from $[0, 1]$ equals the length of $P_\pi = \{x \in [0, 1]:x$ defines $\pi\}$ (as in (1.27)). By ergodicity, the relative frequency of π in an orbit of Λ_v coincides with the length of P_π, except possibly for a set of initial conditions with length zero. For the tent map, the length of the sets P_π can be determined analytically. The simplest case corresponds to the L-pattern $\langle 0, 1, \dots, L-1 \rangle$:

$$P_{\langle 0,1,\dots,L-1 \rangle} = (0, \phi_L(v)),$$

where $\phi_L(v)$ is the leftmost intersection of Λ_v^{L-1} and Λ_v^{L-2}. Therefore, the length of $P_{\langle 0,1,\dots,L-1 \rangle}$, hence the probability that x is of type $\langle 0, 1, \dots, L-1 \rangle$ when drawn uniformly from the interval $[0, 1]$ is $\phi_L(v)$.

In order to calculate $\phi_L(v)$, use

$$\Lambda_v^n(x) = \begin{cases} x/v^n & \text{if } 0 \leq x \leq v^n, \\ (v^{n-1} - x)/v^{n-1}(1 - v) & \text{if } v^n \leq x \leq v^{n-1}. \end{cases}$$

Equating Λ_v^{L-1} and Λ_v^{L-2}, it follows

$$\phi_L(v) = \frac{v^{L-2}}{2 - v}. \tag{2.18}$$

Note that this function is 1-to-1 in the interval $0 \leq v \leq 1$ for $L \geq 2$, with $\phi_2(0) = \frac{1}{2}$, $\phi_L(0) = 0$ for $L \geq 3$, and $\phi_L(1) = 1$ for $L \geq 2$. This fact allows to determine v from $\phi_L(v)$. Furthermore, from the equation

$$\frac{d}{dv}\phi_L(v) = \frac{v^{L-3}}{(2 - v)^2}[2(L - 2) - (L - 3)v] = \begin{cases} 0 & \text{if } v = 0, \\ L - 1 & \text{if } v = 1, \end{cases} \tag{2.19}$$

it follows that $\phi_L(v)$ is a $\cup$-convex map on $0 \leq v \leq 1$ for $L \geq 2$ that converges to 0 on $0 \leq v < 1$ (i.e., it "flattens") as $L \to \infty$. As a result, the higher the L the lower the precision with which v can be numerically read off from $\phi_L(v)$. Consequently, $L = 3, 4$ are the best choices for a quality estimation of v.

In more general terms, suppose that each $f_v \in \mathcal{U}$ is ergodic for $v \in J$ with the same invariant measure μ. Furthermore assume for the time being that $f_v(a) = a$ and $f_v(x) > x$ on a non-empty vicinity of a. Let (a, c) be the maximal interval in (a, x_c) such that $f_v(x) > x$. We claim that the interval

$$I_L^v = (a, c) \cap f_v^{-1}(a, c) \cap \cdots \cap f_v^{-(L-1)}(a, c)$$

coincides with $P_{\langle 0,1,\dots,L-1\rangle}$. Indeed, if $x \in I_L^v$, then $f_v^i(x) \in (a, c)$ for $0 \leq i \leq L - 1$, and

$$x < f_v(x) \Rightarrow f_v(x) < f_v^2(x) \Rightarrow \cdots \Rightarrow f_v^{L-2}(x) < f_v^{L-1}(x).$$

Hence, $I_L^v \subset P_{\langle 0,1,\dots,L-1\rangle}$. Conversely, if $x \in P_{\langle 0,1,\dots,L-1\rangle}$, i.e.,

$$x < f_v(x) < f_v^2(x) < \cdots < f_v^{L-1}(x),$$

then $f_v^i(x) \in (a, c)$ for $0 \leq i \leq L - 1$. Thus, $P_{\langle 0,1,\dots,L-1\rangle} \subset I_L^v$. This proves

$$I_L^v = P_{\langle 0,1,\dots,L-1\rangle}. \tag{2.20}$$

If otherwise $f_v(a) = a$ but $f_v(x) < x$ on a non-empty vicinity of a, then let (a, c) be the maximal interval in (a, x_c) such that $f_v(x) < x$. In this case, a similar reasoning (reversing the inequalities) shows that

$$I_L^v = P_{\langle L-1,L-2,\dots,1,0\rangle}. \tag{2.21}$$

Since the tent map, our workhorse in this section, complies with (2.20), we restrict attention to this case (similar arguments apply mutatis mutandis to case (2.21)). Because of ergodicity, the relative frequency at which a typical trajectory visits I_L^v is $\mu(I_L^v)$. If $\mu(I_L^v)$ happens to be different for each v, then $\mu(I_L^v)$ can be used to determine or estimate the control parameter v. In this case, the relative frequency of the ordinal pattern $\langle 0, 1, \ldots, L-1 \rangle$ in an orbit $\mathcal{O}_{f_v}(x)$ is just the number of times that $f_v^{i+j}(x) \in (a, c)$ for $i \in \mathbb{N}_0$ and $j = 0, 1, \ldots, L-1$.

Figure 2.4 shows the relative frequencies of the ordinal patterns (a) $\langle 0, 1, 2, 3 \rangle$, (b) $\langle 0, 1, 3, 2 \rangle$, (c) $\langle 0, 3, 1, 2 \rangle$, and (d) $\langle 3, 0, 1, 2 \rangle$ found in a numerical simulation with the tent map. As expected, curve (a) approximates the function

$$\phi_4(v) = \frac{v^2}{2 - v}$$

with great precision. Observe that a 1-to-2 functional relation between frequency and v, as it occurs in Fig. 2.4 (b)–(d), can also be acceptable, e.g., for cryptographic applications since it implies a reduction of the secret key space.

So far we have shown the possibility of recovering the control parameter v from the relative frequency of the pattern $\pi = \langle 0, 1, \ldots, L-1 \rangle$ (most conveniently for $L = 3, 4$), in a statistically significant sample of orbits of Λ_v. The ergodicity of Λ_v with respect to the Lebesgue measure on $[0, 1]$ and the 1-to-1 relation between v and

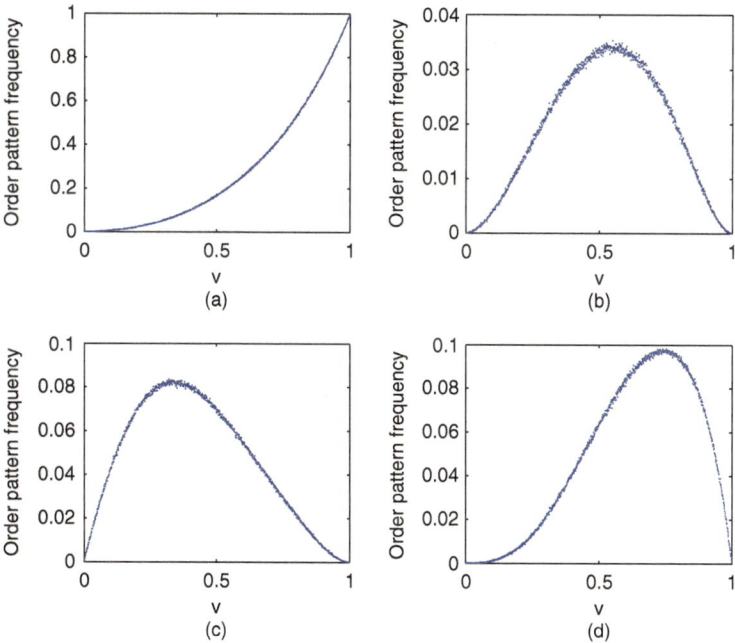

Fig. 2.4 Ordinal pattern frequencies for the tent map family. Here $L = 4$, and (a) $\pi = \langle 0, 1, 2, 3 \rangle$, (b) $\pi = \langle 0, 1, 3, 2 \rangle$, (c) $\pi = \langle 0, 3, 1, 2 \rangle$, and (d) $\pi = \langle 3, 0, 1, 2 \rangle$

the probability $\phi_L(v)$ of observing π were instrumental to achieve that goal. What about if we have access only to coarse-grained orbits $\Phi(\Lambda_v, x)$?

Let $b_0^{M-1} = b_0 b_1 \ldots b_{M-1}$, $b_i \in \{0, 1\}$, be the initial segment of length M of the symbolic sequence $\Phi(f_v, x)$. Take a sliding window of size $W < M$ along b_0^{M-1}. The result is $M - W + 1$ consecutive blocks of length W:

$$b_0^{W-1} = b_0 \ldots b_{W-1}, \ldots, b_i^{i+W-1} = b_i \ldots b_{i+W-1}, \ldots, b_{M-W}^{M-1} = b_{M-W} \ldots b_{M-1}.$$

The blocks b_i^{i+W-1}, $i = 0, 1, \ldots, M - W - L + 1$, define $M - W - L + 2$ ordinal patterns of length L. That is, if

$$b_{i+\pi_0}^{i+\pi_0+W-1} < b_{i+\pi_1}^{i+\pi_1+W-1} < \cdots < b_{i+\pi_{L-1}}^{i+\pi_{L-1}+W-1}, \tag{2.22}$$

then b_i^{i+W-1} is of type $\pi = \langle \pi_0, \pi_1, \ldots, \pi_{L-1} \rangle$. The order for finite sequences in (2.22) is defined the same way as for infinite sequences in (O1)–(O3).

Each block $b_i^{i+W-1} = b_i \ldots b_{i+W-1}$ locates $f_v^i(x)$ up to an uncertainty interval whose length goes to zero when $W, M \to \infty$:

$$f_v^i(x) \in A_{b_i} \cap f_v^{-1} A_{b_{i+1}} \cap \cdots \cap f_v^{-(W-1)} A_{b_{i+W-1}}.$$

This being the case, the ordinal patterns defined by, say, $x, f_v(x), f_v^2(x)$, and b_0^{W-1}, b_1^W, b_2^{W+1} may be different as soon as two of the latter blocks overlap. Otherwise, those ordinal patterns will be the same because of (2.17). In sum, the relative frequencies of an ordinal L-pattern in the finite orbits $(f_v(x))_{i=0}^M$ and $(b_i^{i+W-1})_{i=0}^{M-W-L+1}$ will converge to each other in the limit $M \to \infty$, $W \to \infty$ ($W < M$). In practice, we expect the latter to be a good approximation of the former, at least for $L = 3, 4$, and W large enough, so that a good estimation of the control parameter is feasible.

Figure 2.5 shows the relative frequencies of the same 4-patterns as in Fig. 2.4 for the itineraries of the tent map family. Here $M = 10, 104$ and $W = 100$. Except for $v \simeq 0$ (an uninteresting region for cryptographic applications), the approximation is excellent. Some caveats related to the finite precision of the numerical simulations are discussed in [21]. In practical cases, the error in the estimation of the control parameter ranges between 10^{-3} and 10^{-4}. From the viewpoint of cryptographic applications, this amounts to a strong reduction of the key space, which compromises the security of the cipher.

The tent map family is a specimen of a more general family: unimodal, piecewise linear expanding Markov transformations (Annex A, Definition 9). Each topologically transitive transformation in this family (i.e., some power of its transition matrix is strictly positive) has a unique ergodic invariant measure, which furthermore is absolutely continuous with respect to the Lebesgue measure [134]. This measure can be calculated or numerically estimated by a variety of methods (Perron–Frobenius operator, Ulam's method, or just computation of long time averages) [105]. For the purpose envisaged in this chapter, an exact knowledge of the invariant measures is

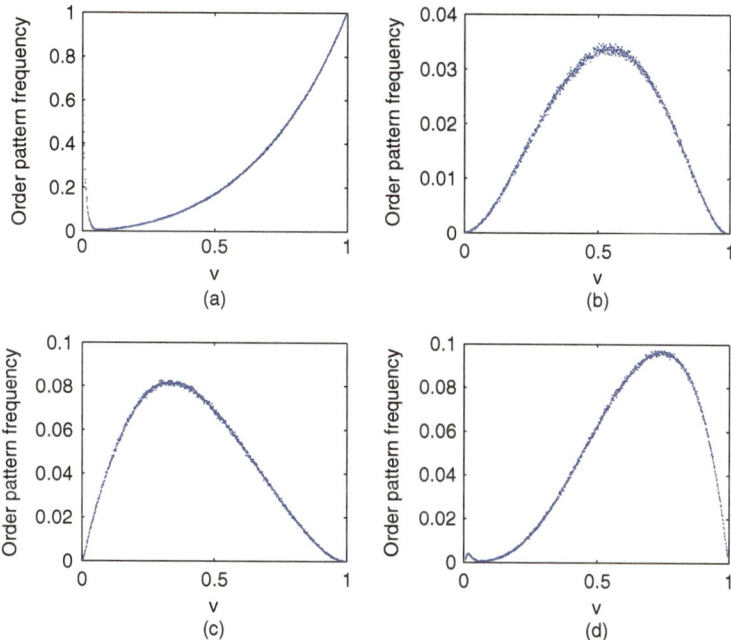

Fig. 2.5 Ordinal pattern frequencies of the tent map family using itineraries. Here $L = 4$, $W = 100$, $M = 10104$, and (**a**) $\pi = \langle 0, 1, 2, 3 \rangle$, (**b**) $\pi = \langle 0, 1, 3, 2 \rangle$, (**c**) $\pi = \langle 0, 3, 1, 2 \rangle$, and (**d**) $\pi = \langle 3, 0, 1, 2 \rangle$

not necessary, since the relative frequencies of the ordinal patterns can be calculated with numerical simulations. The important features are ergodicity, so that the statistical properties of the sharp and coarse-grained orbits do not depend on the initial conditions, and the absolute continuity of the (unique) invariant measure, which guarantees that it is accessible by numerical methods.

2.4 Characterizing Synchronization

As a last application, we are going to summarize the work of R. Monetti et al. [159] on characterizing synchronization in time series using ordinal patterns (therein called "symbols") and some related probability distributions.

Remember that S_L is a group with respect to the product of ordinal patterns (1.25). This being the case, given $\pi, \pi' \in S_L$ there always exists a unique $\tau = \tau(\pi, \pi') \in S_L$, called *transcription* from the *source pattern* π to the *target pattern* π', such that

$$\tau \circ \pi = \pi', \tag{2.23}$$

where (see (1.25))

$$\tau \circ \pi = \left\langle \pi_{\tau_0}, \pi_{\tau_1}, \dots, \pi_{\tau_{L-1}} \right\rangle.$$

It follows that τ is a transcription from π to π' if and only if τ^{-1} is a transcription from π' to π.

As the source pattern π and the target pattern π' vary over $\mathcal{S}_L$, their transcription varies according to $\tau(\pi, \pi') = \pi' \circ \pi^{-1}$. Note that different pairs (π, π') can share the same transcription. As an example in $\mathcal{S}_3$, $\tau(\pi, \pi') = \langle 0, 2, 1 \rangle$ for

$$(\pi, \pi') = (012, 021), (021, 012), (120, 102), (102, 120), (201, 210), (210, 201)$$

(angular parentheses omitted for brevity). More generally, given $\tau \in \mathcal{S}_L$, there exist $L!$ pairs $(\pi, \pi') \in \mathcal{S}_L \times \mathcal{S}_L$ such that τ is the transcript from π to π'.

Another concept we need is that of order of an element. We say that the *order* of $\pi \in \mathcal{S}_L$ is $\mathrm{ord}(\pi) \in \mathbb{N}$ if $\mathrm{ord}(\pi)$ is the minimal number of times we have to multiply π by itself to obtain the identity permutation $\langle 0, 1, \dots, L-1 \rangle$ (this is the only permutation whose order is 1).

The group $\mathcal{S}_L$ can be partitioned into non-overlapping sets of transcriptions according to their order. In mathematical notation, $\mathcal{S}_L = \cup_{1 \le i \le L!} \mathcal{C}_i$, where

$$\mathcal{C}_i = \mathcal{C}_i(L) = \{\tau \in \mathcal{S}_L : \mathrm{ord}(\tau) = i\}.$$

For obvious reasons, the sets $\mathcal{C}_i$ are called *order classes*. From $\mathrm{ord}(\tau^{-1}) = \mathrm{ord}(\tau)$, it follows that $\tau \in \mathcal{C}_i$ if and only if $\tau^{-1} \in \mathcal{C}_i$. Note that $\mathcal{C}_1(L) = \{\langle 0, 1, \dots, L-1 \rangle\}$. The authors of [159] propose to measure the complexity of a transcription between a source and a target pattern by its order.

A permutation of the form $i_1 \mapsto i_2 \mapsto \cdots \mapsto i_n \mapsto i_1$ is called a *cycle* (or cyclic permutation) *of length n* and denoted by $(i_1, i_2, \dots, i_n)$. The order of a cycle of length n is trivially n. It is also trivial that any permutation of $\{0, 1, \dots, L-1\}$ can be written as the product of disjoint cyclic permutations. It follows that the order of any transcription (or any permutation for that matter) is the least common multiple (lcm) of the lengths of its decomposition into cycles. In particular, given L there are ordinal patterns $\tau \in \mathcal{C}_i(L)$ of orders $1 \le i \le L$ (just take $\tau = (0, \dots, i-1)(i)(i+1)\cdots(L-1)$). For $L+1 \le i \le L!$, a hypothetical decomposition $\tau = (i_1, \dots, i_{n_1})(j_1, \dots, j_{n_2}) \cdots (k_1, \dots, k_{n_p})$, $\tau \in \mathcal{C}_i(L)$, has to fulfill the constraints (i) $n_1 + n_2 + \cdots + n_p = L$ and (ii) $\mathrm{lcm}\{n_1, n_2, \dots, n_p\} = i$, which will not be the case in general. For example, for $L = 7$ and $i = 10$ or 12, we can choose $n_1 = 2$ and $n_2 = 5$, or $n_1 = 3$ and $n_2 = 4$, respectively. But for $L = 7$ and $i = 8, 9$, or 11, conditions (i) and (ii) cannot be simultaneously satisfied.

Let us next turn attention to the probability density of transcriptions. Consider source and target ordinal patterns generated by the time series of a coupled dynamics. Due to the symmetry property between source and target patterns pointed out above, it is irrelevant which one refers to which subsystem, any of the two possible assignments being fine. Let $\mathcal{S}_L^s$ and $\mathcal{S}_L^t$ be the state spaces comprising the corresponding admissible source and target patterns of length L, respectively, and let $\Omega_L(\tau)$ be the set of all pairs $(\pi_s, \pi_t) \in \mathcal{S}_L^s \times \mathcal{S}_L^t$ such that $\tau \in \mathcal{S}_L$ is a transcription from π_s to π_t, i.e.,

$$\Omega_L(\tau) = \{(\pi_s, \pi_t) \in \mathcal{S}_L^s \times \mathcal{S}_L^t : \tau \circ \pi_s = \pi_t\}.$$

The probability density of transcriptions $P_L(\tau)$, $\tau \in \mathcal{S}_L$, can be written as

$$P_L(\tau) = \sum_{(\pi_s, \pi_t) \in \Omega_L(\tau)} P^J(\pi_s, \pi_t),$$

where $P^J(\pi_s, \pi_t)$ is the joint probability density. Furthermore, let $P^s(\pi_s)$, $P^t(\pi_t)$ be the marginal probability densities of the patterns $\pi_s \in S_s$ and $\pi_t \in S_t$, respectively. The matrix $M(\pi_s, \pi_t) = P^s(\pi_s)P^t(\pi_t)$ is the probability density matrix of transcriptions for two independent sequences of lengths L. In this case, the corresponding probability density of transcriptions $P_L^{\mathrm{ind}}(\tau)$ can be evaluated as follows:

$$P_L^{\mathrm{ind}}(\tau) = \sum_{(\pi_s, \pi_t) \in \Omega_L(\tau)} M(\pi_s, \pi_t).$$

A natural measure to assess how much $P_L(\tau)$ deviates from $P_L^{\mathrm{ind}}(\tau)$ is provided by the *relative entropy* or *Kullback–Leibler distance* (see Annex B, (B.3))

$$D(P_L \| P_L^{\mathrm{ind}}) = \sum_{\tau \in \mathcal{S}_L} P_L(\tau) \log \frac{P_L(\tau)}{P_L^{\mathrm{ind}}(\tau)}.$$

To circumvent the asymmetry of the relative entropy with respect to its arguments, one can take the harmonic mean of $D(P_L \| P_L^{\mathrm{ind}})$ and $D(P_L^{\mathrm{ind}} \| P_L)$,

$$S_{KL}(L) = \frac{D(P_L \| P_L^{\mathrm{ind}}) \, D(P_L^{\mathrm{ind}} \| P_L)}{D(P_L \| P_L^{\mathrm{ind}}) + D(P_L^{\mathrm{ind}} \| P_L)}.$$

In contrast to the symmetrization via the arithmetic mean, the bound

$$S_{KL}(L) \leq \min\{D(P_L \| P_L^{\mathrm{ind}}), D(P_L^{\mathrm{ind}} \| P_L)\}$$

furnishes more general conditions for the symmetrized Kullback–Leibler distance to be finite. Moreover we shall write $S_{\mathrm{KL}}^{\mathcal{C}}(L)$ when the Kullback–Leibler distance is calculated using the probability densities $P_{\mathcal{C}}$ of the order classes (see Fig. 2.7). Finally, if $P_L(\tau)$ and $P_L^{\mathrm{ind}}(\tau)$ are obtained using only transcriptions from an order class $\mathcal{C}_i(L)$, then the notation will be $S_{\mathrm{KL}}^i(L)$. The point in doing so is that the dynamics of coupled systems may lead to the extinction of order classes, a feature referred to as *saturation* in [159].

Let us apply the method to a bidirectionally coupled Rössler–Rössler system [175] defined by the following set of equations:

$$\dot{x}_{1,2} = -w_{1,2}y_{1,2} - z_{1,2} + k(x_{2,1} - x_{1,2}),$$
$$\dot{y}_{1,2} = w_{1,2}x_{1,2} + 0.165y_{1,2},$$
$$\dot{z}_{1,2} = 0.2 + z_{1,2}(x_{1,2} - 10).$$

Here $w_1 = 0.99$ and $w_2 = 0.95$ are the mismatch parameters and k is the coupling constant. All the time series were generated using a fourth-order Runge–Kutta method with time step $\Delta t = 10^{-3}$ and initial conditions: $x_1(0) = -0.4$, $y_1(0) = 0.6$, $z_1(0) = 5.8$, $x_2(0) = 0.8$, $y_2(0) = -2$, and $z_2(0) = -4$. This chaotic system exhibits a rich synchronization behavior that ranges from phase ($k \approx 0.036$) to lag ($k \approx 0.14$) and finally to complete synchronization as k increases [175]. In [159] the authors only study the x-components of the Rössler subsystems. Specifically, time series of length 2^{19} (about 775 orbits) were sampled with delay $T = 150$ and dimension L, to obtain delay vectors

$$(x(n\Delta t), x((n+T)\Delta t), \dots, x((n+(L-1)T)\Delta t))$$

from either subsystem. Following [80] the delay was chosen so as to minimize the mutual information (Annex B, (B.6)) of the coordinates $x_1(t)$ and $x_1(t + T\Delta t)$ for the uncoupled system ($k = 0$).

Figure 2.6(a) shows the symmetrized Kullback–Leibler distance $S^{\mathcal{C}}_{KL}(L)$ obtained using the probability density $P_{\mathcal{C}}$ of order classes for $L = 6$ and $L = 7$. Figure 2.6 (b)–(d) shows $S_{KL}(L)$ obtained with the probability density of transcriptions in all non-empty order classes for $L = 6$ ($\mathcal{C}_2$–$\mathcal{C}_6$ in subfigure (b)) and $L = 7$ ($\mathcal{C}_7$, $\mathcal{C}_{10}$, and $\mathcal{C}_{12}$ in subfigure (c) and $\mathcal{C}_2$–$\mathcal{C}_6$ in subfigure (d)). We comment first the salient features of $S^{\mathcal{C}}_{KL}(6)$ and $S^{\mathcal{C}}_{KL}(7)$.

The increase of $S^{\mathcal{C}}_{KL}$ at $k \approx 0.036$ is due to the transition from (almost) uncoupled dynamics to phase synchronization. For stronger coupling k, $S^{\mathcal{C}}_{KL}$ increases rather monotonically until $k \approx 0.11$. For $k \in [0.11, 0.145]$, $S^{\mathcal{C}}_{KL}$ displays strong fluctuations revealing the presence of "intermittent-lag synchronization." This particular synchronization regime is characterized by synchronization periods interrupted by bursts of non-synchronized activity [175, 34]. The strong fluctuations sharply vanish at the onset of lag synchronization ($k \approx 0.145$). Lag synchronization is defined by the condition $x_1(t + \delta t) = x_2(t)$, i.e., the coincidence of the time series when shifted in time by a constant time lag δt. Both curves, $S^{\mathcal{C}}_{KL}(6)$ and $S^{\mathcal{C}}_{KL}(7)$, increase monotonically in the interval $k \in [0.145, 0.30]$ reflecting stronger synchronization. This trend is only interrupted within the range $k \in [0.232, 0.256]$, where a period-5 window occurs.

The periodic windows are better observed in Fig. 2.6(b)–(d). In fact, all curves exhibit a peak at $k \approx 0.061$ that corresponds to a period-3 window [175]. $S^6_{KL}(6)$ and $S^{12}_{KL}(7)$ indicate a period-6 window at $k \approx 0.11$. It seems that this window was not reported before [159], probably due to its extremely small size ($k \in [0.1094, 0.1096]$). All curves show clear signatures of periodic behavior in the range $k \in [0.232, 0.256]$. Intermittent-lag synchronization is particularly reflected by the strong fluctuations observed in Fig. 2.6(b) and (c) for $S^6_{KL}(6)$ and $S^{10}_{KL}(7)$, which

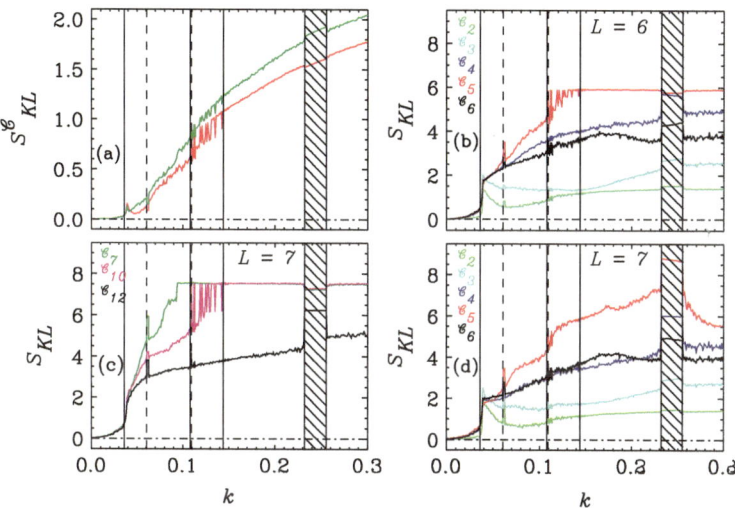

Fig. 2.6 [Reproduced with permission from [159].] **(a)** $S_{\mathrm{KL}}^{\mathcal{C}}$ obtained using the probability density of the order classes for $L = 6$ (*lower curve*) and $L = 7$ (*upper curve*). **(b)**–**(d)** S_{KL} calculated with the probability density of transcriptions and sequence lengths shown in the plots. *Vertical full lines* from left to right locate the transitions to phase synchronization ($k \approx 0.036$), intermittent-lag synchronization ($k \approx 0.11$), and lag synchronization, respectively. The *vertical dashed lines* at $k \approx 0.061$ and $k \approx 0.11$ as well as the *hatched area* ($k \in [0.232, 0.256]$) indicate periodic windows

abruptly disappear at $k = 0.145$. Observe that different order classes provide complementary information of the coupled system. For instance, $S_{\mathrm{KL}}^{5}(6)$ characterizes the intermittent-lag synchronization and the onset of lag synchronization better than $S_{\mathrm{KL}}^{6}(6)$. In any case, these partial pieces of information add altogether to a global picture of the various synchronization stages.

Figure 2.6 also reveals that the Kullback–Leibler distance of some higher order classes saturates when the value of the coupling constant k increases. Indeed, Fig. 2.6(b) and (c) shows that the coupled dynamics lead to the extinction of order classes $\mathcal{C}_5(6)$, $\mathcal{C}_7(7)$, and $\mathcal{C}_{10}(7)$ at $k \approx 0.145$, $k \approx 0.09$, and $k \approx 0.145$, respectively.

Figure 2.7(a) and (b) shows the probability density $P_{\mathcal{C}}$ of the order classes for $L = 6$ and $L = 7$, respectively. Note that Fig. 2.6(a) displays the contrast between probability densities as in Fig. 2.7 and those of the independent processes. In particular, a vanishing contrast as for $k \approx 0.005$ indicates that the corresponding probability density $P_{\mathcal{C}}$ (which is clearly non-uniform) is similar to the probability density of transcriptions generated by two independent Rössler systems. In the vicinity of the transition to phase synchronization, $k \approx 0.039$, $P_{\mathcal{C}}$ deviates from the probability density of independent processes (Fig. 2.6(a)), and higher order classes dominate the coupled dynamics (Fig. 2.7). This trend is reversed when increasing k, and already at $k \approx 0.062$ (resp. $k \approx 0.074$), the class of order 2, $\mathcal{C}_2(6)$ (resp. $\mathcal{C}_2(7)$), is prevalent (except at $k = 0.299$ for $L = 6$, in which case $\mathcal{C}_1(6)$ prevails).

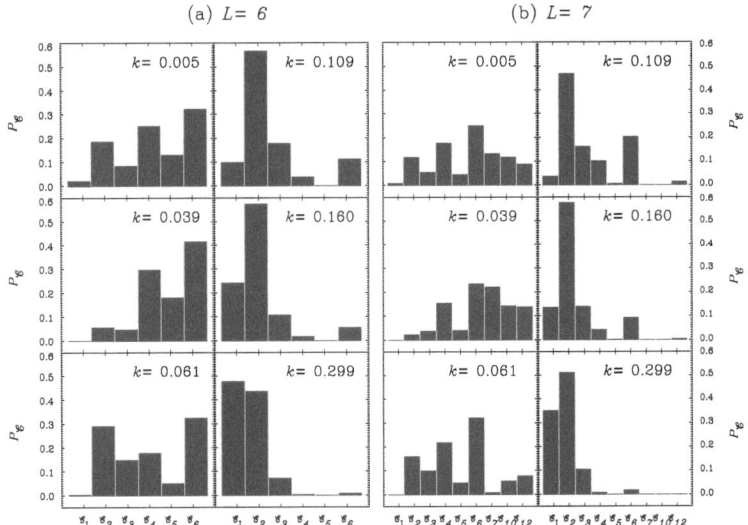

Fig. 2.7 [Reproduced with permission from [159].] (**a**) Probability density P_C of the feasible order classes for $L = 6$ and different values of the coupling constant k. (**b**) Idem for $L = 7$. Classes of orders 8, 9, and 11 are not allowed. Note that $C_1(L) = \{\langle 0, 1, \dots, L-1 \rangle\}$

If following [159] we agree to measure the complexity of a transcription by its order, then the probability density of order classes indicates how complex the relationship between the time series is. Figure 2.7 demonstrates that higher order transcriptions play an important role in the description of complex synchronization states such as phase synchronization ($k \geq 0.036$)—a regimen in which amplitudes remain chaotic and uncorrelated. As k increases, the probability densities of higher order classes decrease and some of them vanish, like $C_5(6)$, $C_7(7)$, and $C_{10}(7)$. In fact, simpler synchronization states such as intermittent-lag and lag synchronizations ($k > 0.11$) are predominantly described by lower order classes ($C_2(L)$ and $C_1(L)$). Clearly, the simplest synchronization state, namely complete synchronization, will only be described by the identity ($C_1(L)$).

Chapter 3
Ordinal Patterns

In this chapter we take a close look to the order relations and their consequences, mostly for dynamical systems defined by self-maps of one-dimensional intervals. More general situations will be considered and studied in detail in the following chapters.

Order has some interesting consequences in discrete-time dynamical systems. Just as one can derive sequences of symbol patterns from such a dynamic via coarse graining of the phase space, so it is also straightforward to obtain sequences of ordinal patterns if the phase space is linearly ordered. As we learnt in Sect. 1.2, not all ordinal patterns can be materialized by the orbits of a given dynamic under some mild mathematical assumptions. Furthermore, if an ordinal pattern of a given length is "forbidden," i.e., cannot occur, its absence pervades all longer patterns in form of more missing ordinal patterns. This cascade of outgrowth forbidden patterns grows super-exponentially (in fact, factorially) with the length, all its patterns sharing a common structure. Of course, forbidden and admissible ordinal patterns can be viewed as permutations; in combinatorial parlance, the admissible patterns are (the inverses of) those permutations avoiding the so-called forbidden root patterns in consecutive positions (see Sect. 3.4.2 for details). Let us mention that permutations avoiding general or consecutive patterns is a popular topic in combinatorics (see, e.g., [25, 74, 75]).

Forbidden *ordinal* patterns should not be mistaken for other sorts of forbidden patterns that may occur in dynamics with constraints. Forbidden patterns in symbol sequences occur, e.g., in Markov subshifts of finite type and, more generally, in random walks on oriented graphs. On the contrary, the existence of forbidden ordinal patterns does not entail necessarily any restriction on the patterns of the corresponding symbolic dynamics; the variability of *symbol* patterns is given by the statistical properties of the dynamics. As a matter of fact, the symbolic dynamics of one-dimensional chaotic maps are used to generate pseudo-random sequences, although all such maps have forbidden ordinal patterns.

J.M. Amigó, *Permutation Complexity in Dynamical Systems,*
Springer Series in Synergetics, DOI 10.1007/978-3-642-04084-9_3,
© Springer-Verlag Berlin Heidelberg 2010

3.1 Symbol Patterns

Before dealing with ordinal patterns, we are going to consider the symbol patterns
defined by a symbolic dynamics with respect to a partition. The scope is to show
that, under general conditions on the map, such symbol patterns have no restrictions,
in contrast with the situation we will encounter when studying ordinal patterns.

Thus, let f be a measure-preserving map from a probability space $(\Omega, \mathcal{B}, \mu)$ to
itself and $\alpha = \{A_0, \ldots, A_{|\alpha|-1}\}$ be a partition of $(\Omega, \mathcal{B}, \mu)$. Recall that the symbolic
dynamics with respect to α, $\mathbf{X}^\alpha = \{X_n^\alpha\}_{n \in \mathbb{N}_0}$ with $X_n^\alpha : \Omega \to S = \{0, \ldots, |\alpha| - 1\}$, is
defined as follows:

$$X_n^\alpha(x) = i_n \quad \text{if } f^n(x) \in A_{i_n}, \quad n \geq 0$$

(see (1.6)). The resulting sequence $(i_n)_{n \in \mathbb{N}_0}$ is called the *coded orbit* (or, sometimes,
the (α, f)-name) of $x \in \Omega$.

In Sect. B.2.2 it is proven that if α is a generating partition for f, then $(\Omega, \mathcal{B}, \mu, f)$
is isomorphic [1] via the *coding map* $\Phi^\alpha : \Omega \to S^{\mathbb{N}_0}$,

$$(\Phi^\alpha(x))_n = X_n^\alpha(x),$$

to the one-sided shift $(S^{\mathbb{N}_0}, \mathcal{B}_\Pi(S), m, \Sigma)$, where $m = \mu \circ (\Phi^\alpha)^{-1}$ and $\Sigma \circ \Phi^\alpha = \Phi^\alpha \circ f$. Here $\mathcal{B}_\Pi(S)$ is the product sigma-algebra generated by the cylinder sets

$$C_{i_0,\ldots,i_n} = \{\mathbf{s} \in S^{\mathbb{N}_0} : s_0 = i_0, \ldots, s_n = i_n\},$$

$i_0, \ldots, i_n \in S$ (see Sect. A.2),

$$m(C_{i_0,i_1,\ldots,i_n}) = \mu(A_{i_0} \cap f^{-1} A_{i_1} \cap \cdots \cap f^{-n} A_{i_n}),$$

and the partition

$$\{\Phi^\alpha(A_i) : i \in S\} = \{C_i : i \in S\}$$

is trivially generating for Σ. It follows that the coded orbits of f contain any arbitrary
pattern. Indeed, given any *symbol pattern* of length $L \geq 1$, $i_0^{L-1} := i_0, i_1, \ldots, i_{L-1}$,
where $i_n \in S$, choose

$$x \in \bigcap_{n=0}^{L-1} f^{-n} A_{i_n} = (\Phi^\alpha)^{-1} C_{i_0,\ldots,i_{L-1}}.$$

---

[1] The general definition of (metric) isomorphy or conjugacy between measure-preserving dynamical systems is given in Definition 12, Sect. A.1. The corresponding concept for continuous
dynamical systems, that usually goes by the name of topological conjugacy, is given in Definition 25, Sect. B.3.

If the pattern has infinite length, $i_0^\infty = i_0, i_1, \ldots$, then there exits a unique point $x \in \Omega$ modulo 0 (i.e., possibly up to sets of measure 0), namely, $x = (\Phi^\alpha)^{-1}(s)$ with $s = (i_0, i_1, \ldots) \in S^{\mathbb{N}_0}$, such that its coded orbit is precisely **s**. Thus,

$$\bigcap_{n=0}^{\infty} f^{-n} A_{i_n} = \{x\}.$$

This means that Φ^α *separates points*: if $x_1 \neq x_2$ then $\Phi^\alpha(x_1) \neq \Phi^\alpha(x_2)$.

We conclude that if α is a generating partition for f, then the coded orbits $\Phi^\alpha(x)$, $x \in \Omega$, define any finite or infinite symbol pattern (in the second case, modulo 0).

If $f : \Omega \to \Omega$ is an automorphism, all the above generalizes to two-sided sequences. Sufficient conditions for f to have a generating partition in such a case are given by Krieger's theorem : ergodicity and a finite entropy.

Example 1 Take $g : [0, 1] \to [0, 1]$ to be the logistic map $g(x) = 4x(1 - x)$ and

$$\alpha = \{A_0 = [0, \tfrac{1}{2}), A_1 = [\tfrac{1}{2}, 1]\}. \tag{3.1}$$

(It is irrelevant whether the midpoint $\frac{1}{2}$ belongs to the left or to the right partition element.) Then α is a generating partition (use, for example, the conjugacy between the logistic map and the symmetric tent map, Example 24). In this case, the coding map $\Phi^\alpha : [0, 1] \to \{0, 1\}^{\mathbb{N}_0}$,

$$(\Phi^\alpha(x))_n = \begin{cases} 0 & \text{if } g^n(x) \in [0, \tfrac{1}{2}), \\ 1 & \text{if } g^n(x) \in [\tfrac{1}{2}, 1], \end{cases}$$

is an isomorphism between $([0, 1], \mathcal{B}, \mu, g)$ and the $(\tfrac{1}{2}, \tfrac{1}{2})$-Bernoulli shift, where $\mathcal{B}$ is the Borel sigma-algebra of $[0, 1]$ and

$$\mu([a, b]) = \int_a^b \frac{dx}{\pi \sqrt{x(1 - x)}},$$

$[a, b] \subset [0, 1]$. For example,

$$m\{C_{0,0}\} = \mu \{x \in [0, 1] : x \in A_0, g(x) \in A_0\} = \int_0^{1/2 - \sqrt{2}/4} \frac{dx}{\pi \sqrt{x(1 - x)}} = \frac{1}{4},$$

$$m\{C_{0,1}\} = \mu \{x \in [0, 1] : x \in A_0, g(x) \in A_1\} = \int_{1/2 - \sqrt{2}/4}^{1/2} \frac{dx}{\pi \sqrt{x(1 - x)}} = \frac{1}{4},$$

$$m\{C_{1,0}\} = \mu \{x \in [0, 1] : x \in A_1, g(x) \in A_0\} = \int_{1/2}^{1/2 + \sqrt{2}/4} \frac{dx}{\pi \sqrt{x(1 - x)}} = \frac{1}{4},$$

$$m\{C_{1,1}\} = \mu \{x \in [0, 1] : x \in A_1, g(x) \in A_1\} = \int_{1/2 + \sqrt{2}/4}^{1} \frac{dx}{\pi \sqrt{x(1 - x)}} = \frac{1}{4}.$$

Exercise 1 Let $E_2 : [0, 1] \to [0, 1]$ be the dyadic map $x \mapsto 2x \bmod 1$ and $\phi : \{0, 1\}^{\mathbb{N}_0} \to [0, 1]$ the map

$$(x_0, x_1, \ldots, x_k, \ldots) \mapsto \sum_{k=0}^{\infty} x_k 2^{-(k+1)}.$$

Check that ϕ is the inverse (modulo 0) of the coding map Φ^α of E_2 with respect to partition (3.1).

Shift transformations have generating partitions (namely, the cylinder sets $C_{i_0,\ldots,i_{n-1}}$ of any given length $n \geq 1$), hence their trajectories realize any possible symbol sequence.

3.2 Order Relations

A relation $\leq$ defined on every pair of elements of a set Ω is said to be a *total* or *linear order* if $\leq$ is reflexive, antisymmetric, and transitive. A set Ω endowed with a total order $\leq$ is called a *totally* or *linearly ordered set* and will be denoted by $(\Omega, \leq)$. As usual, $x < y$ means henceforth $x \leq y$ and $x \neq y$. The product of the totally ordered sets $(\Omega_1, \leq), (\Omega_2, \leq), \ldots, (\Omega_n, \leq)$ is also totally ordered via the *product order* (also called *lexicographical* or *dictionary order*): if $(x^{(1)}, x^{(2)}, \ldots, x^{(n)}) \neq (y^{(1)}, y^{(2)}, \ldots, y^{(n)})$, then $(x^{(1)}, x^{(2)}, \ldots, x^{(n)}) < (y^{(1)}, y^{(2)}, \ldots, y^{(n)})$ if

(i) $x^{(1)} < y^{(1)}$ or
(ii) $x^{(i)} = y^{(i)}$ for $i = 1, \ldots, k \leq n - 1$ and $x^{(k+1)} < y^{(k+1)}$.

Other conventions (e.g., the signed lexicographic order we considered in Sect. 2.3) are of course possible. The product order generalizes straightforwardly to "infinite products" (i.e., sequence spaces).

Suppose now that $(x_n)_{n \in \mathbb{N}_0}$ is a sequence whose elements (symbols, letters,...) x_n belong to a set (state space, alphabet, etc.) S endowed with a total ordering $\leq$. We say that a length-L block (segment, word, etc.) $x_n^{n+L-1} = x_n, x_{n+1}, \ldots, x_{n+L-1}$ defines the *ordinal (L-)pattern* $\pi = \langle \pi_0, \ldots, \pi_{L-1} \rangle$ if

$$x_{n+\pi_0} < x_{n+\pi_1} < \cdots < x_{n+\pi_{L-1}},$$

where in case $x_i = x_j$, we agree to set $x_i < x_j$ if, say, $i < j$.

Alternatively we also say that the block x_n^{n+L-1} is of type π, or that π is realized by x_n^{n+L-1}, and write $\pi = \pi(x_n^{n+L-1})$. As in Sect. 1.2, the set of ordinal L-patterns will be denoted by $\mathcal{S}_L$. Remember from Sect. 1.2 too that $\mathcal{S}_L$ can be promoted to a group of order $L!$ if equipped with the product (1.25). Unlike in Sect. 2.4, the algebraic structure of $\mathcal{S}_L$ will not be exploited in the sequel.

Example 2 Suppose that $S = \{a, b, c\}$ with $a < b < c$, and that we observe the block $x_0^2 = c, a, a$. Then x_0^2 defines the ordinal pattern $\langle 1, 2, 0 \rangle$ since $x_1 = x_2 = a < c = x_0$ and $1 < 2$. Observe that the following blocks of length 3 are also of type $\langle 1, 2, 0 \rangle$: (i) c, b, b, (ii) c, a, b, and (iii) b, a, a.

In other words, $\pi(x_n^{n+L-1})$ is a permutation of $\{0, 1, \ldots, L-1\}$ that encapsulates the ups and downs of the elements $x_n, x_{n+1}, \ldots, x_{n+L-1}$ in the set S; in case that two elements are equal, we take by convention the first one also as the smaller. This qualitative information is shown in Fig. 3.1 for patterns of length 3.

Given the sequence $(x_n)_{n \in \mathbb{N}_0}$, we say that $\pi \in \mathcal{S}_L$ is an *allowed* or *admissible* L-pattern if π is realized by some substring of length L of $(x_n)_{n \in \mathbb{N}_0}$; otherwise, π is called a forbidden L-pattern.

Proposition 1 *1. If $\pi = \langle \pi_0, \ldots, \pi_L \rangle$ is an allowed $(L+1)$-pattern of $(x_n)_{n \in \mathbb{N}_0}$, and $\check{\pi}$ is the L-pattern obtained from π by deleting the entry L, then $\check{\pi}$ is an allowed L-pattern of $(x_n)_{n \in \mathbb{N}_0}$.*

2. If $\pi = \langle \pi_0, \ldots, \pi_{L-1} \rangle$ is a forbidden L-pattern of $(x_n)_{n \in \mathbb{N}_0}$ and $\hat{\pi}$ is the $(L+1)$-pattern obtained from π by adding the entry L at any place, then $\hat{\pi}$ is a forbidden $(L+1)$-pattern of $(x_n)_{n \in \mathbb{N}_0}$.

Proof 1. If $\pi \in \mathcal{S}_{L+1}$ is allowed, this means that there exists a substring $x_n^{n+L} = x_n, x_{n+1}, \ldots, x_{n+L}$ of the sequence $(x_n)_{n \in \mathbb{N}_0}$ such that

$$x_{n+\pi_0} < x_{n+\pi_1} < \cdots < x_{n+\pi_L}. \tag{3.2}$$

Delete then x_{n+L} from (3.2) to show that the substring $x_n, x_{n+1}, \ldots, x_{n+L-1}$ is of type $\check{\pi} \in \mathcal{S}_L$.

2. Suppose by contradiction that the $(L+1)$-pattern $\hat{\pi} = \langle \hat{\pi}_0, \ldots, \hat{\pi}_L \rangle$ is allowed. Then, part 1 implies that the L-pattern obtained by removing from $\hat{\pi}$ the entry L, namely π, is an allowed pattern. $\square$

This general setting will crystallize in different ways as we move on. Let us advance three of them at this point.

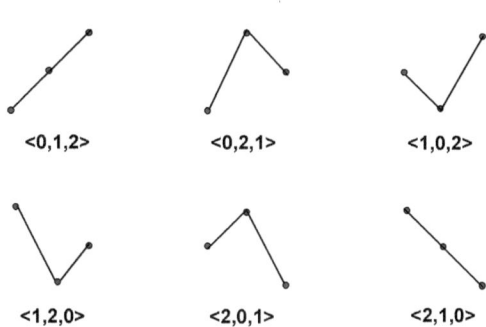

<0,1,2>   <0,2,1>   <1,0,2>

<1,2,0>   <2,0,1>   <2,1,0>

Fig. 3.1 Geometrical illustration of the six ordinal patterns of length 3

- The sequence $(x_n)_{n \in \mathbb{N}_0}$ can be the output of a finite-state stationary stochastic process. This corresponds to the usual information sources emitting a message composed by letters of a finite alphabet.
- The set S can be an interval $I \subseteq \mathbb{R}^q$, $q \geq 1$, and $(x_n)_{n \in \mathbb{N}_0}$ the output of a univariate $(q = 1)$ or multivariate $(q > 1)$ random process taking values in I.
- Still other possibility is that $(x_n)_{n \in \mathbb{N}_0}$ is the orbit of x_0 under a map $f : I \to I$, I being as before a q-dimensional interval or a homeomorphic copy thereof. In this case it is customary to neglect periodic points whose periods are shorter than the pattern length L considered, so as all points in the block x_n^{n+L-1} are different.

In the following sections and chapters we are going to dwell on all these settings.

3.3 Ordinal Patterns Defined by Maps

In Sect. 3.1 we saw that the symbolic dynamics of maps defines any symbol pattern of any length, under rather general assumptions. In this section we shall see that the situation is not quite the same when considering ordinal patterns.

Let $(\Omega, \leq)$ be a totally ordered set and $f : \Omega \to \Omega$ a map. Given $x \in \Omega$, set $x_n = f^n(x)$ for $n \geq 0$. If x is not a periodic point of period less than $L \geq 2$, we can then associate with x an ordinal pattern of length L, as follows. We say that x *defines the ordinal pattern* $\pi = \langle \pi_0, \ldots, \pi_{L-1} \rangle \in \mathcal{S}_L$, if $\pi = \pi(x_0^{L-1})$, i.e.,

$$x_{\pi_0} < x_{\pi_1} < \cdots < x_{\pi_{L-1}},$$

or, equivalently,

$$f^{\pi_0}(x) < f^{\pi_1}(x) < \cdots < f^{\pi_{L-1}}(x). \tag{3.3}$$

Set

$$P_\pi = \{ x \in \Omega : x \text{ defines } \pi \in \mathcal{S}_L \} \tag{3.4}$$

as in Sect. 1.2 and

$$\mathcal{P}_L = \{ P_\pi \neq \emptyset : \pi \in \mathcal{S}_L \}. \tag{3.5}$$

Therefore, $|\mathcal{P}_L|$ is the number of distinct ordinal L-patterns realized by the points of Ω.

Proposition 2 *Let $(\Omega, \mathcal{B}, \mu, f)$ be a measure-preserving dynamical system. Then $\mathcal{P}_L$ is a finite partition of Ω for all $L \geq 2$ if and only if f is aperiodic.*

We say that $f : \Omega \to \Omega$ is *aperiodic*, if

$$\mu\left(\bigcup_{n\geq 1}\{x\in\Omega:f^n(x)=x\}\right)=0. \tag{3.6}$$

Proof In order that $\mathcal{P}_L$ fails to be a finite partition of Ω, it must happen that the complement of the disjoint union $\cup\{P_\pi\in\mathcal{P}_L\}$, which comprises all periodic points of f of period $p\leq L$, has a positive measure. But this possibility is excluded by (3.6). □

In particular, if f is ergodic with respect to μ, then f is aperiodic unless Ω is a finite set modulo 0 [202].

The family of sets $\mathcal{P}_L$ has some elementary properties. In Sect. 1.2 we saw that when going from $\mathcal{P}_L$ to $\mathcal{P}_{L+1}$, each "mother set" $P_{\pi_{\mathrm{mother}}}$, $\pi_{\mathrm{mother}}=\langle\pi_0,\dots,\pi_{L-1}\rangle$, decomposes into several "daughter sets" $P_{\pi_{\mathrm{daughter}}}\in\mathcal{P}_{L+1}$, where

$$\pi_{\mathrm{daughter}}\in\{\langle L,\pi_0,\dots,\pi_{L-1}\rangle,\langle\pi_0,\dots,\pi_k,L,\pi_{k+1},\dots,\pi_{L-1}\rangle,\langle\pi_0,\dots,\pi_{L-1},L\rangle\},$$

$0\leq k\leq L-2$. Therefore, each mother set is the (disjoint) union of her daughter sets. Correspondingly, we speak of mother and daughter patterns. To go back from π_{daughter} to π_{mother}, just delete the entry L. In particular, two different mother intervals cannot give birth to the same daughter interval.

Proposition 3 *(1) $\mathcal{P}_{L+1}$ is a refinement of $\mathcal{P}_L$, i.e., each $P_\pi\in\mathcal{P}_L$ is the union of elements of $\mathcal{P}_{L+1}$.*
(2) For every $P_{\pi'}\in\mathcal{P}_{L+1}$ there is a $P_\pi\in\mathcal{P}_L$ such that $f(P_{\pi'})\subset P_\pi$.

Proof Statement (1) is trivial because

$$P_\pi=\cup\{P_{\pi'}\in\mathcal{P}_{L+1}:\pi'\text{ is a daughter pattern of }\pi\}.$$

To prove (2), let $x\in f(P_{\pi'})$, i.e., $x=f(y)$ where $y\in\Omega$ satisfies

$$f^{\pi'_0}(y)<f^{\pi'_1}(y)<\cdots<f^{\pi'_L}(y). \tag{3.7}$$

Let π'_{n_k}, $0\leq k\leq L-1$, be an order-isomorphic relabeling of those L entries of the ordinal pattern $\pi'\in S_{L+1}$ which are positive. From (3.7) it follows that

$$f^{\pi'_{n_0}-1}(x)<f^{\pi'_{n_1}-1}(x)<\cdots<f^{\pi'_{n_{L-1}}-1}(x),$$

hence $x\in P_\pi$ with $\pi=\langle\pi'_{n_0}-1,\dots,\pi'_{n_{L-1}}-1\rangle\in S_L$. In words, π is obtained from π' after deleting the entry 0 and subtracting 1 from the remaining entries. Therefore, $f(P_{\pi'})\subset P_\pi$. □

Example 3 To illustrate Proposition 3 (1)–(2), consider the logistic map g and the intervals $P_\pi\in\mathcal{P}_3$, (1.29). Then (see Figs. 1.5 and 1.6),

$$g(P_{\langle 0,1,2\rangle}) = g((0,\tfrac{1}{4})) = (0,\tfrac{3}{4}) = P_{\langle 0,1,2\rangle} \cup P_{\langle 0,2,1\rangle} \cup P_{\langle 2,0,1\rangle} = P_{\langle 0,1\rangle},$$

$$g(P_{\langle 0,2,1\rangle}) = g((\tfrac{1}{4},\tfrac{5-\sqrt{5}}{8})) = (\tfrac{3}{4},\tfrac{5+\sqrt{5}}{8}) = P_{\langle 1,0,2\rangle} \subset P_{\langle 1,0\rangle},$$

$$g(P_{\langle 2,0,1\rangle}) = g((\tfrac{5-\sqrt{5}}{8},\tfrac{3}{4})) = (\tfrac{3}{4},1) = P_{\langle 1,0,2\rangle} \cup P_{\langle 1,2,0\rangle} \subset P_{\langle 1,0\rangle},$$

$$g(P_{\langle 1,0,2\rangle}) = g((\tfrac{3}{4},\tfrac{5+\sqrt{5}}{8})) = (\tfrac{5-\sqrt{5}}{8},\tfrac{3}{4}) = P_{\langle 2,0,1\rangle} \subset P_{\langle 0,1\rangle},$$

$$g(P_{\langle 1,2,0\rangle}) = g((\tfrac{5+\sqrt{5}}{8},1)) = (0,\tfrac{5-\sqrt{5}}{8}) = P_{\langle 0,1,2\rangle} \cup P_{\langle 0,2,1\rangle} \subset P_{\langle 0,1\rangle}.$$

Observe that $\mathcal{P}_3$ is a Markov partition for g (i.e., $g(P_\pi) \supset P_\sigma$, whenever $g(P_\pi) \cap P_\sigma \neq \emptyset$, $P_\pi, P_\sigma \in \mathcal{P}_3$) with transition matrix

$$A = \begin{pmatrix} 1 & 1 & 1 & 0 & 0 \\ 0 & 0 & 0 & 1 & 0 \\ 0 & 0 & 0 & 1 & 1 \\ 0 & 0 & 1 & 0 & 0 \\ 1 & 1 & 0 & 0 & 0 \end{pmatrix}$$

(see Definition 9 and (A.2)). Needless to say, the partitions $\mathcal{P}_L$ are not in general Markovian.

Exercise 2 (1) Let $f : [a,b] \to [a,b]$ be a boundary-anchored unimodal map with full range (i.e., $f(a) = f(b) = a$ and $f([a,b]) = [a,b]$). Show that $\mathcal{P}_2$ is a Markov partition for f.

(2) Let Λ be the symmetric tent map. Using the information on $\mathcal{P}_4$ provided in Example 13, Sect. 6.3, shows that $\Lambda(P_{\langle 2,3,0,1\rangle}) \cap P_{\langle 1,2,3,0\rangle} \neq \emptyset$ but $P_{\langle 1,2,3,0\rangle} \not\subset \Lambda(P_{\langle 2,3,0,1\rangle})$.

A plain difference between symbol patterns and ordinal patterns of length L is their cardinality: the former grow exponentially with L (exactly as N^L, where N is the number of symbols) while the latter do superexponentially,

$$|\mathcal{S}_L| = L! \sim e^{L(\ln L - 1) + (1/2)\ln 2\pi L}, \tag{3.8}$$

see (1.34). Although one can construct maps whose orbits realize any possible ordinal pattern (more on this at the end of Sect. 4.2), numerical simulations support the conjecture that the number of ordinal L-patterns realized in the orbits of maps, like symbol patterns, grows only exponentially with L for "well-behaved" maps. In fact, we saw in Sect. 1.2 that if I is a closed interval of $\mathbb{R}$ and $f : I \to I$ is piecewise monotone, then (see (1.33))

$$|\mathcal{P}_L| \sim e^{L h_{\mathrm{top}}(f)}, \tag{3.9}$$

where $h_{\mathrm{top}}(f)$ is the topological entropy of f. From (3.8) and (3.9) we conclude the following result.

Proposition 4 *If f is a piecewise monotone self-map on a finite interval $I \subset \mathbb{R}$, then there exists $L \geq 2$ such that $P_\pi = \emptyset$ for some $\pi \in \mathcal{S}_L$.*

   Ordinal patterns that do not appear in any orbit of f are called *forbidden* (*ordinal*)
patterns for f, at variance with the *admissible* or *allowed patterns*, for which there
are sets of points that realize them.

3.4 Properties of the Ordinal Patterns

We examine in this section three basic properties of ordinal patterns: invariance
under order isomorphism, superexponential growth of the forbidden patterns with
the length, and robustness against noise.

3.4.1 Invariance Under Order Isomorphism

Since ordinal patterns are not related to measure-theoretical or topological properties, metrically or topologically conjugate dynamical systems need not have the
same allowed (and hence forbidden) patterns, unless the conjugacy preserves linear
order—supposing that both state spaces are linearly ordered. In general, this will not
be the case.

   For instance, we saw in Sect. 1.2 that the logistic map has the forbidden 3-pattern
$\langle 2, 1, 0 \rangle$, i.e., there are no three consecutive points in any orbit of the logistic map,
forming a strictly decreasing trio (see Fig. 1.6). However, Fig. 3.2 shows that the
dyadic map $E_2 : x \mapsto 2x \pmod 1$, $0 \le x \le 1$, has no forbidden patterns of length
3, despite being isomorphic to the logistic map. The reason is simple: the isomorphism between these two maps is proved via the semi-conjugacy[2] $\varphi : [0, 1] \to [0, 1]$,
$\varphi(x) = \sin^2 \pi x$, which does not preserve order on account of being increasing on
$(0, \frac{1}{2})$ and decreasing on $(\frac{1}{2}, 1)$.

Definition 1 Given two totally ordered sets $(\Omega_1, \le_1)$ and $(\Omega_2, \le_2)$, two maps
$f_1 : \Omega_1 \to \Omega_1$ and $f_2 : \Omega_2 \to \Omega_2$, and an invertible map $\phi : \Omega_1 \to \Omega_2$ such that
$\phi \circ f_1 = f_2 \circ \phi$, we say that f_1 and f_2 are *order isomorphic* if ϕ is order-preserving
(i.e., $x \le_1 y$ implies $\phi(x) \le_2 \phi(y)$). The map ϕ is called an order isomorphism.

   It is trivial that if $\phi : \Omega_1 \to \Omega_2$ is an order isomorphism, then $x \in \Omega_1$ and
$\phi(x) \in \Omega_2$ define the same ordinal L-patterns, for all $L \ge 2$, under the f_1- and
f_2-dynamics, respectively. In other words, order-isomorphic maps have the same
allowed and forbidden patterns of any length. We conclude that ordinal patterns are
not invariants of metric nor topological conjugacy, but of order isomorphy.

Example 4 (1) The logistic map g (1.19) and the symmetric tent map Λ (1.17) are
   not only isomorphic but also order isomorphic. Indeed, the isomorphism

$$\phi : x \mapsto \sin^2 \left(\tfrac{\pi}{2} x \right),$$

---

[2] Definition 25.

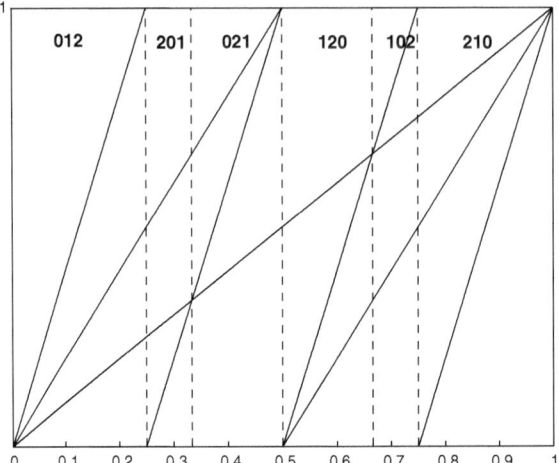

Fig. 3.2 All six 3-patterns are allowed for the shift map E_2:$x \mapsto 2x$ (mod 1): $P_{012} = \left(0, \frac{1}{4}\right)$, $P_{201} = \left(\frac{1}{4}, \frac{1}{3}\right)$, $P_{021} = \left(\frac{1}{3}, \frac{1}{2}\right)$, $P_{120} = \left(\frac{1}{2}, \frac{2}{3}\right)$, $P_{102} = \left(\frac{2}{3}, \frac{3}{4}\right)$, $P_{210} = \left(\frac{3}{4}, 1\right)$. A pattern $\langle \pi_0, \pi_1, \pi_2 \rangle$ has been shorthanded as $\pi_0 \pi_1 \pi_2$. Note that ordinal patterns are mirrored with respect to the central line $x = \frac{1}{2}$

(see Example 24) is strictly increasing and, hence, order preserving. This entails that allowed patterns for f correspond to allowed patterns for Λ in a one-to-one way.

(2) The same happens with the dyadic map $E_2 : x \mapsto 2x$ (mod 1), $0 \leq x \leq 1$, and the $(\frac{1}{2}, \frac{1}{2})$-Bernoulli shift, since the isomorphism (modulo 0) $\phi_2 : \{0, 1\}^{\mathbb{N}_0} \rightarrow [0, 1]$,

$$\phi_2 : (x_1, x_2, \dots) \mapsto \sum_{k=1}^{\infty} x_k 2^{-k}$$

is order-preserving ($\{0, 1\}^{\mathbb{N}}$ endowed with the lexicographical order).

(3) The logistic map is isomorphic but not order isomorphic to the $(\frac{1}{2}, \frac{1}{2})$-Bernoulli shift. Indeed, the corresponding isomorphy (actually, the coding map of Example 1) $\Phi^{\alpha} : [0, 1] \rightarrow \{0, 1\}^{\mathbb{N}_0}$ is not order preserving; e.g.,

$$\Phi^{\alpha} \left(\frac{1}{4}\right) = (0, 1^{\infty}) < \Phi^{\alpha}(\frac{3}{4}) = (1^{\infty}),$$

where binary strings are ordered lexicographically, while

$$\Phi^{\alpha} \left(\frac{1}{2}\right) = (1, 1, 0^{\infty}) > \Phi^{\alpha}(1) = (1, 0^{\infty}).$$

The forbidden ordinal patterns of the shift systems will be studied in Chap. 4.

On the other hand, if $\phi : \Omega_1 \rightarrow \Omega_2$ is order preserving but not one-to-one, then ordinal patterns are not necessarily invariant under ϕ. Let, for example, $\Omega_1 = \Omega_2 = $

$[0, 1] \times [0, 1] =: [0, 1]^2$ endowed with lexicographical order, $f : [0, 1]^2 \rightarrow [0, 1]^2$, $\phi : [0, 1]^2 \rightarrow [0, 1]$ the projection onto the first coordinate, $\mathbf{x}_0 = (x_0^{(1)}, x_0^{(2)}) \mapsto x_0^{(1)}$, and

$$(x_0^{(1)}, x_0^{(2)}) = \mathbf{x}_0 > f(\mathbf{x}_0) = (x_1^{(1)}, x_1^{(2)}),$$

so that $\mathbf{x}_0$ is of type $\langle 1, 0 \rangle$. If $x_0^{(1)} > x_1^{(1)}$, then $\phi(\mathbf{x}_0)$ is also of type $\langle 1, 0 \rangle$. But if $x_0^{(1)} = x_1^{(1)}$ (and $x_0^{(2)} > x_1^{(2)}$), then $\phi(\mathbf{x}_0)$ will be of type $\langle 0, 1 \rangle$ in virtue of the lexicographical convention (Sect. 3.2).

Proposition 5 *Given $\Omega_1, \Omega_2 \subset \mathbb{R}$, let $f_i : \Omega_i \rightarrow \Omega_i$, $i = 1, 2$, be topologically conjugate via a (continuous) map $\phi : \Omega_1 \rightarrow \Omega_2$. If f_1 is topologically transitive and, for all $x \in \Omega_1$, both x and $\phi(x)$ define the same ordinal pattern, then ϕ is order preserving.*

Proof Pick $x, x' \in \Omega_1$ such that $x < x'$. We must prove that $\phi(x) < \phi(x')$.

Because of continuity, for all $\varepsilon > 0$ there exists $0 < \delta < \frac{x'-x}{2}$ such that $|y - x| < \delta \Rightarrow |\phi(y) - \phi(x)| < \varepsilon$ and $|y' - x'| < \delta \Rightarrow |\phi(y') - \phi(x')| < \varepsilon$. Moreover, topological transitiveness implies that, given x, x' and δ as above, there exists $x_0 \in \Omega_1$, and positive integers $N = N(x, \delta)$, $N' = N'(x', \delta)$ such that $\left| f_1^N(x_0) - x \right| < \delta$ and $\left| f_1^{N'}(x_0) - x' \right| < \delta$. Suppose without restriction $N < N' = N + k$, $k > 0$, and set $f_1^N(x_0) = y$, $f_1^{N'}(x_0) = y'$, hence $y' = f_1^k(y)$. By assumption, $y \in \Omega_1$ and $\phi(y) \in \Omega_2$ define the same ordinal $(k + 1)$-pattern, i.e.,

$$f_1^{\pi_0}(y) < \cdots < f_1^{\pi_k}(y) \Leftrightarrow f_2^{\pi_0}(\phi(y)) < \cdots < f_2^{\pi_k}(\phi(y)), \tag{3.10}$$

where $0 \leq \pi_i \leq k$, and $\pi_i \neq \pi_j$ for $i \neq j$. Since $|y - x| < \delta$, $\left| f_1^k(y) - x' \right| < \delta$, and $\delta < \frac{x'-x}{2}$, we have $y < f_1^k(y) = y'$. From (3.10) it follows

$$\phi(y) < f_2^k(\phi(y)) = \phi(f_1^k(y)) = \phi(y').$$

By continuity, $\phi(y)$ and $\phi(y')$ can be made to lie arbitrarily close to $\phi(x)$ and $\phi(x')$. It follows $\phi(x) < \phi(x')$. $\qquad \square$

Finally, observe that the setting we are considering is more general than the setting of kneading theory [150] since our functions need not be continuous, but only piecewise continuous. Under some assumptions, the so-called kneading invariants completely characterize the order isomorphy of continuous, one-dimensional interval maps.

3.4.2 Growth of Forbidden Patterns with Length: Outgrowth Patterns

Forbidden ordinal patterns come in two flavors: *outgrowth* patterns and *root* patterns.

Outgrowth forbidden patterns appeared already in Sect. 1.2 when discussing the ordinal patterns of the logistic map: they are the patterns on the "trail" of a given forbidden pattern (see (1.36)). Consider now a general map $f : \Omega \to \Omega$. That $\pi = \langle \pi_0, \ldots, \pi_{L-1} \rangle$ is forbidden for f means that the order relations

$$f^{\pi_0}(x) < f^{\pi_1}(x) < \cdots < f^{\pi_{L-1}}(x) \tag{3.11}$$

cannot occur. This implies that the following $2(L+1)$ patterns of length $L+1$ are also forbidden for f:

Group I: $\langle L, \pi_0, \ldots, \pi_{L-1} \rangle, \langle \pi_0, L, \pi_1, \ldots, \pi_{L-1} \rangle, \ldots, \langle \pi_0, \ldots, \pi_{L-1}, L \rangle,$
Group II: $\langle 0, \pi_0 + 1, \ldots, \pi_{L-1} + 1 \rangle, \langle \pi_0 + 1, 0, \pi_1 + 1, \ldots, \pi_{L-1} + 1 \rangle,$
$\ldots, \langle \pi_0 + 1, \ldots, \pi_{L-1} + 1, 0 \rangle.$
$$\tag{3.12}$$

For suppose by contradiction that the pattern $\langle \pi_0, \ldots, \pi_i, L, \pi_{i+1}, \ldots, \pi_{L-1} \rangle$ is allowed. Then the inequalities

$$f^{\pi_0}(x) < \cdots < f^{\pi_i}(x) < f^L(x) < f^{\pi_{i+1}}(x) < \cdots < f^{\pi_{L-1}}(x)$$

would hold for some $x \in I$, hence (3.11) would occur for the same $x \in I$, contradicting the assumption that π is forbidden. Analogously, if $x \in I$ would realize the pattern $\langle \pi_0 + 1, \ldots, \pi_i + 1, 0, \pi_{i+1} + 1, \ldots, \pi_{L-1} + 1 \rangle$, then $f(x)$ would realize the pattern π—again a contradiction.

A weak form of the converse holds also true: if $\langle L, \pi_0, \ldots, \pi_{L-1} \rangle, \langle \pi_0, L, \ldots, \pi_{L-1} \rangle, \ldots, \langle \pi_0, \ldots, \pi_{L_0-1}, L \rangle \in S_{L+1}$ are forbidden, then $\langle \pi_0, \ldots, \pi_{L-1} \rangle \in S_L$ is also forbidden.

Assume for the time being that the forbidden patterns (3.12), belonging to the "first generation," are all different. Then, proceeding similarly as before, we would find

$$2(L+1) \times 2(L+2) = 2^2(L+1)(L+2)$$

forbidden patterns of length $L+2$ in the second generation and, in general,

$$2^m(L+1) \cdots (L+m) = 2^m \frac{(L+m)!}{L!}$$

forbidden patterns of length $L+m$ in the mth generation, provided that all forbidden patterns up to (and including) the mth generation are different. Observe that all these forbidden patterns generated by π have the form

$$\langle *, \pi_0 + n, *, \pi_1 + n, *, \ldots, *, \pi_{L-1} + n, * \rangle \in \mathcal{S}_M. \tag{3.13}$$

Here $n = 0, 1, \ldots, M - L$, where $M - L \geq 1$ is the number of wildcards $* \in \{0, 1, \ldots, n - 1, L + n, \ldots, M - 1\}$ (with $* \in \{L, \ldots, M - 1\}$ if $n = 0$ and $* \in \{0, \ldots, M - L - 1\}$ if $n = M - L$). Forbidden M-patterns of the form (3.13), where $\pi = \langle \pi_0, \ldots, \pi_{L-1} \rangle$ is a forbidden pattern for f and $M > L$, are called *outgrowth (forbidden) patterns* of π. It is straightforward that if π' is an outgrowth pattern of π and π'' is an outgrowth pattern of π', then π'' is an outgrowth pattern of π.

A better upper bound on the number of outgrowth forbidden patterns of length M of π is obtained using the following reasoning. For fixed n, the number of outgrowth patterns of π of form (3.13) is $M!/L!$. This is because out of all possible permutations of the numbers $\{0, 1, \ldots, M - 1\}$, we only count those that have the entries $\{\pi_0 + n, \pi_1 + n, \ldots, \pi_{L-1} + n\}$ in the required order. Next, note that we have $M - L + 1$ choices for the value of n. Each choice generates a set of $M!/L!$ outgrowth patterns. These sets are not necessarily disjoint, but an upper bound on the size of their union, i.e., the set of all outgrowth forbidden patterns of length M of π, is given by

$$(M - L + 1)\frac{M!}{L!}.$$

Forbidden patterns that are not outgrowth patterns of other forbidden patterns of shorter length are called *root forbidden patterns*. They can be viewed as the root of the tree of forbidden patterns spanned by the outgrowth patterns they generate, branching taking place when going from one length (or generation) to the next. Therefore, they are instrumental in the study of the ordinal structure defined by a transformation—the remaining patterns, whether forbidden or allowed, follow from them. In view of (3.12), for proving that a forbidden L-pattern is a root pattern it suffices to show that it does not belong to group I nor to group II of a forbidden $(L - 1)$-pattern.

Example 5 Figure 3.3 depicts the graphs of the identity (main diagonal), the map $E_2 : x \rightarrow 2x \bmod 1, 0 \leq x \leq 1$, and its second and third iterates. The vertical dashed lines rise at the endpoints of the intervals $P_\pi \neq \emptyset$ of points x defining the allowed patterns $\pi \in \mathcal{S}_4$. We conclude that E_2 has 18 allowed 4-patterns, all consisting of a single component, and hence 6 forbidden 4-patterns, namely

$$\langle 0, 2, 3, 1 \rangle, \langle 1, 0, 2, 3 \rangle, \langle 1, 3, 2, 0 \rangle, \langle 2, 0, 1, 3 \rangle, \langle 3, 1, 0, 2 \rangle, \langle 3, 2, 0, 1 \rangle. \tag{3.14}$$

Since E_2 has no forbidden 3-patterns (see Fig. 3.2), we deduce that all these six forbidden 4-patterns are root patterns.   □

Given a permutation

$$\sigma = \begin{pmatrix} 0 & 1 & \ldots & M - 1 \\ \sigma_0 & \sigma_1 & \ldots & \sigma_{M-1} \end{pmatrix} = [\sigma_0, \ldots, \sigma_{M-1}],$$

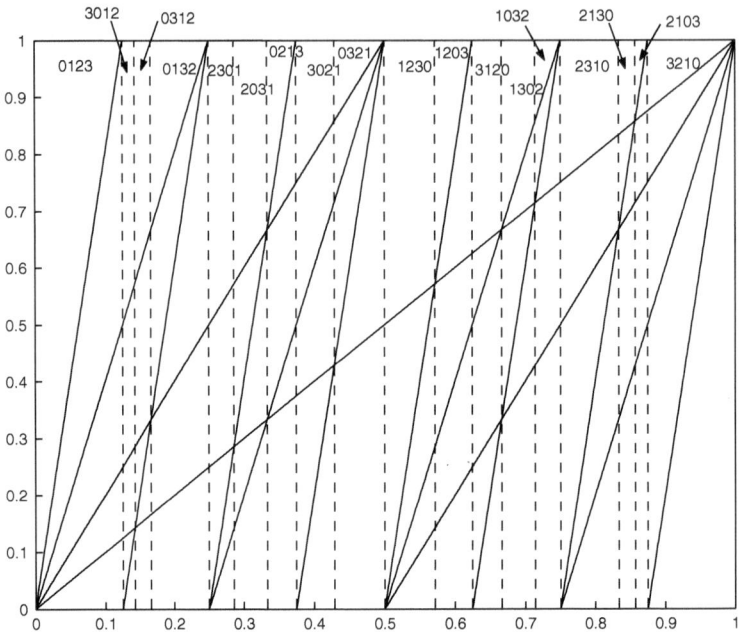

Fig. 3.3 The eighteen allowed 4-patterns of the map $E_2 : x \mapsto 2x \pmod 1$. For clarity, the allowed patterns have been written without angular parentheses nor separating commas. Note that the intervals P_π and the allowed patterns are mirrored with respect to the central line $x = 1/2$

we say that σ contains the *consecutive pattern* $\tau = [\tau_0, \ldots, \tau_{L-1}]$, $L < M$, if the sequence $\sigma_0, \ldots, \sigma_{M-1}$ contains a *consecutive* subsequence order isomorphic to the sequence $\tau_0, \ldots, \tau_{L-1}$. Alternatively, we say that σ avoids the *consecutive pattern* τ if it contains no consecutive subsequence order isomorphic to τ [74]. For instance, $\sigma = [5, 2, 0, 1, 4, 3]$ contains the consecutive pattern $\tau = [0, 2, 1]$ because σ contains the consecutive subsequence $1, 4, 3$ which is order isomorphic to $0, 2, 1$.

In order to apply results on pattern avoidance in combinatorics to forbidden ordinal patterns, recall from Sect. 1.2 that any ordinal pattern $\pi = \langle \pi_0, \ldots, \pi_{L-1} \rangle$ corresponds to the permutation $[\pi]^{-1} : \pi_i \mapsto i$, $0 \le i \le L-1$, (1.23). Suppose furthermore that $\pi' = \langle \pi'_0, \ldots, \pi'_{M-1} \rangle$, $L < M$, is an outgrowth pattern of π, i.e., π' has the form (3.13). The permutation $[\pi'_0, \ldots, \pi'_{M-1}]^{-1} =: [\pi']^{-1}$ performs the substitutions

$$\ldots \quad \pi_0 + n \mapsto i_0, \ldots \quad \pi_1 + n \mapsto i_1, \ldots \quad \pi_{L-1} + n \mapsto i_{L-1}, \ldots,$$

where $n \in \{0, 1, \ldots, M - L\}$ and $0 \le i_0 < i_1 < \cdots < i_{L-1} \le M - 1$. Thus the sequence $i_0, i_1, \ldots, i_{L-1}$ is order isomorphic to $0, 1, \ldots, L-1$. Note, furthermore, that $\pi_0, \ldots, \pi_{L-1}$ is a rearrangement of the consecutive sequence $0, \ldots, L-1$, hence $\pi_0 + n, \ldots, \pi_{L-1} + n$ is a rearrangement of the consecutive sequence $n, \ldots, n + L - 1$. It follows that $\langle \pi'_0, \ldots, \pi'_{M-1} \rangle$ is an outgrowth pattern of $\langle \pi_0, \ldots, \pi_{L-1} \rangle$ if and only if the permutation $[\pi']^{-1}$ contains the permutation $[\pi]^{-1}$ as a consecutive pattern. Therefore the allowed patterns for f are the permutations that avoid all such consecutive subsequences for every forbidden root pattern of f.

Example 6 Take $\pi = \langle 2, 0, 1 \rangle$ to be a forbidden pattern for a certain function f. Then $\pi' = \langle 4, 2, 1, 5, 3, 0 \rangle$ is an outgrowth pattern of π because it contains the subsequence $4, 2, 3$ ($n = 2$). Equivalently, the permutation $[4, 2, 1, 5, 3, 0]^{-1} = [5, 2, 1, 4, 0, 3]$ contains the consecutive pattern $1, 4, 0$, which is order isomorphic to $[2, 0, 1]^{-1} = [1, 2, 0]$.

Let $\mathcal{S}^{\mathrm{out}}(\pi)$ denote the family of outgrowth patterns of the forbidden pattern π,

$$\mathcal{S}_M^{\mathrm{out}}(\pi) = \mathcal{S}^{\mathrm{out}}(\pi) \cap \mathcal{S}_M$$
$$= \{\pi' \in \mathcal{S}_M : [\pi']^{-1} \text{ contains } [\pi]^{-1} \text{ as a consecutive pattern}\},$$

and

$$\mathcal{S}_M^{\mathrm{avoid}}(\pi) = \mathcal{S}_M \backslash \mathcal{S}_M^{\mathrm{out}}(\pi)$$
$$= \{\pi' \in \mathcal{S}_M : [\pi']^{-1} \text{ avoids } [\pi]^{-1} \text{ as a consecutive pattern}\}.$$

where $\backslash$ stands for set difference. The fact that some of the outgrowth patterns of a given length will be the same and that this depends on π makes the analytical calculation of $\left|\mathcal{S}_M^{\mathrm{out}}(\pi)\right|$ extremely complicated. Yet, from [74] we know that there are constants $0 < c, d < 1$ such that

$$c^M M! < \left|\mathcal{S}_M^{\mathrm{avoid}}(\pi)\right| < d^M M!$$

(for the first inequality, $L \geq 3$ is needed). This implies that

$$(1 - d^M)M! < \left|\mathcal{S}_M^{\mathrm{out}}(\pi)\right| < (1 - c^M)M!. \tag{3.15}$$

This factorial growth with M is one of the mechanisms that make forbidden patterns a practical tool for detection of determinism in noisy time series. This topic will be addressed in detail in Chap. 9.

3.4.3 Robustness Against Noise in Deterministic Time Series

Determinism means functional dependence between the "current" value of a univariate or multivariate time series and some of its past values. In some theoretical models this dependence can involve infinitely many values, but we shall limit our attention to the more realistic processes with a finite number of dependent variables. Multivariate time series appear not only when the data source is vectorial but also when scalar deterministic processes are modeled as dynamical systems. Consider, for instance, a time process $y_{n+1} = g(y_n, y_{n-1}, \ldots, y_{n-M+1})$, where g is a scalar self-map, the "memory" $M \geq 2$, and $(y_0, y_{-1}, \ldots, y_{-M+1}) \in \mathbb{R}^M$ is the initial condition. This process can be modeled as an M-dimensional dynamical system (or a multivariate process with memory one) via the change of variables

$$x_n^{(1)} = y_n, \ x_n^{(2)} = y_{n-1}, \ \ldots, \ x_n^{(M)} = y_{n-M+1},$$

so as

$$y_{n+1} = g(y_n, y_{n-1}, \ldots, y_{n-M+1}) \Leftrightarrow \mathbf{x}_{n+1} = \mathbf{f}(\mathbf{x}_n),$$

where

$$\mathbf{x}_n = (x_n^{(1)}, x_n^{(2)}, \ldots, x_n^{(M)}) \in \mathbb{R}^M,$$

and $\mathbf{f} : \mathbb{R}^M \to \mathbb{R}^M$ with

$$\mathbf{f}(\mathbf{x}_n) = (g(\mathbf{x}_n), x_n^{(1)}, x_n^{(2)}, \ldots, x_n^{(M-1)}) \in \mathbb{R}^M.$$

A similar strategy works out for vectorial maps. In sum, any deterministic time series can be considered as the orbit of a dynamical system of adequate dimensionality.

Exercise 3 Write the evolution process

$$x_{n+1} = f(x_n, x_{n-1}, y_{n-1}),$$
$$y_{n+1} = g(x_{n-1}, y_{n-2}),$$

as a five-dimensional dynamical system.

The perturbations that distort a deterministic time series during generation, transmission, observation, and/or measurement are generically referred to as *noise*. We elaborate next on the persistence of admissible and forbidden patterns when the observed data are "noisy," a property called robustness against noise. This property is essential for the applications of ordinal analysis since noise is ubiquitous in real data.

When modeling noise, there are two basic approaches:

- *Dynamical noise* is due to errors in the determination of the initial state and propagates with the dynamic. Thus, if we observe $y_0 = x_0 + \eta_0 \in \Omega \subset \mathbb{R}^q$ instead of the true initial state x_0, then the dynamical noise $(\eta_n)_{n \in \mathbb{N}_0}$ is defined as

$$y_n = f^n(y_0) = f^n(x_0 + \eta_0) = x_n + \eta_n,$$

where $\eta_n = f^n(x_0 + \eta_0) - f^n(x_0)$ depends on x_0 and η_0. Dynamical noise is detrimental to the predictability of the sequence $(f^n(x_0))_{n \in \mathbb{N}_0}$ when f exhibits sensitivity to initial conditions. This sensitivity is measured by its Lyapunov exponent(s) with respect to the natural invariant measure.

- *Observational* (or *additive*) *noise* adds a random fluctuation to the true value $x_n = f^n(x_0)$ in each iteration, that is, the observed value at "time" n is

$$z_n = x_n + \zeta_n,$$

where $\zeta = (\zeta_n)_{n \in \mathbb{N}_0}$ is an $\mathbb{R}^q$-valued random process that accounts for the different macroscopic and/or microscopic factors affecting the true value $f^n(x_0)$. If the random variables ζ_n are independent, then one says that ζ is *white noise*, otherwise the noise is *colored*. Since ordinal patterns depend only on arithmetical differences between observations close in time, the mean of the noise probability distribution is irrelevant. By the same token, we also expect that observational noises with similar variances and finite supports (or possibly thin-tailed distributions) will produce a similar structure of admissible and forbidden patterns. In numerical simulations, the support of the random variables ζ_n will be certainly bounded. White and colored noise are random time series, so random sequences can be viewed as consisting only of noise.

Dynamical noise belongs only to deterministic time series and is important in numerical simulations, whereas observational noise corrupts actual observations of experimental deterministic and random sequences.

Given a deterministic or random time series $\mathbf{x} = (x_n)_{n \in \mathbb{N}_0}$, we say that an ordinal pattern $\pi = \langle \pi_0, \pi_1, \ldots, \pi_{L-1} \rangle$ is *observable* or *visible* in $\mathbf{x}$ if $\mathbf{x}$ contains a length-L block $x_k^{k+L-1} = x_k, \ldots, x_{k+L-1}$ of type π, i.e., if $x_{k+\pi_0} < x_{k+\pi_1} < \cdots < x_{k+\pi_{L-1}}$. Otherwise, π is said to be *unobservable* or *missing* in $\mathbf{x}$. If $\mathbf{x}$ has been deterministically generated by f, then visible patterns are necessarily admissible for f, while forbidden patterns for f cannot be visible in $\mathbf{x}$ (nor in any other orbit of f for that matter). On the other hand, if π is missing in $\mathbf{x}$, this does not necessarily mean that π is forbidden for f—it might be visible in other orbit of f. Thus, forbidden patterns are a subset of the missing patterns. The same considerations apply to real, finite-length sequences.

We say that a visible (correspondingly, missing) ordinal L-pattern π in a deterministic time series $x_n = f^n(x_0)$ is *unconditionally robust* against dynamical or observational noise, if π is also visible (correspondingly, missing) in any perturbed time series $x_n + \xi_n$, $n \in \mathbb{N}_0$, where ξ_n is dynamical or observational noise, respectively. Likewise, we say that a visible (correspondingly, missing) ordinal L-pattern π in a deterministic time series $x_n = f^n(x_0)$ is *conditionally robust* against dynamical or observational noise, if π is also visible (correspondingly, missing) in any perturbed *finite* time series (or initial segment) $x_n + \xi_n$, $0 \le n \le N$, where ξ_n is, respectively, dynamical or observational noise with sufficiently small *amplitude* $A = A(x_0, N) = \max_{0 \le n \le N} \|\xi_n\|$.

Lemma 1 *Consider time series generated by a continuous self-map f of a closed interval $I \subset \mathbb{R}$.*

(1) Forbidden patterns are unconditionally robust against dynamical noise. (This is also true if f is not continuous.)

(2) Visible patterns are conditionally robust against dynamical noise.
(3) Visible and missing patterns are conditionally robust against observational noise.

Proof (1) If $\pi = \langle \pi_0, \ldots, \pi_{L-1} \rangle$ is a forbidden pattern for f, then π will be not visible in the sequence $(f^n(x_0))_{n \in \mathbb{N}_0}$ nor in the perturbed sequence $(f^n(x_0 + \eta_0))_{n \in \mathbb{N}_0}$ for any η_0 such that $x_0 + \eta_0 \in I$.

(2) An ordinal L-pattern π visible in $(f^n(x_0))_{n \in \mathbb{N}_0}$ will remain visible in a finite noisy sequence $y_n = f^n(x_0 + \eta_0) = x_n + \eta_n$, $0 \le n \le N$, only if $|\eta_0|$ is small enough. The size of $|\eta_0|$ will depend on the Lyapunov exponent (with respect to the natural invariant measure) of f.

(3) Consider the segment $x_n = f^n(x_0)$, $0 \le n \le N$, of the time series $(f(x_0))_{n \in \mathbb{N}_0}$, and suppose that

$$f^{\pi_0}(x_k) < f^{\pi_1}(x_k) < \cdots < f^{\pi_{L-1}}(x_k)$$

for some $k \in \{0, 1, , \ldots, N - L + 1\}$. Then

$$f^{\pi_0}(x_k) + \zeta_0 < f^{\pi_1}(x_k) + \zeta_1 < \cdots < f^{\pi_{L-1}}(x_k) + \zeta_{L-1}$$

holds also true as long as the perturbations ζ_i satisfy

$$\zeta_i < f^{\pi_{i+1}}(x_k) - f^{\pi_i}(x_k) + \zeta_{i+1}$$

for $i = 0, 1, \ldots, L - 2$.

From the result that visible patterns are robust against small observational noise, it follows that missing patterns (in particular, forbidden patterns) are likewise robust against small observational noise. $\qquad\square$

We conclude from Lemma 1 that visible patterns in univariate time series are conditionally robust however the kind of noise, whereas forbidden patterns are unconditionally robust against dynamical noise but conditionally robust against observational noise.

In case of multivariate time sequences ($I \subset \mathbb{R}^q$ with $q \ge 2$), property (1) of Lemma 1 remains the same, since the dimensionality of I does not enter in the proof. The situation is different with the conditional robustness. For example, suppose that $\mathbb{R}^q$ is lexicographically ordered, $x_k^{(1)} = x_{k+1}^{(1)}$ and $x_k^{(2)} < x_{k+1}^{(2)}$, hence $\mathbf{x}_k := (x_k^{(1)}, x_k^{(2)}) < (x_{k+1}^{(1)}, x_{k+1}^{(2)}) =: \mathbf{x}_{k+1}$. Then $\mathbf{x}_k + \zeta_k, \mathbf{x}_{k+1} + \zeta_{k+1}$ will not define the pattern $\pi = \langle 0, 1 \rangle$ if $\zeta_k^{(1)} > \zeta_{k+1}^{(1)}$, however, small their sizes are. A corresponding result holds for dynamical noise if, in the example above, the first component of $f^k(\mathbf{x}_0)$ can be made to increase or decrease by varying $\mathbf{x}_0$. In real cases though, in which time series are finite and maps have random-like properties, the coincidence of components is highly unlikely, at least if real numbers are represented with a high enough precision. We may infer that, although visible and missing patterns in multivariate sequences are not, in general, robust against observational nor dynamical noise, in practice they may be considered conditionally robust (as in the univariate case).

Conditional robustness has to do with the amplitude of the perturbation. What about the dependence of visible and missing patterns on the length N of the initial segment $(z_n)_{n=0}^{N-1}$ of a noisy time series $(z_n)_{n \in \mathbb{N}_0}$ of either type? Since an increase of N eventually transforms missing patterns of length $L < N$ into visible L-patterns, while visible patterns remain visible, it is clear that the number of missing L-patterns in time series contaminated by dynamical or observational noise will decrease with N. In other words, the longer the sequence, the higher the odds that some block z_n^{n+L-1} defines π. In the case of white noise only, $(z_n)_{n \in \mathbb{N}_0} = (\zeta_n)_{n \in \mathbb{N}_0}$, one can show that the decrease of missing ordinal L-patterns goes exponentially with N (see also Fig. 9.7).

If forbidden patterns were not robust against noise, they would be not useful in time-series analysis. The sort of applications we have in mind belong in the detection of determinism in univariate and multivariate time-series analysis, since (unconstrained) random real-valued time series have no forbidden patterns with probability 1. These and related issues will be discussed in Chap. 9.

Chapter 4
Ordinal Structure of the Shifts

Shift systems are dynamical systems which are used as universal models in information theory and stochastic processes. Besides they are interesting on its own because, in spite of their conceptual simplicity, they exhibit some of the intricacies of low-dimensional chaos, like sensitivity to initial conditions, strong mixing, and a dense set of periodic points.

In the last chapter we studied some general properties of the allowed and forbidden patterns associated with a dynamical system whose state space is linearly ordered. In this chapter we will be more specific and study the ordinal structure of the shift transformations. By ordinal structure we mean such properties as the length and number of the root forbidden patterns. Contrary to the generality of maps, we shall see that these issues can be ascertained with great detail for the shifts.

4.1 Ordinal Patterns and the Shift Maps

Let $E_N : [0, 1] \to [0, 1]$, $N \in \{2, 3, \dots\}$, be the shift or sawtooth map

$$E_N(x) = Nx \pmod 1 \tag{4.1}$$

(Fig. 4.1). Observe that if

$$x = \sum_{n=0}^{\infty} x_n \cdot N^{-(n+1)} =: 0.\, x_0\, x_1 \dots x_n \dots,$$

$0 \le x_n \le N - 1$, is an N-ary expansion of $x \in [0, 1]$, then

$$Nx = \sum_{n=0}^{\infty} x_n \cdot N^{-n} = x_0 + \sum_{n=1}^{\infty} x_n \cdot N^{-n} = x_0 .\, x_1\, x_2 \dots x_{n+1} \dots$$

and

J.M. Amigó, *Permutation Complexity in Dynamical Systems*,
Springer Series in Synergetics, DOI 10.1007/978-3-642-04084-9_4,
© Springer-Verlag Berlin Heidelberg 2010

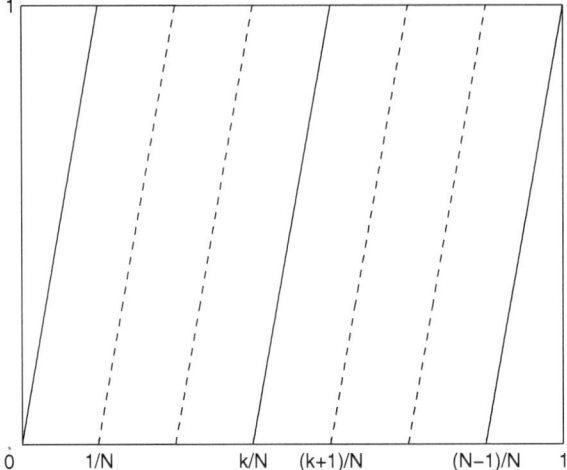

Fig. 4.1 The function $E_N(x) = Nx \bmod 1$. The figure shows only the first, the kth, and the last laps of the graph

$$E_N (0. x_0 x_1 \ldots x_n \ldots) = 0. x_1 x_2 \ldots x_{n+1} \ldots . \tag{4.2}$$

In other words, if we write $E_N (0. x_0 x_1 \ldots x_n \ldots) = 0. y_0 y_1 \ldots y_n \ldots$, then $y_n = x_{n+1}$ for $n \in \mathbb{N}_0$. This justifies the name "shift map" for E_N since it shifts the digits of the representation of x in base N, one position to the left (the first digit is deleted). Let us recall an N-ary expansion is not unique for some $x \in [0, 1]$ since

$$0. x_0 \ldots x_{n-1} 1 0^\infty = 0. x_0 \ldots x_{n-1} 0 (N-1)^\infty,$$

where (as in Sect. 1.1.2) the upper symbol "∞" stands for indefinite repetition. But the set of points $x \in [0, 1]$ whose N-ary expansion ends with 10^∞ or $0 (N-1)^\infty$ has zero Lebesgue measure, so such points can eventually be thought to have been removed from $[0, 1]$.

If we identify now an N-ary expansion $0. x_0 x_1 \ldots x_n \ldots$ of $x \in [0, 1]$ with the one-sided sequence $(x_0, x_1, \ldots, x_n, \ldots) \in S^{\mathbb{N}_0}$, $S = \{0, \ldots, N-1\}$, then action (4.2) translates into the action of the one-sided shift Σ on $S^{\mathbb{N}_0}$. Formally, if $\phi_N : S^{\mathbb{N}_0} \to [0, 1]$ is the map defined by

$$\phi_N : (x_n)_{n \in \mathbb{N}_0} \mapsto \sum_{n=0}^{\infty} x_n N^{-(n+1)}, \tag{4.3}$$

then ϕ_N is an order isomorphism modulo 0 between E_N and the one-sided shift Σ on $S^{\mathbb{N}_0}$, i.e.,

$$\phi_N \circ \Sigma = E_N \circ \phi_N, \tag{4.4}$$

the order of $S^{\mathbb{N}_0}$ being given by the lexicographical rule:

$$\mathbf{x} < \mathbf{x}' \Leftrightarrow \begin{cases} x_0 < x'_0, \\ \text{or} \\ x_0 = x'_0, \dots, x_{n-1} = x'_{n-1} \text{ and } x_n < x'_n \ (n \geq 1), \end{cases} \tag{4.5}$$

where $\mathbf{x} = (x_n)_{n \in \mathbb{N}_0}$ and $\mathbf{x}' = (x'_n)_{n \in \mathbb{N}_0}$. Observe that ϕ_N maps the cylinder set $C_{i_0 \dots i_n} = \{(x_n)_{n \in \mathbb{N}_0} : x_0 = i_0, \dots, x_n = i_n\}$ $(i_0, \dots, i_n \in S)$ to the interval

$$\left[\frac{i_0 N^n + \dots + i_n}{N^{n+1}}, \frac{i_0 N^n + \dots + i_n + 1}{N^{n+1}} \right].$$

Exercise 4 Let $\mathcal{B}$ be the Borel sigma-algebra on $[0, 1]$, λ the corresponding Lebesgue measure, and E_N the sawtooth map (4.1). Prove that the dynamical system $([0, 1], \mathcal{B}, \lambda, E_N)$ and the $(\frac{1}{N}, \dots, \frac{1}{N})$-Bernoulli one-sided shift are isomorphic (modulo 0).

Once we know that E_N and the one-sided shift Σ on N symbols are order-isomorphic (up to sets measure 0), it follows that they have the same forbidden patterns (see Sect. 3.4.1).

In general it is very difficult to work out the specifics of the forbidden patterns of a given map; the graphical methods can only help for small values of L. But we shall see next that the shifts and the signed shifts (to be defined in Chap. 5.) are an important exception. In particular, owing to the simple structure of one-sided and two-sided shifts, the structure of their admissible and forbidden patterns can be analyzed with great detail. By order isomorphy these conclusions will hold also for the sawtooth map family E_N (one-sided shifts), the baker map (two-sided shifts), and the logistic and symmetric tent maps (one-sided signed shifts), among others.

4.2 Forbidden Patterns for One-Sided Shifts

In Sect. 1.1.2 we saw that one-sided shifts Σ are continuous maps on the compact metric spaces $(\{0, 1, \dots, N-1\}^{\mathbb{N}_0}, d)$, $N \geq 2$. Furthermore, if $\{0, 1, \dots, N-1\}^{\mathbb{N}_0}$ is lexicographically ordered (see (4.5)), then Σ is order-isomorphic (modulo 0) to E_N via map (4.3).

What is the structure of the allowed ordinal patterns for Σ? It is easy to convince oneself (see Example 7) that, given $\mathbf{x} = (x_0, \dots, x_{L-1}, \dots) \in \{0, 1, \dots, N-1\}^{\mathbb{N}_0}$ of type $\pi \in \mathcal{S}_L$, π can be decomposed into at most N blocks (separated by semicolons),

$$\langle \pi_0, \dots, \pi_{k_0-1}; \pi_{k_0}, \dots, \pi_{k_0+k_1-1}; \dots; \pi_{k_0+\dots+k_{N-2}}, \dots, \pi_{k_0+\dots+k_{N-2}+k_{N-1}-1} \rangle, \tag{4.6}$$

where $k_s \geq 0$ is the number of times the symbol $s \in \{0, 1, \ldots, N-1\}$ appears in the segment $x_0^{L-1} = x_0, \ldots, x_{L-1}$ of $\mathbf{x}$ ($k_s = 0$ if none, with the corresponding block empty) and $k_0 + \cdots + k_{N-1} = L$. The entries $\pi_0, \ldots, \pi_{k_0-1}$ are the locations of the symbol 0 in x_0^{L-1}, the entries $\pi_{k_0}, \ldots, \pi_{k_0+k_1-1}$ are the locations of the symbol 1 in x_0^{L-1}, etc. For this reason, the first block will also be called the 0-block, and, in general, the $(s+1)$th block,

$$\pi_{k_0+\cdots+k_{s-1}}, \ldots, \pi_{k_0+\cdots+k_{s-1}+k_s-1}, \tag{4.7}$$

$1 \leq s \leq N-1$, will also be called the s-block. Decomposition (4.6) is sometimes called the *decomposition* of an allowed ordinal pattern $\pi \in \mathcal{S}_L$ in s-blocks.

A (finite) subsequence of components of π of the form $\pi_i, \ldots, \pi_i + 1, \ldots, \pi_i + 2, \ldots$ (respectively, $\pi_i, \ldots, \pi_i - 1, \ldots, \pi_i - 2, \ldots$) will be called an *increasing* (respectively, *decreasing*) *subsequence*. Increasing or decreasing subsequences will be collectively called *monotone*. Observe that we use these concepts in a restrictive way.

We will see next that from the fact that allowed patterns for the one-sided shift must be decomposable as in (4.6), it is possible to deduce their structure.

Lemma 2 *The blocks in decomposition (4.6) obey the following basic restrictions.*

R1 *The first (leftmost) block, $\pi_0, \ldots, \pi_{k_0-1}$, contains the locations of the 0's in x_0^{L-1}. Each 0-run (i.e., a segment of two or more consecutive 0's contained in or intersected by x_0^{L-1}), if any, contributes an increasing subsequence of the same length as the 0-run. Solitary symbols 0's in x_0^{L-1}, if any, contribute components to the first block that do not form monotone subsequences.*

R2 *The last (rightmost) block, $\pi_{k_0+\cdots+k_{N-2}}, \ldots, \pi_{k_0+\cdots+k_{N-2}+k_{N-1}-1}$, contains the locations of the $(N-1)$'s in x_0^{L-1}. Each $(N-1)$-run contained in or intersected by x_0^{L-1}, if any, contributes a decreasing subsequence of the same length as the $(N-1)$-run. Solitary symbols 1's in x_0^{L-1}, if any, contribute components to the last block that do not form monotone subsequences.*

R3 *Every intermediate block, $\pi_{k_0+\cdots+k_{j-1}}, \ldots, \pi_{k_0+\cdots+k_{j-1}+k_j-1}$, $1 \leq j \leq N-2$, contains the locations of the j's in x_0^{L-1}. Each j-run contained in or intersected by x_0^{L-1}, if any, contributes a subsequence of the same length as the j-run that is increasing if the run is followed by a symbol $> j$, or decreasing if the run is followed by a symbol $< j$. Isolated symbols j's in x_0^{L-1}, if any, contribute components to the corresponding block that do not form monotone subsequences.*

R4 *If the entries $\pi_m \leq L-2$ and $\pi_n \leq L-2$ belong to the same block of $\pi \in \mathcal{S}_L$ and π_m appears on the left of π_n (i.e., $0 \leq m < n \leq L-1$), then $\pi_m + 1$ appears also on the left of $\pi_n + 1$ (i.e., $\pi_m + 1 = \pi_{m'}$, $\pi_n + 1 = \pi_{n'}$ and $0 \leq m' < n' \leq L-1$), not necessarily in the same block.*

Proof **R1)** Consider a 0-run of length l in **x**:

$i =$	$n-1$	n	$n+1$	$\ldots$	$n+l-1$	$n+l$
$\mathbf{x} =$	a	0	0	$\ldots$	0	b

with $0 \leq n, n+l \leq L$, and $a, b > 0$. Hence the 0-block of $\pi(\mathbf{x})$ contains the increasing subsequence

$$\ldots, n, \ldots, n+1, \ldots, n+l-1, \ldots,$$

The "..." stands for entries proceeding from other 0-runs in **x**.

R2) Consider an $(N-1)$-run of length l in **x**:

$i =$	$n-1$	n	$n+1$	$\ldots$	$n+l-1$	$n+l$
$\mathbf{x} =$	c	$N-1$	$N-1$	$\ldots$	$N-1$	d

with $0 \leq n, n+l \leq L$, and $c, d < N-1$. Hence the $(N-1)$-block of $\pi(\mathbf{x})$ contains the decreasing subsequence

$$\ldots, n+l-1, \ldots, n+1, \ldots, n, \ldots,$$

The "..." allows for entries proceeding from other $(N-1)$-runs in **x**.

R3) This restriction follows similarly to R1 for s-runs, $0 < s < N-1$, terminated with $b > s$ (increasing subsequences), and similarly to R2 for s-runs terminated with $d < s$ (decreasing subsequences).

R4) Since π_m and π_n belong to the same block and $\Sigma^{\pi_m}(\mathbf{x}) < \Sigma^{\pi_n}(\mathbf{x})$ for some $\mathbf{x} \in \{0, 1, \ldots, N-1\}^{\mathbb{N}_0}$, there exists $k \in \{0, 1, \ldots, N-1\}$ such that

$$\Sigma^{\pi_m}(\mathbf{x}) = (k, x_{\pi_m+1} \ldots) < (k, x_{\pi_n+1}, \ldots) = \Sigma^{\pi_n}(\mathbf{x}).$$

By the definition of lexicographical order, there are two possibilities: (i) $x_{\pi_m+1} < x_{\pi_n+1}$ and (ii) $x_{\pi_m+\kappa} = x_{\pi_n+\kappa}$ for $1 \leq \kappa \leq l-1$, $l \geq 2$, and $x_{\pi_m+l} < x_{\pi_n+l}$. In both cases,

$$\Sigma^{\pi_m+1}(\mathbf{x}) = (x_{\pi_m+1} \ldots) < (x_{\pi_n+1}, \ldots) = \Sigma^{\pi_n+1}(\mathbf{x})$$

and, hence, the entry $\pi_m + 1$ appears on the left of $\pi_n + 1$.   □

Example 7 Consider in $\{0, 1, 2\}^{\mathbb{N}_0}$ the sequence

$$\mathbf{x} = (2_0, 1_1, 1_2, 1_3, 2_4, 2_5, 0_6, 0_7, 1_8, 1_9, 0_{10}, 0_{11}, 2_{12}, 2_{13}, 2, 1, \ldots), \qquad (4.8)$$

where a_k indicates that the entry $a \in \{0, 1, 2\}$ is at place k. Then **x** defines the ordinal pattern

$$\pi = \langle 6, 10, 7, 11; 9, 8, 1, 2, 3; 5, 0, 4, 13, 12 \rangle \in \mathcal{S}_{14}.$$

The 0-block, $\pi_0^3 = 6, 10, 7, 11$, codifies the $k_0 = 4$ times the symbol 0 appears in x_0^{13}, grouped in two runs, x_6^7 and x_{10}^{11} (note the two increasing subsequences 6, 7 and 10, 11 in this block). The order results from

$$\Sigma_3^6(\mathbf{x}) = (0, 0, 1, \dots) < \Sigma_3^{10}(\mathbf{x}) = (0, 0, 2, \dots)$$
$$< \Sigma_3^7(\mathbf{x}) = (0, 1, 1, \dots) < \Sigma_3^{11}(\mathbf{x}) = (0, 2, \dots).$$

The 1-block, $\pi_4^8 = 9, 8, 1, 2, 3$, codifies the $k_1 = 5$ times the symbol 1 appears in x_0^{13}, grouped also in two runs: x_1^3, followed by the symbol $2 > 1$, and x_8^9, followed by the symbol $0 < 1$ (note the corresponding increasing subsequence $1, 2, 3$, and decreasing subsequence 9, 8, in this block). The order results from

$$\Sigma_3^9(\mathbf{x}) = (1, 0, 0, \dots) < \Sigma_3^8(\mathbf{x}) = (1, 1, 0, \dots) < \Sigma_3^1(\mathbf{x}) = (1, 1, 1, \dots) < \cdots$$

etc. Finally, the 2-block $\pi_9^{13} = 5, 0, 4, 13, 12$ codifies the $k_2 = 5$ appearances of the symbol 2 in x_0^{13}. The decreasing subsequences $5, 4$ and $13, 12$ come from the runs x_4^5 and x_{12}^{13}, respectively, where x_{12}^{13} is the intersection within x_0^{13} of a longer 2-run. The order results from

$$\Sigma_3^5(\mathbf{x}) = (2, 0, 0, \dots) < \Sigma_3^0(\mathbf{x}) = (2, 1, 1, \dots) < \Sigma_3^4(\mathbf{x}) = (2, 2, 0, \dots) < \cdots.$$

The restriction R4 is easily checked to be fulfilled.

Observe that two sequences $\mathbf{x}, \mathbf{x}'$ with $x_0^{L-1} \neq x_0'^{L-1}$ may define the same ordinal L-pattern, while two sequences $\mathbf{y}, \mathbf{y}'$ with $y_0^{L-1} = y_0'^{L-1}$ may define different ordinal L-patterns (depending on $y_{L-2}, \dots$ and $y_{L-2}', \dots$).

The restriction R4 implies some simple consequences for the relative locations of increasing and decreasing subsequences within the same block and their continuations (if any) outside the block.

Corollary 1 *In an allowed ordinal pattern $\pi \in \mathcal{S}_L$, the following relations among its components hold.*

(A) *If $\pi_i, \pi_i + 1, \dots, \pi_i + l - 1$, $1 \le l \le L - 1$, is an increasing subsequence within the same block of $\pi \in \mathcal{S}_L$ with $\pi_i + l < L$, then $\pi_i + l$ is on the right of $\pi_i + l - 1$ (i.e., $\pi_i + l - 1 = \pi_m$, $\pi_i + l = \pi_n$, and $m < n$).*

(B) *If $\pi_i, \pi_i - 1, \dots, \pi_i - l + 1$, $1 \le l \le L - 1$, is a decreasing subsequence within the same block of $\pi \in \mathcal{S}_L$ with $\pi_i < L - 1$, then $\pi_i + 1$ is on the left of π_i (i.e., $\pi_i + 1 = \pi_j$ with $j < i$).*

(C) *If $\pi_i, \pi_i \pm 1, \dots, \pi_i \pm l \mp 1$ and $\pi_j, \pi_j \pm 1, \dots, \pi_j \pm h \mp 1$, $1 \le l, h \le L - 1$, are two subsequences with the same monotonicity (upper signs for increasing, lower signs for decreasing subsequences) within the same block of $\pi \in \mathcal{S}_L$, then they are fully separated or, if intertwined, then it may not happen that two or more entries of one of them are between two entries of the other.*

The proof is left as an easy exercise.

Theorem 1 *The one-sided shift on $N \geq 2$ symbols has no forbidden patterns of length $L \leq N + 1$.*

Proof If $L \leq N$ and $\pi = \langle \pi_0, \pi_1, \ldots, \pi_{L-1} \rangle$, then any "point" $\mathbf{x} \in \{0, 1, \ldots, N - 1\}^{\mathbb{N}_0}$ with $x_{\pi_n} = n$, $0 \leq n \leq L - 1 \leq N - 1$, is trivially of type π:

$$\Sigma^{\pi_0}(\mathbf{x}) = (0, \ldots) < \Sigma^{\pi_1}(\mathbf{x}) = (1, \ldots) < \cdots < \Sigma^{\pi_{L-1}}(\mathbf{x}) = (L - 1, \ldots).$$

Thus, suppose $L = N + 1$ and note if $\mathbf{x} = (x_0, x_1, x_2, \ldots)$ is of type $\pi = \langle \pi_0, \pi_1, \ldots, \pi_N \rangle$, then the sequence $\bar{\mathbf{x}} = (N - 1 - x_0, N - 1 - x_1, N - 1 - x_2, \ldots)$ is of type $\pi_{\text{mirrored}} = \langle \pi_N, \pi_{N-1}, \ldots, \pi_1, \pi_0 \rangle$.

Given $\pi = \langle \pi_0, \pi_1, \ldots, \pi_N \rangle$, we can therefore assume, without loss of generality, that $\pi_0 < \pi_N$. Consider two cases.

- If $\pi_N \neq N$, then there is some $l \in \{1, 2, \ldots, N - 1\}$ such that $\pi_l = N$. In this case, the point $\mathbf{x} = (x_0, x_1, \ldots) \in \{0, 1, \ldots, N - 1\}^{\mathbb{N}_0}$, where

$$x_{\pi_0} = 0, \ x_{\pi_1} = 1, \ \ldots, \ x_{\pi_{l-1}} = l - 1, \ x_{\pi_l} = l - 1, \ x_{\pi_{l+1}} = l, \ \ldots,$$
$$x_{\pi_{N-1}} = N - 2, \ x_{\pi_N} = N - 1, \ x_{N+1} = x_{N+2} = N - 1$$

is of type π. Indeed, it is enough to note that

$$\Sigma^{\pi_{l-1}}(\mathbf{x}) = (l - 1, x_{\pi_{l-1}+1}, \ldots) < (l - 1, N - 1, N - 1, \ldots)$$
$$= \Sigma^N(\mathbf{x}) = \Sigma^{\pi_l}(\mathbf{x}).$$

- If $\pi_N = N$, let us first assume that $\pi_0 \neq 0$. Then there is $k \in \{1, 2, \ldots, N - 1\}$ such that $\pi_k + 1 = \pi_0$. In this case, the point $\mathbf{x} = (x_0, x_1, \ldots) \in \{0, 1, \ldots, N - 1\}^{\mathbb{N}_0}$, where

$$x_{\pi_0} = 0, \ \ldots, \ x_{\pi_k} = k, \ x_{\pi_{k+1}} = k, \ x_{\pi_{k+2}} = k + 1, \ \ldots,$$
$$x_{\pi_{N-1}} = N - 2, \ x_{\pi_N} = N - 1, \ x_{N+1} = N - 1$$

is of type π. This is clear because

$$\Sigma^{\pi_k}(\mathbf{x}) = (k, 0, \ldots) < (k, x_{\pi_{k+1}+1}, \ldots) = \Sigma^{\pi_{k+1}}(\mathbf{x}).$$

In the case that $\pi_0 = 0$, then there is $l \in \{1, 2, \ldots, N - 1\}$ such that $\pi_l = N - 1$. Now the sequence $\mathbf{x} = (x_0, x_1, \ldots) \in \{0, 1, \ldots, N - 1\}^{\mathbb{N}_0}$, where

$$x_{\pi_0} = 0, \ x_{\pi_1} = 1, \ \ldots, \ x_{\pi_{l-1}} = l - 1, \ x_{\pi_l} = l - 1, \ x_{\pi_{l+1}} = l, \ \ldots,$$
$$x_{\pi_{N-1}} = N - 2, \ x_{\pi_N} = N - 1$$

is of type π, since

$$\Sigma^{\pi_{l-1}}(\mathbf{x}) = (l - 1, x_{\pi_{l-1}+1}, \dots)$$
$$< (l - 1, N - 1, \dots) = \Sigma^{N-1}(\mathbf{x}) = \Sigma^{\pi_l}(\mathbf{x}). \ \square$$

Next we are going to show that the one-sided shift on N symbols has forbidden patterns (more specifically, forbidden *root* patterns) of any length $L \geq N + 2$. In order to construct explicit instances, we need first to introduce some notation and definitions.

Consider a partition of the sequence $0, 1, \dots, L - 1$ of the form

$$\vec{\mathbf{p}_1}, \vec{\mathbf{p}_2}, \dots, \vec{\mathbf{p}_d}, \dots, \vec{\mathbf{p}_D}, \tag{4.9}$$

where

$$\vec{\mathbf{p}_d} = e_d, e_d + 1, \dots, e_d + h_d - 1, \tag{4.10}$$

$1 \leq d \leq D, D \geq 2$, with (i) $h_d \geq 1$, $h_1 + \dots + h_D = L$, (ii) $e_1 = 0$, $e_D + h_D - 1 = L - 1$, and (iii) $e_d + h_d = e_{d+1}$ for $1 \leq d \leq D - 1$, i.e., the *follower* of $\vec{\mathbf{p}_d}$, $e_d + h_d$, $d \leq D - 1$, is the first element of p_{d+1}, namely, e_{d+1}. We call (4.9) a partition of $0, 1, \dots, L - 1$ in D segments, (4.10) being an *increasing segment*, and denote by $\overleftarrow{\mathbf{p}_d}$ the *decreasing* or *reversed segment*

$$\overleftarrow{\mathbf{p}_d} = e_d + h_d - 1, \dots, e_d + 1, e_d.$$

We also call e_d the first element of $\overleftarrow{\mathbf{p}_d}$ and e_{d+1} the follower of $\overleftarrow{\mathbf{p}_d}$.

Since increasing and decreasing segments are nothing else but special cases of increasing and decreasing subsequences, respectively, the consequences (A)–(C) of restriction R4 apply as well. In the proof of the existence of forbidden root patterns below (Lemmas 3 and 4 and Theorem 2) we are going to use (A) and (B) in the following, particularized version (that will be also referred to as R4): *the follower (if any) of an increasing segment $\vec{\mathbf{p}_n}$ (correspondingly, decreasing segment $\overleftarrow{\mathbf{p}_n}$) in an allowed pattern π appears always to the right of $\vec{\mathbf{p}_n}$ (correspondingly, to the left of $\overleftarrow{\mathbf{p}_n}$).*

Definition 2 Consider partition (4.9) of $0, 1, \dots, L - 1$ in segments.

1. We call

$$\pi = \langle \vec{\mathbf{p}_1}, \vec{\mathbf{p}_3}, \dots, \overleftarrow{\mathbf{p}_4}, \overleftarrow{\mathbf{p}_2} \rangle \quad \text{and} \quad \pi_{\text{mirrored}} = \langle \vec{\mathbf{p}_2}, \vec{\mathbf{p}_4}, \dots, \overleftarrow{\mathbf{p}_3}, \overleftarrow{\mathbf{p}_1} \rangle \tag{4.11}$$

   a tent *pattern of length L.*
2. We call

$$\pi = \langle \dots, \overleftarrow{\mathbf{p}_3}, \overleftarrow{\mathbf{p}_1}, \vec{\mathbf{p}_2}, \vec{\mathbf{p}_4}, \dots \rangle \quad \text{and} \quad \pi_{\text{mirrored}} = \langle \dots, \overleftarrow{\mathbf{p}_4}, \overleftarrow{\mathbf{p}_2}, \vec{\mathbf{p}_1}, \vec{\mathbf{p}_3}, \dots \rangle \tag{4.12}$$

   a spiraling *pattern of* length L.

Observe that the relation between partitions of $0, 1, \ldots, L-1$ in segments and spiraling patterns of length L is one-to-one except when $\overrightarrow{\mathbf{p_1}} = 0$ ($h_1 = 1$). In this case, $\overleftarrow{\mathbf{p_1}}, \overrightarrow{\mathbf{p_2}} = 0, 1, \ldots, e_2 + h_2 - 1$ can be taken for $\overrightarrow{\mathbf{p_1'}} := 0, 1, \ldots, e_2 + h_2 - 1$ ($h_1' = h_2 + 1$).

Lemma 3 *If $N \geq 2$ is the number of symbols and π is a tent pattern with D segments, then π is forbidden if and only if $D \geq N + 2$.*

Proof Consider the tent pattern $\pi = \langle \overrightarrow{\mathbf{p_1}}, \overrightarrow{\mathbf{p_3}}, \ldots, \overleftarrow{\mathbf{p_4}}, \overleftarrow{\mathbf{p_2}} \rangle$. To begin with, the last entry $h_1 - 1$ of $\overrightarrow{\mathbf{p_1}}$ and the first entry e_3 of $\overrightarrow{\mathbf{p_3}}$ may not be in the same block, otherwise the R4 would be violated ($e_2 = h_1$ should be on the left of $e_3 + 1$ if $h_3 \geq 2$ or on the left of e_4 if $h_3 = 1$). Thus we separate them with a first semicolon:

$$\pi = \langle \overrightarrow{\mathbf{p_1}}; \overrightarrow{\mathbf{p_3}}, \ldots, \overleftarrow{\mathbf{p_4}}, \overleftarrow{\mathbf{p_2}} \rangle.$$

Observe that the resulting leftmost block, $\overrightarrow{\mathbf{p_1}}$, complies with R1. Consider now the followers of $\overleftarrow{\mathbf{p_2}}$ and $\overleftarrow{\mathbf{p_4}}$ to conclude similarly that we need to separate these segments by a second semicolon:

$$\pi = \langle \overrightarrow{\mathbf{p_1}}; \overrightarrow{\mathbf{p_3}}, \ldots, \overleftarrow{\mathbf{p_4}}; \overleftarrow{\mathbf{p_2}} \rangle.$$

The resulting rightmost block satisfies R2.

The procedure continues along the same lines. In the kth step, R4 requires a kth semicolon between the segments $\overrightarrow{\mathbf{p_k}}$ and $\overrightarrow{\mathbf{p_{k+2}}}$, so that, if $D \geq N + 1$, the $(N-1)$th semicolon will separate $\overrightarrow{\mathbf{p_{N-1}}}$ and $\overrightarrow{\mathbf{p_{N+1}}}$. All these intermediary blocks trivially fulfill R3.

In the particular case $D = N + 1$, the "central" block $\overrightarrow{\mathbf{p_N}}\overleftarrow{\mathbf{p_{N+1}}}$ (N odd) or $\overrightarrow{\mathbf{p_{N+1}}}\overleftarrow{\mathbf{p_N}}$ (N even) complies with R3 and R4, and hence π is allowed. A further segment $\overrightarrow{\mathbf{p_{N+2}}}$ would require an Nth semicolon to separate $\overrightarrow{\mathbf{p_N}}$ and $\overrightarrow{\mathbf{p_{N+1}}}$ in order not to violate R4.

The proof for π_{mirrored} is completely analogous. $\qquad \square$

Lemma 4 *If $N \geq 2$ is the number of symbols, π is a spiraling pattern with D segments, and $h_1 \geq 2$ (i.e., $\overrightarrow{\mathbf{p_1}} = 0, 1, \ldots$), then*

1. *π is forbidden if and only if (a) $D = N$ and $h_D \geq 2$ or (b) $D \geq N + 1$;*
2. *π is allowed if and only if (a') $D < N$ or (b') $D = N$ and $h_D = 1$.*

Part 2 of Lemma 4, which is the logical negation of part 1, has been explicitly formulated for further references.

Proof Consider the spiraling pattern (4.12). To begin with, the entries $h_1 - 1$ and $h_1 - 2$ of $\overleftarrow{\mathbf{p_1}} = h_1 - 1, \ldots, 1, 0$ may not be in the same block, otherwise R4 would be violated (e_2 should be on the left of $h_1 - 1$). Thus we separate them with a first semicolon:

$$\pi = \langle \ldots, \overleftarrow{\mathbf{p_3}}, h_1 - 1; h_1 - 2, \ldots, 1, 0, \overrightarrow{\mathbf{p_2}}, \overrightarrow{\mathbf{p_4}}, \ldots \rangle.$$

From here on, three possibilities can occur that we illustrate in a general step of even order. (i) If $\overrightarrow{\mathbf{p}_{2v}}$ consists of more than one element (i.e., $h_{2v} \geq 2$), then we apply R4 to $\overrightarrow{\mathbf{p}_{2v}}$ to conclude that we need a semicolon between $e_{2v} + h_{2v} - 2$ and $e_{2v} + h_2 - 1$ (since the follower of $\overrightarrow{\mathbf{p}_{2v}}$, i.e., the first entry of $\overrightarrow{\mathbf{p}_{2v+1}}$, is on the wrong side). (ii) If $\overrightarrow{\mathbf{p}_{2v}}$ consists of one element ($h_{2v} = 1$) and $\overrightarrow{\mathbf{p}_{2v-2}}$ consists of more than one element ($h_{2v-2} \geq 2$), then we apply R4 to the pair $\overrightarrow{\mathbf{p}_{2v}} = e_{2v}$ and $e_{2v-2} + h_{2v-2} - 1$, the last element of $\overrightarrow{\mathbf{p}_{2v-2}}$, which has been separated with a semicolon from the rest of elements in $\overrightarrow{\mathbf{p}_{2v-2}}$ two steps earlier. (iii) If both $\overrightarrow{\mathbf{p}_{2v}}$ and $\overrightarrow{\mathbf{p}_{2v-2}}$ consist of a single element ($h_{2v} = h_{2v-2} = 1$), apply R4 to the pair $\overrightarrow{\mathbf{p}_{2v-2}} = e_{2v-2} < \overrightarrow{\mathbf{p}_{2v}} = e_{2v}$ to infer the need for a semicolon separating them (since $e_{2v-2} + 1 = e_{2v-1}$, the first element of $\overleftarrow{\mathbf{p}_{2v-1}}$, is on the right of $e_{2v} + 1 = e_{2v+1}$, the first element of $\overleftarrow{\mathbf{p}_{2v+1}}$). As a general rule, we need one semicolon per segment $\overrightarrow{\mathbf{p}_{2v}}$ or $\overleftarrow{\mathbf{p}_{2v+1}}$ as long as there are still a posterior segment $\overrightarrow{\mathbf{p}_{2v+1}}$ or $\overrightarrow{\mathbf{p}_{2v+2}}$, respectively, on the "wrong" side. Note that all (intermediary) blocks ensued so far comply with R3.

Following this way, we run out of the $N - 1$ semicolons we may use (corresponding to the N symbols), after having considered the segment $\overrightarrow{\mathbf{p}_{N-1}}$. Yet if $D = N$ and $h_N \geq 2$, then $\overrightarrow{\mathbf{p}_N}$ will violate R1 if N is odd or R2 if N is even. If $D \geq N + 1$, then the segment $\overrightarrow{\mathbf{p}_{N+1}}$ will be on the wrong side of $\overrightarrow{\mathbf{p}_N}$ and the pattern will not comply with R4.

The proof for π_{mirrored} is completely analogous.   □

The constructive, stepwise procedure used in the proofs of Lemmas 3 and 4 can be used mutatis mutandis in general to decompose any ordinal pattern into well-formed (i.e., complying with R1–R4) blocks. For instance, one could start from the leftmost entry and move on rightward one entry at a time, inserting a semicolon between the current and the previous entry whenever necessary to enforce the restrictions R1–R4. Reciprocally, given a decomposition of an ordinal pattern π in s-blocks, one can easily construct a sequence $\mathbf{x} \in \{0, \ldots, N - 1\}^{\mathbb{N}_0}$ of type π.

Theorem 2 *The following patterns of length $L \geq N + 2$, together with their corresponding mirrored patterns, are forbidden root patterns.*

1. *The tent patterns with $N + 2$ segments*

$$\langle 0, \overrightarrow{\mathbf{p}_3}, \ldots, \overrightarrow{\mathbf{p}_N}, L - 1, \overleftarrow{\mathbf{p}_{N+1}}, \ldots, \overleftarrow{\mathbf{p}_2} \rangle \tag{4.13}$$

if N is odd or

$$\langle 0, \overrightarrow{\mathbf{p}_3}, \ldots, \overrightarrow{\mathbf{p}_{N+1}}, L - 1, \overleftarrow{\mathbf{p}_N}, \ldots, \overleftarrow{\mathbf{p}_2} \rangle \tag{4.14}$$

if N is even. Here $\overrightarrow{\mathbf{p}_1} = 0$ and $\overrightarrow{\mathbf{p}_{N+2}} = L - 1$.
2. *The spiraling pattern with $N + 1$ segments*

$$\langle L - 2, \overleftarrow{\mathbf{p}_{N-2}}, \ldots, \overleftarrow{\mathbf{p}_3}, 1, 0, \overrightarrow{\mathbf{p}_2}, \ldots, \overrightarrow{\mathbf{p}_{N-1}}, L - 1 \rangle \tag{4.15}$$

if N is odd or

$$\langle L-1, \overleftarrow{\mathbf{p}_{N-1}}, \ldots, \overleftarrow{\mathbf{p}_3}, 1, 0, \overrightarrow{\mathbf{p}_2}, \ldots, \overrightarrow{\mathbf{p}_{N-2}}, L-2 \rangle, \tag{4.16}$$

if N is even. Here $\overrightarrow{\mathbf{p}_1} = 0, 1$, $\overrightarrow{\mathbf{p}_N} = L-2$, and $\overrightarrow{\mathbf{p}_{N+1}} = L-1$.

3. The spiraling pattern with N segments

$$\langle L-1, L-2, \overleftarrow{\mathbf{p}_{N-2}}, \ldots, \overleftarrow{\mathbf{p}_3}, 1, 0, \overrightarrow{\mathbf{p}_2}, \ldots, \overrightarrow{\mathbf{p}_{N-1}} \rangle \tag{4.17}$$

if N is odd or

$$\langle \overleftarrow{\mathbf{p}_{N-1}}, \ldots, \overleftarrow{\mathbf{p}_3}, 1, 0, \overrightarrow{\mathbf{p}_2}, \ldots, \overrightarrow{\mathbf{p}_{N-2}}, L-2, L-1 \rangle, \tag{4.18}$$

if N is even. Here $\overrightarrow{\mathbf{p}_1} = 0, 1$, and $\overrightarrow{\mathbf{p}_N} = L-2, L-1$.

Of course, cases 2 and 3 are related to the two possibilities in Lemma 4.

Proof First of all, remember from Sect. 3.4.2, (3.12), that given a forbidden pattern

$$\langle \pi_0, \ldots, \pi_{L-2} \rangle \in \mathcal{S}_{L-1},$$

its outgrowth patterns of length L have the form (*group I*)

$$\langle L-1, \pi_0, \ldots, \pi_{L-2} \rangle, \langle \pi_0, L-1, \ldots, \pi_{L-2} \rangle, \ldots, \langle \pi_0, \ldots, \pi_{L-2}, L-1 \rangle$$

or the form (*group II*)

$$\langle 0, \pi_0 + 1, \ldots, \pi_{L-2} + 1 \rangle, \langle \pi_0 + 1, 0, \ldots, \pi_{L-2} + 1 \rangle, \ldots, \langle \pi_0 + 1, \ldots, \pi_{L-2} + 1, 0 \rangle.$$

1. This case is trivial. Any tent pattern made out of $N+2$ segments is forbidden according to Lemma 3. Moreover, since the entries $L-1$ and 0 in patterns (4.13) and (4.14) are segments on their own, the number of segments D of these tent patterns will fall below the threshold value $D = N+2$ once $L-1$ (group I) or 0 (group II) are deleted.

2. Only (4.15) will be considered here, the proof for (4.16) and their mirrored patterns being completely analogous. That (4.15) is forbidden follows readily from Lemma 4 (b). To prove that π is also a root pattern, we need to show that it is not the outgrowth of any forbidden pattern of shorter length.

There are two possibilities. Suppose first that π is an outgrowth forbidden pattern of group I. Deletion of the entry $L-1$ yields then the spiraling pattern

$$\langle L-2, \overleftarrow{\mathbf{p}_{N-2}}, \ldots, \overleftarrow{\mathbf{p}_3}, 1, 0, \overrightarrow{\mathbf{p}_2}, \ldots, \overrightarrow{\mathbf{p}_{N-1}} \rangle,$$

which is allowed on account of having N segments, $h_1 = 2$, and a last segment $\overrightarrow{\mathbf{p}_N} = L-2$ of length 1 (Lemma 4 (b')).

Thus, suppose that π is an outgrowth forbidden pattern of group II. In this case, after removing the entry 0 and subtracting 1 from the remaining entries we are left with the pattern

$$\langle L - 3, \overleftarrow{\mathbf{p}'_{N-2}}, \ldots, \overleftarrow{\mathbf{p}'_3}, 0, \overrightarrow{\mathbf{p}'_2}, \ldots, \overrightarrow{\mathbf{p}'_{N-1}}, L - 2 \rangle, \tag{4.19}$$

where $\overrightarrow{\mathbf{p}'_d} = e_d - 1, \ldots, e_d + h_d - 2, 2 \le d \le N+1$. Since $\overrightarrow{\mathbf{p}'_1} = 0$ ($h'_1 = h_1 - 1 = 1$) and $\overrightarrow{\mathbf{p}'_2} = 1, \ldots$ ($h'_2 = h_2 \ge 1$), we can merge $\overrightarrow{\mathbf{p}'_1}$ and $\overrightarrow{\mathbf{p}'_2}$ into the new segment $\overrightarrow{\mathbf{p}''_1} := 0, 1, \ldots$, so that (4.19) is a spiraling pattern with $h''_1 \ge 2$ and the following N segments: $\overrightarrow{\mathbf{p}''_1}, \overrightarrow{\mathbf{p}'_3}, \ldots, \overrightarrow{\mathbf{p}'_{N-1}}, \overrightarrow{\mathbf{p}_N} = L - 3, \overrightarrow{\mathbf{p}_{N+1}} = L - 2$. According to Lemma 4 (b′), the ordinal pattern (4.19) is allowed.

3. This case uses Lemma 4 (a)–(a′) instead. The proof proceeds similar to case 2.   □

Example 8 For $N = 2n+1$, Theorem 2 provides the following six forbidden patterns of minimal length $L = N + 2$:

$$\langle 0, 2, \ldots, 2n, 2n + 2, 2n + 1, \ldots, 3, 1 \rangle,$$
$$\langle 2n + 1, 2n - 1, \ldots, 1, 0, 2, \ldots, 2n, 2n + 2 \rangle,$$
$$\langle 2n + 2, 2n + 1, \ldots, 1, 0, 2, \ldots, 2n - 2, 2n \rangle,$$

and their mirrored patterns. For $N = 2n$, the six forbidden patterns of minimal length $L = N + 2$ provided by Theorem 2 are

$$\langle 0, 2, \ldots, 2n, 2n + 1, \ldots, 3, 1 \rangle,$$
$$\langle 2n + 1, 2n - 1, \ldots, 1, 0, 2, \ldots, 2n - 2, 2n \rangle,$$
$$\langle 2n - 1, 2n - 3 \ldots, 1, 0, 2, \ldots, 2n, 2n + 1 \rangle,$$

and their mirrored patterns. In particular, for $N = 2$ we obtain the following minimal-length forbidden patterns:

$$\langle 0, 2, 3, 1 \rangle \quad \langle 1, 3, 2, 0 \rangle,$$
$$\langle 3, 1, 0, 2 \rangle \quad \langle 2, 0, 1, 3 \rangle,$$
$$\langle 1, 0, 2, 3 \rangle \quad \langle 3, 2, 0, 1 \rangle.$$

Needless to say, these are the six 4-patterns we got in (3.14) by graphical means.

It was proven in [76] that the shift Σ_N has exactly six root forbidden L-patterns for each $L \ge N + 2$, namely, those delivered by Theorem 2 after setting $\overrightarrow{\mathbf{p}_k} = k - 1$ (respectively, $\overrightarrow{\mathbf{p}_k} = k$) in those segments not explicitly given in the tent patterns (4.13) and (4.14) (respectively, in the spiraling patterns (4.15), (4.16), (4.17), and (4.18)).

Corollary 2 *For every $K \ge 2$ there are self-maps on the interval $[0, 1]$ without forbidden patterns of length $L \le K$.*

Proof Let $E_N : [0, 1] \to [0, 1]$ be the shift map $x \mapsto Nx \pmod 1$, $N = 2, 3, \ldots$. We know that E_N and Σ have the same allowed and forbidden patterns because they are order isomorphic (see (4.4)). Therefore if $N + 1 \leq K$, then E_N has no forbidden patterns of length $L \leq K$ because of Theorem 1. $\square$

It follows that *there exist n-dimensional interval maps without forbidden patterns*. For example, see Fig. 4.2, one can decompose $[0, 1]$ in infinite many half-open intervals (of vanishing length), $[0, 1] = \cup_{N=2}^{\infty} I_N$ and define on each I_N a properly scaled version of E_N, $\tilde{E}_N : I_N \to I_N$. In $\mathbb{R}^2$ one can repeat the said decomposition along the 1-axis and define on $I_N \times [0, 1]$ the function $(\tilde{E}_N, \mathrm{Id})$, where Id denotes the identity. Proposition 4 shows that adding some natural assumption, like piecewise monotonicity, can make all the difference.

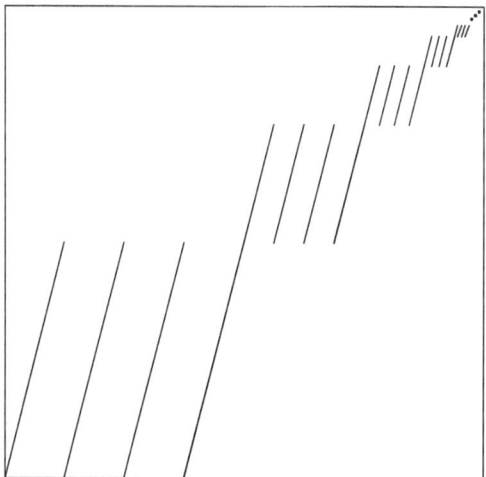

Fig. 4.2 A map with infinitely many monotonicity intervals and no forbidden patterns

4.3 Forbidden Patterns for Two-Sided Shifts

Consider now the bisequence space, $\{0, 1, \ldots, N-1\}^{\mathbb{Z}}$, equipped with the following lexicographical order. With the notation $\mathbf{x}^-$ for the *left sequence* $(x_{-n})_{n \in \mathbb{N}}$ of $\mathbf{x} \in \{0, 1, \ldots, N-1\}^{\mathbb{Z}}$ and $\mathbf{x}^+$ for the *right sequence* $(x_n)_{n \in \mathbb{N}_0}$, we set

$$\mathbf{x} < \mathbf{x}' \Leftrightarrow \begin{cases} \mathbf{x}^+ < \mathbf{x}'^+ \\ \text{or} \\ \mathbf{x}^- < \mathbf{x}'^- \quad \text{if } \mathbf{x}^+ = \mathbf{x}'^+ \end{cases}, \tag{4.20}$$

where $\mathbf{x} = (\mathbf{x}^-, \mathbf{x}^+)$, $\mathbf{x}' = (\mathbf{x}'^-, \mathbf{x}'^+)$, and $<$ between right (respectively, left) sequences denote lexicographical order in $\{0, 1, \ldots, N-1\}^{\mathbb{N}_0}$ (respectively, $\{0, 1, \ldots, N-1\}^{\mathbb{N}}$). If we map $\{0, 1, \ldots, N-1\}^{\mathbb{Z}}$ onto $[0, 1] \times [0, 1] \equiv [0, 1]^2$ via

$$(\mathbf{x}^-, \mathbf{x}^+) \mapsto \left(\sum_{n=1}^{\infty} x_{-n} N^{-n}, \sum_{n=0}^{\infty} x_n N^{-(n+1)} \right), \qquad (4.21)$$

we find that the lexicographical order (4.20) in $\{0, 1, \ldots, N-1\}^{\mathbb{Z}}$ corresponds to the usual lexicographical order in $[0, 1]^2$. In order for this map to be one-to-one, we have to dispose of the usual ambiguities in either direction.

In relation with the ordinal patterns defined by the orbits of two-sided sequences,

$$\Sigma^i(\mathbf{x}) < \Sigma^j(\mathbf{x})$$
$$\Leftrightarrow \begin{cases} (x_i, x_{i+1}, \ldots) < (x_j, x_{j+1}, \ldots) \\ \text{or} \\ (x_{i-1}, x_{i-2}, \ldots) < (x_{j-1}, x_{j-2}, \ldots) \text{ if } (x_i, x_{i+1}, \ldots) = (x_j, x_{j+1}, \ldots), \end{cases}$$

where $i, j \geq 0$, $i \neq j$. It follows that the "exceptional" condition $(x_i, x_{i+1}, \ldots) = (x_j, x_{j+1}, \ldots)$ occurs if and only if $\Sigma^{|i-j|}(\mathbf{x}^+) = \mathbf{x}^+$, i.e., when the right sequence $\mathbf{x}^+$ of $\mathbf{x} \in \{0, 1, \ldots, N-1\}^{\mathbb{Z}}$ is periodic from the entry $\min\{i, j\}$ on with period $p = |i - j|$.

Lemma 5 *One-sided and two-sided shifts on N symbols have the same admissible and forbidden ordinal patterns.*

Proof (i) Suppose that the one-sided sequence $\mathbf{x}^+ \in \{0, 1, \ldots, N-1\}^{\mathbb{N}_0}$ defines an ordinal L-pattern π, i.e.,

$$\Sigma^{\pi_0}(\mathbf{x}^+) < \Sigma^{\pi_1}(\mathbf{x}^+) < \cdots < \Sigma^{\pi_{L-1}}(\mathbf{x}^+).$$

Then, the two-sided sequences $\mathbf{x} = (\mathbf{x}^-, \mathbf{x}^+)$, with $\mathbf{x}^- \in \{0, 1, \ldots, N-1\}^{\mathbb{N}}$ arbitrary, define the same ordinal pattern.

(ii) Suppose now that the two-sided sequence $\mathbf{x} = (\mathbf{x}^-, \mathbf{x}^+) \in \{0, 1, \ldots, N-1\}^{\mathbb{Z}}$ defines an ordinal L-pattern π,

$$\Sigma^{\pi_0}(\mathbf{x}) < \Sigma^{\pi_1}(\mathbf{x}) < \cdots < \Sigma^{\pi_{L-1}}(\mathbf{x}). \qquad (4.22)$$

If $\mathbf{x}^+$ is not eventually periodic, then (4.22) implies

$$\Sigma^{\pi_0}(\mathbf{x}^+) < \Sigma^{\pi_1}(\mathbf{x}^+) < \cdots < \Sigma^{\pi_{L-1}}(\mathbf{x}^+),$$

hence the pattern π is realized by the one-sided sequence $\mathbf{x}^+$. If $\mathbf{x}^+$ is eventually periodic, say

$$\mathbf{x}^+ = (x_0, \ldots, x_{k-1}, (x_k, \ldots, x_{k+p-1})^{\infty}),$$

i.e., $(\mathbf{x}^+)_{k+np} = (\mathbf{x}^+)_k$ for $k \geq 0$ and every $n \in \mathbb{N}$, then there are two subcases.

(ii-a) If $L \leq k + 2p$, then the periodicity of $\mathbf{x}^+$ is not visible in the segment x_0^{L-1}, so the pattern π is realized by the one-sided pattern $\mathbf{x}^+$.

(ii-b) If $L = k+np+v$ with $n \geq 2$ and $v \geq 1$, then $\Sigma^{k+p+i}(\mathbf{x}) = \cdots = \Sigma^{k+np+i}(\mathbf{x})$ for $i = 0, \ldots, v-1$, so their negative sequences $(\Sigma^{k+p+i}(\mathbf{x}))^-, \ldots, (\Sigma^{k+np+i}(\mathbf{x}))^-$ have to be compared before ordering them. In this case, the pattern π is realized by the one-sided sequence

$$\tilde{\mathbf{x}}^+ = (x_0, \ldots, x_{k+np+v-1}, (\Sigma^{k+np+v-1}(\mathbf{x}))^-)$$
$$= (x_0, \ldots, x_{k+np+v-1}, x_{k+np+v-2}, \ldots, x_0, x_{-1}, \ldots).$$

From (i) and (ii) we deduce that one-sided and two-sided shifts on $N \geq 2$ symbols have the same admissible ordinal patterns, hence they have also the same forbidden patterns. $\square$

As a corollary of Lemma 5, together with Theorems 1 and 2, we obtain the following result.

Theorem 3 *The two-sided shift on N symbols has no forbidden patterns of length $L \leq N+1$ and has forbidden root patterns for $L \geq N+2$.*

Example 9 Let $I^2 = [0,1] \times [0,1]$ endowed with the Lebesgue measure, and let $B : I^2 \to I^2$ be the *baker map*,

$$B(\xi, \eta) = \begin{cases} (2\xi, \frac{1}{2}\eta), & 0 \leq \xi < \frac{1}{2}, \\ (2\xi - 1, \frac{1}{2}\eta + \frac{1}{2}), & \frac{1}{2} \leq \eta \leq 1. \end{cases}$$

A generating partition of B is $A_0 = [0, \frac{1}{2}) \times [0,1]$ and $A_1 = [\frac{1}{2}, 1] \times [0,1]$. For Σ take the two-sided $\left(\frac{1}{2}, \frac{1}{2}\right)$-Bernoulli shift. Then B and Σ are isomorphic (mod 0) via the coding map $\Phi : I^2 \to \{0,1\}^{\mathbb{Z}}$, given by

$$\Phi(\xi, \eta) = (\ldots, x_{-1}, x_0, x_1, \ldots),$$

where $x_n = i_n$ if $B^n(\xi, \eta) \in A_{i_n}$, $n \in \mathbb{Z}$. Since Φ preserves order (in fact, Φ is the inverse of the order-preserving map $(\mathbf{x}^-, \mathbf{x}^+) \mapsto (\sum_{n=0}^{\infty} x_{-n} 2^{-(n+1)}, \sum_{n=1}^{\infty} x_n 2^{-n}))$, we conclude that the baker transformation has no forbidden patterns of length ≤ 3. The forbidden 4-patterns of the baker map are the same as those of the one-sided shift, see (3.14).

Chapter 5
Ordinal Structure of the Signed Shifts

Shift transformations are a special case of a more general family: signed shift transformations—a sort of state-dependent shifts. The tent map is the simplest and perhaps most popular representative of the signed shifts. In this chapter we are going to show that most of the results on the ordinal structure of the shifts can be generalized to the signed shifts. By order isomorphy, these results apply also to more interesting cases, like the signed sawtooth maps.

5.1 Ordinal Patterns and the Tent Map

In this section we mimic the strategy used in the previous chapter, in order to get a handle on the ordinal patterns of the symmetric tent map. We will also address an issue pointed out in Fig. 1.7, namely, the interval structure of the sets P_π defining the allowed ordinal patterns of the logistic map.

5.1.1 A State-Dependent Shift Approach to the Tent Map

Just as some important dynamical properties of the sawtooth map E_N (like density of periodic points, sensitivity to initial conditions, topological transitivity, and the structure of its admissible and forbidden ordinal patterns) can be easily studied in the sequence space with the help of the relevant order isomorphisms, the same happens with the symmetric tent map. Remember from Sect. 1.1.3 that the symmetric tent map $\Lambda:[0, 1] \to [0, 1]$ is given by

$$\Lambda(x) = 1 - |1 - 2x| = \begin{cases} 2x & 0 \leq x \leq \frac{1}{2} \\ 2(1 - x) & \frac{1}{2} \leq x \leq 1 \end{cases}. \tag{5.1}$$

For $x \in [0, 1]$, write

$$x = \sum_{n=0}^{\infty} x_n 2^{-(n+1)} = 0.x_0 x_1 \ldots x_n \ldots,$$

J.M. Amigó, *Permutation Complexity in Dynamical Systems*,
Springer Series in Synergetics, DOI 10.1007/978-3-642-04084-9_5,
© Springer-Verlag Berlin Heidelberg 2010

$x_n \in \{0, 1\}$. If $0 \leq x < 1/2$, then

$$\Lambda(x) = 2x = 0.x_1x_2 \ldots x_{n+1} \ldots,$$

hence the action of Λ coincides with the action of the sawtooth map E_2. Otherwise, if $1/2 \leq x \leq 1$, then

$$\Lambda(x) = 2 - 2x \equiv 1 - 2x \bmod 1$$
$$= 1 - 0.x_2x_3 \ldots x_{n+1} \ldots$$

Introducing the *dual bit*

$$x^* = 1 - x = \begin{cases} 1 & \text{if } x = 0 \\ 0 & \text{if } x = 1 \end{cases} \tag{5.2}$$

(thus, $(x^*)^* = x$), we have

$$\Lambda(x) = 0.x_1^* x_2^* \ldots x_{n+1}^* \ldots$$

because

$$0.x_1x_2 \ldots x_{n+1} + \cdots + 0.x_1^* x_2^* \ldots x_{n+1}^* \cdots = 0.11 \ldots 1 \ldots = 1.$$

All in all,

$$\Lambda(0.x_0x_1 \ldots x_n \ldots) = \begin{cases} 0.x_1x_2 \ldots x_{n+1} \ldots & \text{if } x_0 = 0, \\ 0.x_1^* x_2^* \ldots x_{n+1}^* \ldots & \text{if } x_0 = 1. \end{cases} \tag{5.3}$$

Identify now the binary representation $0. x_0 x_1 \ldots x_n \ldots$, $x_n \in \{0, 1\}$, of a number $x \in [0, 1]$, with the sequence

$$(x_0, x_1, \ldots, x_n, \ldots) \in \{0, 1\}^{\mathbb{N}_0},$$

via the map $\phi_2 : \{0, 1\}^{\mathbb{N}_0} \to [0, 1]$ defined as in (4.3) with $N = 2$. Then action (5.3) translates into the following zeroth-state-dependent shift on $\{0, 1\}^{\mathbb{N}_0}$:

$$\Sigma_{(+,-)}(x_0, x_1, \ldots, x_n, \ldots) = \begin{cases} (x_1, x_2, \ldots, x_{n+1}, \ldots) & \text{if } x_0 = 0 \\ (x_1^*, x_2^*, \ldots, x_{n+1}^*, \ldots) & \text{if } x_0 = 1 \end{cases} \tag{5.4}$$

(the subscripts $(+, -)$ will be explained later). Observe that if we write

$$\mathbf{x}^* = (x_0^*, x_1^*, \ldots, x_n^*, \ldots),$$

then

$$\Sigma_{(+,-)}(\mathbf{x}) = \begin{cases} \Sigma_2(\mathbf{x}) & \text{if } x_0 = 0, \\ \Sigma_2(\mathbf{x}^*) & \text{if } x_0 = 1, \end{cases}$$

where Σ_2 is the usual one-sided shift on sequences of two symbols.

A method of visualizing how the orbits of $\mathbf{x}$ are generated by $\Sigma_{(+,-)}$ is the following. Take as way of illustration

$$\mathbf{x} = (0, 1, 1, 0, 0, 0, 1, 0, 1, 1, 0, 0, 1, 1, 1, 0, 1, 0, 0, 1, 1, 0, 1, \ldots), \qquad (5.5)$$

so as

$$
\begin{array}{llllllllllllllll}
\Sigma^1_{(+,-)}(\mathbf{x}) & = & (1 & 1 & 0 & 0 & 0 & 1 & 0 & 1 & 1 & 0 & 0 & 1 \ldots) & = & \Sigma^1_2(\mathbf{x}) \\
\Sigma^2_{(+,-)}(\mathbf{x}) & = & (0 & 1 & 1 & 1 & 0 & 1 & 0 & 0 & 1 & 1 & 0 & 0 \ldots) & = & \Sigma^2_2(\mathbf{x}^*) \\
\Sigma^3_{(+,-)}(\mathbf{x}) & = & (1 & 1 & 1 & 0 & 1 & 0 & 0 & 1 & 1 & 0 & 0 & 0 \ldots) & = & \Sigma^3_2(\mathbf{x}^*) \\
\Sigma^4_{(+,-)}(\mathbf{x}) & = & (0 & 0 & 1 & 0 & 1 & 1 & 0 & 0 & 1 & 1 & 1 & 0 \ldots) & = & \Sigma^4_2(\mathbf{x}) \\
\Sigma^5_{(+,-)}(\mathbf{x}) & = & (0 & 1 & 0 & 1 & 1 & 0 & 0 & 1 & 1 & 1 & 0 & 1 \ldots) & = & \Sigma^5_2(\mathbf{x}) \\
\Sigma^6_{(+,-)}(\mathbf{x}) & = & (1 & 0 & 1 & 1 & 0 & 0 & 1 & 1 & 1 & 0 & 1 & 0 \ldots) & = & \Sigma^6_2(\mathbf{x}) \\
\Sigma^7_{(+,-)}(\mathbf{x}) & = & (1 & 0 & 0 & 1 & 1 & 0 & 0 & 0 & 1 & 0 & 1 & 1 \ldots) & = & \Sigma^7_2(\mathbf{x}^*) \\
\Sigma^8_{(+,-)}(\mathbf{x}) & = & (1 & 1 & 0 & 0 & 1 & 1 & 1 & 0 & 1 & 0 & 0 & 1 \ldots) & = & \Sigma^8_2(\mathbf{x}) \\
\Sigma^9_{(+,-)}(\mathbf{x}) & = & (0 & 1 & 1 & 0 & 0 & 0 & 1 & 0 & 1 & 1 & 0 & 0 \ldots) & = & \Sigma^9_2(\mathbf{x}^*) \\
\Sigma^{10}_{(+,-)}(\mathbf{x}) & = & (1 & 1 & 0 & 0 & 0 & 1 & 0 & 1 & 1 & 0 & 0 & 1 \ldots) & = & \Sigma^{10}_2(\mathbf{x}^*) \\
\end{array}
$$

etc., that is,

$$\Sigma^i_{(+,-)}(\mathbf{x}) = \begin{cases} \Sigma^i_2(\mathbf{x}) & \text{for } i = 0, 1, 4, 5, 6, 8, \ldots, \\ \Sigma^i_2(\mathbf{x}^*) & \text{for } i = 2, 3, 7, 9, 10, \ldots. \end{cases}$$

Write now $\mathbf{x}^*$ directly under $\mathbf{x}$, and mark (for example, with an underline) the initial digit of $\Sigma^i_{(+,-)}(\mathbf{x}), i \geq 0$:

$i =$	0	1	2	3	4	5	6	7	8	9	10	11	12
$\mathbf{x} =$	$\underline{0}$	$\underline{1}$	1	0	0	0	$\underline{1}$	0	$\underline{1}$	$\underline{1}$	0	0	$\underline{1}$
$\mathbf{x}^* =$	1	0	$\underline{0}$	$\underline{1}$	$\underline{1}$	$\underline{1}$	0	$\underline{1}$	0	0	$\underline{1}$	$\underline{1}$	0

(5.6)

That is, we set out from x_0, which is always underlined. If $x_0 = 0$, then go over to x_1 and underline it. If $x_0 = 1$, then go down to x_1^* and underline it. In general, if $\underline{x_i} = 0$ or $x_i^* = 0$, go one step rightward on the same row and underline x_{i+1} or x_{i+1}^*, respectively. On the other hand, if $\underline{x_i} = 1$ or $x_i^* = 1$, we go one step rightward on the other row and underline x_{i+1}^* or x_{i+1}, respectively. The L-pattern π defined by $\mathbf{x}$ can be found now by ordering all the sequences on the $\mathbf{x}$-row and $\mathbf{x}^*$-row starting with an underlined bit, for $0 \leq i \leq L - 1$.

If $\mathbf{x}$ is sequence (5.5), then the ordinal L-patterns of $\mathbf{x}$ under $\Sigma_{(+,-)}$ are obtained by comparing the shifts $\Sigma^i(\mathbf{x})$ for $i = 0, 1, 4, 5, 6, 8, \ldots$ with the shifts $\Sigma^j(\mathbf{x}^*)$ for $j \neq i$. In particular, $\mathbf{x}$ is of type

$$\pi = \langle 4, 5, 9, 0, 2; 7, 6, 10, 1, 8, 3 \rangle \in \mathcal{S}_{11} \tag{5.7}$$

under the action of $\Sigma_{(+,-)}$.

Rather than deriving at this point the structure of the allowed ordinal patterns for $\Sigma_{(+,-)}$ (or the tent map Λ for this matter), which follows from the general results of the next section, let us prove here a particular property of the allowed patterns for $\Sigma_{(+,-)}$.

Lemma 6 *The subsequence $n+2, \ldots, n+1, \ldots, n$ ($0 \le n \le L-3$) cannot appear in the entries of an allowed L-pattern for $\Sigma_{(+,-)}$. Thus, the allowed ordinal patterns of $\Sigma_{(+,-)}$ cannot contain decreasing subsequences of length 3.*

Proof We prove by contradiction that the order relation

$$\Sigma^2_{(+,-)}(\mathbf{x}) < \Sigma_{(+,-)}(\mathbf{x}) < \mathbf{x} \tag{5.8}$$

cannot hold true. If $x_0 = 0$ there is no way that $\Sigma_{(+,-)}(\mathbf{x}) \equiv \Sigma_2(\mathbf{x}) < \mathbf{x}$. Hence $\mathbf{x} = (1, x_1, x_2, \ldots)$ and

$$\Sigma_{(+,-)}(\mathbf{x}) \equiv \Sigma_2(\mathbf{x}^*) = (x_1^*, x_2^*, \ldots).$$

By the same token, if $x_1^* = 0$ there is no way that $\Sigma_{(+,-)}(\Sigma_{(+,-)}(\mathbf{x})) \equiv \Sigma^2_{(+,-)}(\mathbf{x}) < \Sigma_{(+,-)}(\mathbf{x})$. Hence

$$\mathbf{x} = (1, 0, x_2, \ldots), \quad \Sigma_{(+,-)}(\mathbf{x}) = (1, x_2^*, x_3^*, \ldots), \quad \Sigma^2_{(+,-)}(\mathbf{x}) = (x_2, x_3, \ldots).$$

From $\Sigma_{(+,-)}(\mathbf{x}) < \mathbf{x}$ it follows $x_2^* = 0$. In turn, from $\Sigma^2_{(+,-)}(\mathbf{x}) = (1, x_3, \ldots) < \Sigma_{(+,-)}(\mathbf{x}) = (1, 0, x_3^*, \ldots)$ it follows $x_3 = 0$. So far, we found that $\mathbf{x} = (1, 0, 1, 0, x_4, \ldots)$ (thus $\Sigma_{(+,-)}(\mathbf{x}) = (1, 0, 1, x_4^*, \ldots)$ and $\Sigma^2_{(+,-)}(\mathbf{x}) = (1, 0, x_4, \ldots)$).

A straightforward induction along these lines yields

$$\mathbf{x} = (1, 0, 1, 0, \ldots, 1, 0, \ldots) = ((1, 0)^\infty),$$

which is the binary expansion of the rational number $2/3$. Since $\Sigma^2_{(+,-)}(\mathbf{x}) = \Sigma_{(+,-)}(\mathbf{x}) = \mathbf{x}$ for this particular sequence (in other words, $2/3$ is a fixed point of $\Sigma_{(+,-)}$), the statement follows by contradiction. $\square$

Exercise 5 Prove, using representation (5.4) that the symmetric tent map has dense periodic points, sensitive dependence on initial conditions, and is topologically transitive.

5.1.2 The Interval Structure of the Sets P_π

The points in state space Ω defining an ordinal L-pattern π under the action of a map $f:\Omega \to \Omega$ build the set P_π, (3.4). The sets $P_\pi \neq \emptyset$, $\pi \in S_L$, build in turn the set $\mathcal{P}_L$, which build a finite partition of Ω under the condition set by Proposition 2. In this section we examine the "topology" of $P_\pi \in \mathcal{P}_L$ for some one-dimensional interval maps. For continuous maps, those sets are clearly open sets (hence, an enumerable union of disjoint open intervals), but no further dissection can be made. For the sawtooth map family $x \mapsto Nx \bmod 1$, $N \geq 2$, it is easy to convince oneself that P_π consists of a single open or half-open interval for all admissible patterns $\pi \in S_L$, $L \geq 2$ (see Figs. 3.2 and 3.3). For the logistic map, Figs. 1.5 and 1.6 show that all $P_\pi \in \mathcal{P}_L$ with $L = 2, 3$ consist of a single open interval, but from Fig. 1.7 it can be read that

$$P_{\langle 0,3,1,2 \rangle} \approx (0.09549, 0.11698) \cup (0.18826, 0.25),$$
$$P_{\langle 2,0,3,1 \rangle} \approx (0.34549, 0.41318) \cup (0.61126, 0.65451),$$
$$P_{\langle 1,2,3,0 \rangle} \approx (0.93301, 0.95048) \cup (0.96985, 1).$$

We claim the following.

Proposition 6 *For the logistic map and the symmetric tent map, all $P_\pi \neq \emptyset$ consist of one or two components.*

As stated in Example 4 (1), the logistic map g and the symmetric tent map Λ are order isomorphic. Specifically, $g(\phi(x)) = \phi(\Lambda(x))$, where $\phi(x) = \sin^2(\frac{\pi}{2}x)$, $0 \leq x \leq 1$, so that

$$g^n(\phi(x)) = g^m(\phi(x)) \Leftrightarrow \phi(\Lambda^n(x)) = \phi(\Lambda^m(x)) \Leftrightarrow \Lambda^n(x) = \Lambda^m(x).$$

Thus, the curves $y = g^n(x)$ and $y = g^m(x)$ cross at x_0 if and only if the piecewise straight lines $y = \Lambda^n(x)$ and $y = \Lambda^m(x)$ cross at $\phi^{-1}(x_0)$. Moreover, the iterates of Λ have not only a simple graphical representation (triangular waves with frequencies increasing as powers of 2) but also a scaling property that makes Λ handier for the proof of Proposition 6:

$$\begin{aligned} \Lambda^n(x) &= \Lambda^{n-1}(2x), & 0 \leq x \leq \tfrac{1}{2}, \\ \Lambda^n(x) &= \Lambda^{n-1}(2(1-x)), & \tfrac{1}{2} \leq x \leq 1. \end{aligned} \tag{5.9}$$

Therefore, the left-half part of the graphs $(x, \Lambda^0(x)), (x, \Lambda^1(x)), \ldots, (x, \Lambda^L(x))$ is a "squeezed" copy of the graphs $(x, \frac{x}{2}), (x, \Lambda^0(x)), \ldots, (x, \Lambda^{L-1}(x))$ on the interval $0 \leq x \leq \frac{1}{2}$; indeed, upon rescaling the X-axis by a factor $\frac{1}{2}$, we have $(x, \frac{x}{2}) \mapsto (\frac{x}{2}, \frac{x}{2})$ and $(x, \Lambda^l(x)) \mapsto (\frac{x}{2}, \Lambda^l(x)) = (\frac{x}{2}, \Lambda^{l+1}(\frac{x}{2}))$. The corresponding right-half parts require the squeezed copy of the graphs $(x, 1 - \frac{x}{2}), (x, \Lambda^0(x)), \ldots, (x, \Lambda^{L-1}(x))$ on $0 \leq x \leq \frac{1}{2}$ to be further mirrored with respect to the line $x = \frac{1}{2}$ (this is the transformation $(x, y) \mapsto (x, 1 - x)$); see Fig. 5.1 for further insights.

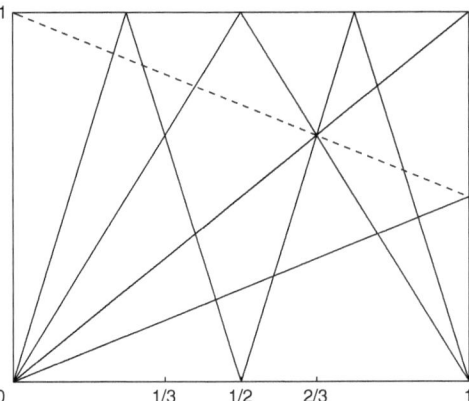

Fig. 5.1 If this figure is "opened" at the right side as a book put upside down, with the line $y = x/2$ only on the left page, the (*dashed*) line $y = 1 - x/2$ only on the right page, and the triangular waves $y = \Lambda(x)$, $y = \Lambda^2(x)$ on both, and the resulting graph is shrunk by a factor $1/2$ along the X-axis, then we get the graphs of $y = \Lambda^n(x)$, $0 \le n \le 3$. Alternatively, we can go from $\mathcal{P}_3^*$ to $\mathcal{P}_4^*$ just by going first rightward on the bottom page (containing $y = x/2$) of the closed book and then leftward on the top page (containing $y = 1 - x/2$)

Proof Proposition 6 follows from the considerations prior to Proposition 3 (remember the terminology mother and daughter intervals, here shortened to *mother* and *daughter*), together with the following facts.

The decomposition of a mother $P_{\pi_{\text{mother}}} \in \mathcal{P}_L$ into several daughters including two or more *twins* (disjoint subintervals with the same ordinal label) can only happen in intervals containing "vertex" or "bouncing-off" points x_v. As their name indicates, these points correspond to projections onto the X-axis of points at the bottom ($y = 0$) or at the ceiling ($y = 1$) of the unit square at which incoming (left) and outgoing (right) lines $y = \Lambda^l(x)$ meet, like $(\frac{1}{2}, 0)$ and $(\frac{1}{4}, 1)$ in Fig. 5.1. Possibly the most intuitive way to follow the growth of twins around vertex points uses the scaling property (5.9). If $0 < x_v < \frac{1}{2}$, consider the graphs of $y = \frac{x}{2}, y = \Lambda^0(x), \dots, y = \Lambda^{L-1}(x)$ around $x = 2x_v$. If $2x_v \in P_{\langle \pi_0, \dots, \pi_{L-1} \rangle}$, then the straight line $y = \frac{x}{2}$ generates (left to right) daughters of $P_{\pi_{\text{mother}}}$ (after squeezing) with labels $\pi_{\text{left}} = \langle \pi_0 + 1, \dots, 0, \pi_k + 1, \dots, \pi_{L-1} + 1 \rangle$, $\pi_{\text{central}} = \langle \pi_0 + 1, \dots, \pi_k + 1, 0, \dots, \pi_{L-1} + 1 \rangle$ and $\pi_{\text{right}} = \langle \pi_0 + 1, \dots, 0, \pi_k + 1, \dots, \pi_{L-1} + 1 \rangle = \pi_{\text{left}}$, with $x_v \in P_{\pi_{\text{central}}} \in \mathcal{P}_{L+1}$. Here k depends on the number of lines meeting at $(x_v, 0)$; if $k = 0$ or $L - 1$, then 0 is the first or last entry of the label, respectively. Hence, the set $P_{\pi_{\text{left}}} \cup P_{\pi_{\text{right}}} \in \mathcal{P}_{L+1}$ ($\pi_{\text{left}} = \pi_{\text{right}}$) consists of two disjoint interval components, one on each side of $P_{\pi_{\text{central}}}$. If, on the other hand, $\frac{1}{2} < x_v < 1$, consider the graphs of $y = 1 - \frac{x}{2}, y = \Lambda^0(x), \dots, y = \Lambda^{L-1}(x)$ around $x = 2(1 - x_v)$. If $2(1 - x_v) \in P_{\langle \pi_0, \dots, \pi_{L-1} \rangle}$, then the straight line $y = 1 - \frac{x}{2}$ generates daughters of $P_{\pi_{\text{mother}}}$ (after squeezing and mirroring) with labels $\pi_{\text{left}} = \langle \pi_0 + 1, \dots, \pi_k + 1, 0, \dots, \pi_{L-1} + 1 \rangle$, $\pi_{\text{central}} = \langle \pi_0 + 1, \dots, 0, \pi_k + 1, \dots, \pi_{L-1} + 1 \rangle$, and $\pi_{\text{right}} = \langle \pi_0 + 1, \dots, \pi_k + 1, 0, \dots, \pi_{L-1} + 1 \rangle = \pi_{\text{left}}$. As before, $P_{\pi_{\text{left}}} \cup P_{\pi_{\text{right}}}$ consists of two disjoint interval components, one on each side of $P_{\pi_{\text{central}}}$. Finally, for

$x_v = \frac{1}{2}$ the first set of graphs produces π_{left} and π_{central}, while the second produces π_{central} and $\pi_{\text{right}} = \pi_{\text{left}}$, with $x_v \in P_{\pi_{\text{central}}}$. This mechanism repeats again and again over all generations. After the step $L \to L + 1$, only the one-component daughters $P_{\pi_{\text{central}}}$, all of which contain some x_v, can in turn generate twins (two-component grand daughters); the corresponding two-component sisters $P_{\pi_{\text{left}}} \cup P_{\pi_{\text{right}}}$ cannot generate twins because they contain no vertex point. As a result, only one- or two-component intervals are possible, the latter forming a nested structure around some vertex points. From Fig. 5.1 it is clear that all such vertex points originate from $x = \frac{1}{2}, 1$ by squeezing and from $x = \frac{1}{4}$ by squeezing and mirroring.   □

Exercise 6 Discuss the interval structure of the sets P_π for the map $E_{-2}: x \mapsto -2x$ mod 1.

5.2 Ordinal Patterns and the Signed Shifts

The results of Sect. 5.1.1 can be generalized to a particular case of piecewise linear maps. Partition the unit interval $[0, 1]$ in $N \geq 2$ equal subintervals,

$$I_k = \left[\frac{k}{N}, \frac{k+1}{N} \right), \quad 0 \leq k \leq N - 2 \quad \text{and} \quad I_{N-1} = \left[\frac{N-1}{N}, 1 \right]$$

(other choices regarding the endpoints are of course possible), and raise over I_k a "/-lap" of slope $+N$,

$$f(x) = Nx - k, \; x \in I_k,$$

or a "\-lap" of slope $-N$,

$$f(y) = k + 1 - Nx, \; x \in I_k.$$

A map of the unit interval whose graph consists of /-laps and \-laps of slopes $\pm N$, respectively, over the intervals I_k, $0 \leq k \leq N - 1$, will be called a *signed saw-tooth map*, the term "signed" referring to the fact that its laps can have positive or negative slope (see Fig. 5.2). We say that a signed sawtooth map f has *signature* $\sigma = (\sigma_0, \sigma_1, \ldots, \sigma_{N-1})$, where $\sigma_k \in \{+, -\}$, $0 \leq k \leq N - 1$, to summarize that (the graph of) f has a /-lap over I_k whenever $\sigma_k = +$ and a \-lap whenever $\sigma_k = -$. In other words, the kth component of the signature gives the slope sign of the kth lap.

We have already met two important representatives of the signed sawtooth map family: the sawtooth map $E_N: x \mapsto Nx$ mod 1 ($\sigma = (+, \ldots, +)$) and the symmetric tent map Λ ($\sigma = (+, -)$).

Given a signature σ, define the *signed shift* $\Sigma_\sigma: \{0, \ldots, N-1\}^{\mathbb{N}_0} \to \{0, \ldots, N-1\}^{\mathbb{N}_0}$ as follows:

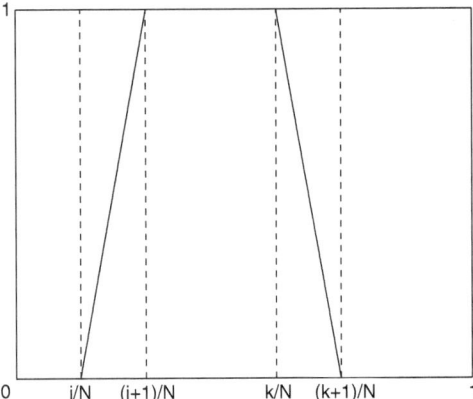

Fig. 5.2 The graph of a generic signed sawtooth map with slopes $\pm N$. The figure only depicts the jth lap, with positive slope, and the kth slope, with negative slope

$$\Sigma_\sigma(x_0, \ldots, x_n, \ldots) = \begin{cases} (x_1, \ldots, x_{n+1}, \ldots) & \text{if } x_0 = k, \sigma_k = +, \\ (N - 1 - x_1, \ldots, N - 1 - x_{n+1}, \ldots) & \text{if } x_0 = k, \sigma_k = -. \end{cases}$$

Therefore, if we define the *dual digit* of $k \in \{0, 1, \ldots, N - 1\}$ as

$$k^* = N - 1 - k, \tag{5.10}$$

(thus $(k^*)^* = k$), then

$$\Sigma_\sigma(\mathbf{x}) = \begin{cases} \Sigma_N(\mathbf{x}) & \text{if } x_0 - k \text{ and } \sigma_k = +, \\ \Sigma_N(\mathbf{x}^*) & \text{if } x_0 = k \text{ and } \sigma_k = -, \end{cases} \tag{5.11}$$

where

$$\mathbf{x}^* = (x_0^*, \ldots, x_n^*, \ldots) = (N - 1 - x_0, \ldots, N - 1 - x_n, \ldots)$$

is the *dual sequence* to $\mathbf{x} = (x_0, \ldots, x_n, \ldots) \in \{0, 1, \ldots, N - 1\}^{\mathbb{N}_0}$. In particular, if

$$N = 2\nu + 1,$$

then $\nu = (N - 1)/2$ is "self-dual": $\nu^* = \nu$. Note that (5.10) generalizes the definition of dual bit, (5.2).

Important for us is that if f is a signed sawtooth map with signature σ, then f and Σ_σ are order isomorphic via the map $\phi_N:\{0, 1, \ldots, N - 1\}^{\mathbb{N}_0} \to [0, 1]$ defined in (4.3). Observe that $\phi_N(0^\infty) = 0$, $\phi_N(1^\infty) = 1$, and

$$\frac{k}{N} \leq \phi_N(\mathbf{x}) \leq \frac{k + 1}{N} \quad \text{iff} \quad x_0 = k.$$

The technique described in Sect. 5.1 to keep track of the orbits of $\mathbf{x}$ under $\Sigma_{(+,-)}$ can be used for Σ_σ too. The number of symbols N goes in the definition of $\mathbf{x}^*$, while σ_k tells whether we have to jump from the current entry $x_i = k$ to x^*_{i+1} or from the current entry $x^*_i = k$ to x_{i+1} ($\sigma_k = -$), instead of remaining on the same line ($\sigma_k = +$), when underlining the entries of $\mathbf{x}$ in table (5.6).

Exercise 7 Check that

$$f(\phi_N(\mathbf{x})) = \phi_N(\Sigma_\sigma \mathbf{x}) = \begin{cases} \sum_{n=1}^{\infty} x_n N^{-n} & \text{if } x_0 = k \text{ and } \sigma_k = +, \\ 1 - \sum_{n=1}^{\infty} x_n N^{-n} & \text{if } x_0 = k \text{ and } \sigma_k = -. \end{cases}$$

We turn now to the ordinal patterns realized by a signed shift Σ_σ. Completely analogous to the case $\Sigma_{(+,\dots,+)} \equiv \Sigma_N$, Chap. 4, the allowed ordinal patterns for Σ_σ can also be decomposed into s-blocks, (4.6), where now the s-block (4.7) contains the locations of the symbol $s \in \{0, \dots, N-1\}$ in the segments $x_0^{L-1} := x_0, \dots, x_{L-1}$ of $\mathbf{x}$ and $(x^*)_0^{L-1} := x_0^*, \dots, x_{L-1}^*$ of $\mathbf{x}^*$, such that the zeroth component of $\Sigma_\sigma^i \mathbf{x}$, $0 \le i \le L-1$, is s (i.e., the locations of the symbol s which are underlined in the $\mathbf{x}$- or $\mathbf{x}^*$-row of table (5.6)). We shall presently see that each s-block consists basically of two kinds of subsequences: monotone ($\sigma_s = +$) or spiraling ($\sigma_s = -$), eventually intertwined by other subsequences of the same kind. Entries in an s-block not belonging to a subsequence will be referred to as solitary or single components or entries.

Theorem 4 *The non-empty blocks $\pi_{k_0+\cdots+k_{s-1}}, \dots, \pi_{k_0+\cdots+k_{s-1}+k_s-1}$, $0 \le s \le N-1$, of $\pi(\mathbf{x}) \in \mathcal{S}_L$ fulfill the following basic restrictions:*

R*1   *If $\sigma_s = +$, $0 < s < N-1$, then the s-block is built by increasing subsequences,*

$$n, \dots, n+1, \dots, n+l-1 \tag{5.12}$$

($l \ge 2$) and/or decreasing subsequences,

$$n+l-1, \dots, n+1, \dots, n \tag{5.13}$$

($l \ge 2$) and/or solitary components ($l = 1$). If $\sigma_0 = +$, then the 0-block consists of increasing subsequences (5.12) and/or solitary components. If $\sigma_{N-1} = +$, then the $(N-1)$-block consists of decreasing subsequences (5.13) and/or solitary components.

R*2   *If $\sigma_s = -$, $0 < s < N-1$, then the s-block is built by even-length spiraling subsequences*

$$n+2l-2, \dots, n+2, \dots, n, \dots, n+1, \dots, n+3, \dots, n+2l-1 \tag{5.14}$$

with the entry $n+2l$ on an anterior block (if $n+2l \le L-1$) and/or the mirrored subsequences

$$n + 2l - 1, \ldots, n + 3, \ldots, n + 1, \ldots, n, \ldots, n + 2, \ldots, n + 2l - 2 \quad (5.15)$$

with the entry $n+2l$ on a posterior block (if $n+2l \le L-1$) and/or odd-length spiraling subsequences

$$n + 2l, \ldots, n + 2l - 2, \ldots, n + 2, \ldots, n, \ldots, n + 1, \ldots, n + 3, \ldots, n + 2l - 1$$
$$(5.16)$$

with the entry $n + 2l + 1$ on a posterior block (if $n + 2l + 1 \le L - 1$) and/or the mirrored subsequences

$$n + 2l - 1, \ldots, n + 3, \ldots, n + 1, \ldots, n, \ldots, n + 2, \ldots, n + 2l - 2, \ldots, n + 2l$$
$$(5.17)$$

with the entry $n + 2l + 1$ on an anterior block (if $n + 2l + 1 \le L - 1$) and/or solitary components. If $\sigma_0 = -$, then the first block consists of spiraling subsequences of the form (5.15) and/or (5.16) and/or solitary components. If $\sigma_{N-1} = -$, then the last block consists of spiraling subsequences of the form (5.14) and/or (5.17) and/or solitary components.

R*3   *If (i) $\sigma_s = +$, (ii) the entries $m, n \le L - 2$ belong to the s-block of $\pi(\mathbf{x})$, and (iii) m appears on the left of n, then $m + 1$ appears also on the left of $n + 1$ (not necessarily in the same block). If, on the other hand, (i) $\sigma_s = -$, (ii) the entries $m, n \le L - 2$ belong to the s-block of $\pi(\mathbf{x})$, and (iii) m appears on the left of n, then $m + 1$ appears on the right of $n + 1$ (not necessarily in the same block).*

Proof **R*1)** Let $s \in \{0, 1, \ldots, N - 1\}$ and consider an s-run of length $l \ge 2$ in the segment x_0^{L-1} of $\mathbf{x}$:

$i =$	$\ldots$	n	$n + 1$	$\ldots$	$n + l - 1$	$n + l$	$\ldots$
$\mathbf{x} =$	$\ldots$	s	s	$\ldots$	s	r	$\ldots$
$\mathbf{x}^* =$	$\ldots$	$N - 1 - s$	$N - 1 - s$	$\ldots$	$N - 1 - s$	$N - 1 - r$	$\ldots$

where $r \in \{0, 1, \ldots, N - 1\}$ and $r \ne s$. If (i) $s < N - 1$ and (ii) $x_{n+l} = r > s$, then this s-run contributes the increasing subsequence

$$n, \ldots, n + 1, \ldots, n + l - 1 \quad (5.18)$$

to the s-block of $\pi(\mathbf{x})$. If, on the other hand, (i) $s > 0$ and (ii) $x_{n+l} = r < s$, then the s-run contributes the decreasing subsequence

$$n + l - 1, \ldots, n + 1, \ldots, n. \quad (5.19)$$

The "$\ldots$" between the entries of these subsequences allow for entries eventually proceeding from other s-runs in $\mathbf{x}$ or $\mathbf{x}^*$ (see Example 7).

It follows that the 0-block can contain only increasing subsequences (and single entries not belonging to subsequences in the block), whereas the $(N - 1)$-block can contain only decreasing subsequences (and single entries not belonging to subsequences in the block).

R*2) Consider an s-run of even length $2l$ in the segment x_0^{L-1} of $\mathbf{x}$. Thus,

$i =$	n	$n + 1$	$\ldots$	$n + 2l - 2$	$n + 2l - 1$	$n + 2l$
$\mathbf{x} =$	s	$N - 1 - 1$	$\ldots$	s	$N - 1 - s$	r
$\mathbf{x}^* =$	$N - 1 - s$	s	$\ldots$	$N - 1 - s$	s	$N - 1 - r$

where $r \in \{0, 1, \ldots, N - 1\}$ and $r \neq s$. Therefore, if (i) $s > 0$ and (ii) $x_{n+2l-1} = N - 1 - s < x_{n+2l}^* = N - 1 - r$, i.e., $r < s$, then the s-block of $\pi(\mathbf{x})$ will contain the spiraling subsequence

$$n + 2l - 2, \ldots, n + 2, \ldots, n, \ldots, n + 1, \ldots, n + 3, \ldots, n + 2l - 1. \tag{5.20}$$

Hence the entry $n + 2l$ will appear in the r-block (provided $n + 2l \leq L - 1$), which precedes the s-block in $\pi(\mathbf{x})$ because $r < s$. If, on the other hand, (i) $s < L - 1$ and (ii) $x_{n+2l-1} = N - 1 - s > x_{n+2l}^* = N - 1 - r$, i.e., $r > s$, then we obtain the mirrored, spiraling subsequence

$$n + 2l - 1, \ldots, n + 3, \ldots, n + 1, \ldots, n, \ldots, n + 2, \ldots, n + 2l - 2, \tag{5.21}$$

with the symbol $n + 2l$ in a posterior block (provided $n + 2l \leq L - 1$), namely, on the r-block.

Consider now an s-run of odd length $2l + 1$ in the segment x_0^{L-1} of $\mathbf{x}$. Thus,

$i =$	n	$n + 1$	$\ldots$	$n + 2l - 1$	$n + 2l$	$n + 2l + 1$
$\mathbf{x} =$	s	$N - 1 - s$	$\ldots$	$N - 1 - s$	s	$N - 1 - r$
$\mathbf{x}^* =$	$N - 1 - s$	s	$\ldots$	s	$N - 1 - s$	r

where $r \in \{0, 1, \ldots, N - 1\}$ and $r \neq s$. Therefore, if (i) $s > 0$ and (ii) $x_{n+2l}^* = N - 1 - s < x_{n+2l+1} = N - 1 - r$, i.e., $r < s$, then the s-block of $\pi(\mathbf{x})$ will contain the spiraling subsequence

$$n + 2l - 1, \ldots, n + 3, \ldots, n + 1, \ldots, n, \ldots, n + 2, \ldots, n + 2l - 2, \ldots, n + 2l.$$

The entry $n + 2l + 1$ will appear on the r-block (provided $n + 2l + 1 \leq L - 1$), which is on the left of the s-block because $r < s$. If, on the other hand, (i) $s < L - 1$ and (ii) $x_{n+2l}^* = N - 1 - s > x_{n+2l+1} = N - 1 - r$, i.e., $r > s$, then we obtain the mirrored, spiraling subsequence

$$n + 2l, \ldots, n + 2l - 2, \ldots, n + 2, \ldots, n, \ldots, n + 1, \ldots, n + 3, \ldots, n + 2l - 1,$$

with the entry $n + 2l + 1$ in a block on the right of the s-block (provided $n + 2l + 1 \leq L - 1$).

The corresponding results for the first ($s = 0$) and last ($s = N - 1$) blocks follow readily from these general results.

R*3) If m and n belong to the s-block, $\sigma_s = +$, and $\Sigma_\sigma^m(\mathbf{x}) < \Sigma_\sigma^n(\mathbf{x})$ for $\mathbf{x} \in \{0, 1, \ldots, N - 1\}^{\mathbb{N}_0}$, then

$$\Sigma_\sigma^m(\mathbf{x}) = (s, x_{m+1}, \ldots) < (s, x_{n+1}, \ldots) = \Sigma_\sigma^n(\mathbf{x}).$$

By the definition of lexicographical order, there are two possibilities: (i) $x_{m+1} < x_{n+1}$ or (ii) $x_{m+k} = x_{n+k}$ for $1 \leq k \leq l - 1$, $l \geq 2$, and $x_{m+l} < x_{n+l}$. In both cases,

$$\Sigma_\sigma^{m+1}(\mathbf{x}) = (x_{m+1}, \ldots) < (x_{n+1}, \ldots) = \Sigma_\sigma^{n+1}(\mathbf{x})$$

and, hence, the entry $m + 1$ appears on the left of $n + 1$ in $\pi(\mathbf{x})$.

If, on the other hand, m and n belong to the s-block, $\sigma_s = -$, and $\Sigma_\sigma^m(\mathbf{x}) < \Sigma_\sigma^n(\mathbf{x})$, then

$$\Sigma_\sigma^m(\mathbf{x}) = (s, x_{m+1}, \ldots) < (s, x_{n+1}, \ldots) = \Sigma_\sigma^n(\mathbf{x}).$$

As before, there are two possibilities: (i) $x_{m+1} < x_{n+1}$ and (ii) $x_{m+k} = x_{n+k}$ for $1 \leq k \leq l - 1$, $l \geq 2$, and $x_{m+l} < x_{n+l}$. In both cases,

$$\Sigma_\sigma^{m+1}(\mathbf{x}) = (N - 1 - x_{m+1}, \ldots) > (N - 1 - x_{n+1}, \ldots) = \Sigma_\sigma^{n+1}(\mathbf{x})$$

and, hence, the entry $m + 1$ appears on the right of $n + 1$ in $\pi(\mathbf{x})$. $\square$

Conditions R*1–R*3 are not only necessary for an ordinal pattern to be allowed for Σ_σ, $\sigma = (\sigma_0, \ldots, \sigma_{N-1})$, but also sufficient. Indeed, given the s-block decomposition of $\pi \in S_L$ with each block satisfying the pertinent restrictions, then it is a simple matter to construct sequences $\mathbf{x} \in \{0, \ldots, N - 1\}^{\mathbb{N}_0}$ of type π. Furthermore, it is obvious that all L-patterns with $L \leq N$ are allowed for Σ_σ.

Corollary 3 *If $\pi = \langle \pi_0, \pi_1, \ldots, \pi_{L-1} \rangle$ is allowed (correspondingly, forbidden) for Σ_σ, $\sigma = (\sigma_0, \sigma_1, \ldots, \sigma_{N-1})$, then $\pi_{mirrored} = \langle \pi_{L-1}, \pi_{L-2}, \ldots, \pi_0 \rangle$ is allowed (correspondingly, forbidden) for $\Sigma_{\sigma_{mirrored}}$, where*

$$\sigma_{mirrored} := (\sigma_{N-1}, \sigma_{N-2}, \ldots, \sigma_0).$$

In the particular case $\sigma = \sigma_{mirrored}$, it follows that π is allowed (correspondingly, forbidden) for Σ_σ, iff $\pi_{mirrored}$ is also allowed (correspondingly, forbidden) for Σ_σ. These statements hold also true if "forbidden pattern" is replaced by "root forbidden pattern."

Proof The s-block structure of an allowed ordinal pattern is preserved under the transformation $\pi \mapsto \pi_{mirrored}$. Indeed, monotone subsequences transform into monotone subsequences (in particular, increasing subsequences of the 0-block transform in decreasing subsequences of the $(N - 1)$-block and vice versa), and spiraling subsequences go over to spiraling subsequences.

By the same token, mirrored outgrowth forbidden patterns for Σ_σ will be outgrowth forbidden patterns for $\Sigma_{\sigma_{\mathrm{mirrored}}}$. It follows that $\pi \in \mathcal{S}_L$ is a root forbidden pattern for Σ_σ in the case $\sigma = \sigma_{\mathrm{mirrored}}$, iff π_{mirrored} is also a root forbidden pattern for Σ_σ. □

Remark 1 If the first or last element of a monotone subsequence appearing in an s-block is assigned to the anterior or posterior block, respectively (if any), then the remaining subsequence preserves its increasing or decreasing character—or it becomes a single entry. If the leftmost or the rightmost element of a spiraling subsequence is assigned to the anterior or posterior block (if any), then the remaining subsequence preserves its spiraling character, eventually appearing also a new single entry in the same block. This implies that, when carrying out a decomposition of an ordinal L-pattern into s-blocks, $L \geq N$, we may assume without loss of generality that all s-blocks are non-empty.

For $\sigma_k = +, 0 \leq k \leq N - 1$, we recover from Theorem 4 the restrictions fulfilled by the allowed patterns for Σ_N (Lemma 2). In the case $\sigma = (+, -)$, considered in Sect. 5.1.1, there are only two symbols and two blocks in the decomposition of the ordinal patterns. Restrictions R*1 and R*2 entail then that $\pi = \langle 2, 1, 0 \rangle$ is forbidden for $\Sigma_{(+,-)}$ (Lemma 6). Indeed, $\pi_0, \pi_1 = 2, 1$ cannot occur in the 0-block because it is a decreasing sequence (R*1), hence $\pi = \langle 2;1,0 \rangle$; but then the entry 2 should appear on the right of $\pi_1, \pi_2 = 1, 0$ in order to form a spiraling subsequence (R*2); the restriction R*3 is also violated.

The five root forbidden 4-patterns for the logistic map (hence, for Λ and $\Sigma_{(+,-)}$) were found graphically in Sect. 1.2, (1.38). We check here that they do fail to satisfy the restrictions R*1–R*3:

- $\langle 0;2,3,1 \rangle$ violates R*2; $\langle 0,2;3,1 \rangle$ and $\langle 0,2,3;1 \rangle$ violate R*3.
- $\langle 1;0,2,3 \rangle$ violates R*3; $\langle 1,0;2,3 \rangle$ and $\langle 1,0,2;3 \rangle$ violate R*1.
- $\langle 1;0,3,2 \rangle$ violates R*3; $\langle 1,0;3,2 \rangle$ and $\langle 1,0,3;2 \rangle$ violate R*1.
- $\langle 1;3,0,2 \rangle$ violates R*3; $\langle 1,0;3,2 \rangle$ and $\langle 1,0,3;2 \rangle$ violate R*1.
- $\langle 3;1,2,0 \rangle$ violates R*2; $\langle 3,1;2,0 \rangle$ violates R*3 and $\langle 3,1,2;0 \rangle$ violates R*1.

Exercise 8 Check that the allowed patterns for the logistic map, Fig. 1.7, comply with the restrictions (R*1)–(R*4).

Finally, let us prove that $\Sigma_{(+,-)}$ has root forbidden L-patterns for $L \geq 5$.

Theorem 5 *The patterns*

$$\pi = \langle 3, \ldots, L - 2, 0, 1, 2, L - 1 \rangle \in \mathcal{S}_L, \tag{5.22}$$

$L \geq 5$, *are root forbidden patterns for* $\Sigma_{(+,-)}$.

Proof Let us check that (5.22) is a forbidden pattern. First of all, $\pi_{L-5}, \pi_{L-4} = L - 2, 0$ cannot belong to the 0-block because $\pi_{L-5} + 1 = L - 1$ is not on the left of $\pi_{L-4} + 1 = 1$ (R*3). Hence

$$\pi = \langle 3, \ldots, L - 2; 0, 1, 2, L - 1 \rangle.$$

But $\pi_{L-4}, \pi_{L-3}, \pi_{L-2} = 0, 1, 2$ is not a spiraling subsequence, hence it violates R*2.

Furthermore, we claim that (5.22) is a root forbidden pattern. Otherwise, see (3.12), (i) π would be an outgrowth pattern of group I, i.e., the $(L-1)$-pattern obtained from π after removing the entry $L-1$,

$$\langle 3, \ldots, L-2, 0, 1, 2 \rangle \in \mathcal{S}_{L-1}, \qquad (5.23)$$

would be forbidden or (ii) π would be an outgrowth pattern of group II, i.e., the $(L-1)$-pattern obtained from π after removing the entry 0 and subtracting 1 from each remaining entry,

$$\langle 2, \ldots, L-3, 0, 1, L-2 \rangle \in \mathcal{S}_{L-1}, \qquad (5.24)$$

would be forbidden. But (5.23) admits the s-block decompositions

$$\langle 3, \ldots, L-2, 0; 1, 2 \rangle \text{ and } \langle 3, \ldots, L-2, 0, 1; 2 \rangle,$$

while (5.24) admits the decomposition

$$\langle 2, \ldots, L-3; 0, 1, L-2 \rangle.$$

$\square$

Exercise 9 Consider the eight cylinder sets $C_{i_0 i_1 i_2}$ of $\{0, 1\}^{\mathbb{N}_0}$. Check that the sequences of these sets are of the following types under $\Sigma_{(+,-)}$:

   (i) The sequences of C_{000} are of type $\langle 0, 1, 2 \rangle$.
   (ii) The sequences of C_{001} are also of type $\langle 0, 1, 2 \rangle$.
 (iii) The sequences $(0, 1, 0, 0, \ldots) \in C_{010}$ are of type $\langle 0, 1, 2 \rangle$, while the sequences $(0, 1, 0, 1, \ldots) \in C_{010}$ are of type $\langle 0, 2, 1 \rangle$.
  (iv) The sequences of C_{011} are of type $\langle 0, 2, 1 \rangle$ or $\langle 2, 0, 1 \rangle$.
   (v) The sequences of C_{100} are of type $\langle 2, 0, 1 \rangle$.
  (vi) The sequences of C_{101} are of type $\langle 1, 0, 2 \rangle$ or $\langle 2, 0, 1 \rangle$.
 (vii) The sequences of C_{110} are of type $\langle 1, 0, 2 \rangle$ or $\langle 1, 2, 0 \rangle$.
(viii) The sequences of C_{111} are of type $\langle 1, 2, 0 \rangle$.

Among the signed sawtooth maps, those with signatures of alternating signs (we call them *alternating signatures*) have the special property of being continuous. The tent map is one of the two possibilities for $N = 2$. The next theorem generalizes the result that the tent map has a forbidden pattern already for $L = 3$.

Theorem 6 *Let Σ_σ be a shift with alternating signature $\sigma = (\sigma_0, \ldots, \sigma_{N-1})$.*

1. *If N is even, then Σ_σ has forbidden L-patterns for $L \geq N + 1$.*
2. *If N is odd and $\sigma = (+, -, \ldots, -, +)$, then Σ_σ has forbidden L-patterns for $L \geq N + 1$.*

3. *If N is odd and $\sigma = (-, +, \ldots, +, -)$, then (i) all ordinal $(N + 1)$-patterns are allowed for Σ_σ and (ii) Σ_σ has forbidden L-patterns for $L \geq N + 2$.*

In cases 2 and 3, along with a forbidden pattern $\pi \in \mathcal{S}_L$, π_{mirrored} will also be a forbidden pattern (Corollary 3).

Proof Remember that if Σ_σ has a forbidden pattern of length L_0, then its outgrowth patterns provide forbidden L-patterns for every $L \geq L_0$. Hence, we need only to exhibit forbidden patterns of the minimal lengths claimed in each case of Theorem 6.

1. Let $N \geq 2$ be even. There are two possibilities: (a) $\sigma_0 = +$ and $\sigma_{N-1} = -$ and (b) $\sigma_0 = -$ and $\sigma_{N-1} = +$. Since the signatures of these cases are mirrored from each other, we need to consider only one of them (Corollary 3), say (b).

A forbidden pattern of length $L = N + 1$ can be constructed attending to the positive signs of σ, together with the first and last negative signs, as follows. Take the entry $\pi_0 = 0$ for $\sigma_0 = -$,

$$\pi = \langle 0, \ldots \rangle ,$$

the decreasing subsequence $\pi_{2k-1}, \pi_{2k} = 2k, 2k - 1$ for $\sigma_{2k-1} = +$, $1 \leq k \leq N/2 - 1$,

$$\pi = \langle 0, 2, 1, \ldots, 2k, 2k - 1, \ldots, N - 2, N - 3, \ldots \rangle ,$$

and the increasing subsequence $\pi_{N-1}, \pi_N = N - 1, N$ for $\sigma_{N-1} = -$,

$$\pi = \langle 0, 2, 1, \ldots, 2k, 2k - 1, \ldots, N - 2, N - 3, N - 1, N \rangle \in \mathcal{S}_{N+1}.$$

(For $N = 2$, $\pi = \langle 0, 1, 2 \rangle \in \mathcal{S}_3$.) Then R*3 requires a first semicolon between $\pi_0 = 0$ and $\pi_1 = 2$, a second semicolon between $\pi_1 = 2$ and $\pi_2 = 1$, ..., and an $(N - 1)$th semicolon (the maximal number allowed) between $\pi_{N-2} = N - 3$ and $\pi_{N-1} = N - 1$. Still the increasing subsequence $\pi_{N-1}, \pi_N = N - 1, N$ in the last block ($\sigma_{N-1} = +$) violates R*1.

2. Let $N \geq 3$ be odd and $\sigma_0 = \sigma_{N-1} = +$. A forbidden pattern of length $L = N + 1$ can then be constructed attending to positive signs of σ. Take the decreasing subsequence $\pi_0, \pi_1 = 1, 0$ for $\sigma_0 = +$,

$$\pi = \langle 1, 0, \ldots \rangle ,$$

the decreasing subsequence $\pi_{2k}, \pi_{2k+1} = 2k+1, 2k$ for $\sigma_{2k} = +$, $1 \leq k \leq (N-1)/2$,

$$\pi = \langle 1, 0, 3, 2, \ldots, 2k + 1, 2k, \ldots, N - 2, N - 3, \ldots \rangle ,$$

and the increasing subsequence $\pi_{N-1}, \pi_N = N - 1, N$ for $\sigma_{N-1} = +$,

$$\pi = \langle 1, 0, 3, 2, \ldots, 2k + 1, 2k, \ldots, N - 2, N - 3, N - 1, N \rangle \in \mathcal{S}_{N+1}.$$

(For $N = 3$, $\pi = \langle 1, 0, 2, 3 \rangle \in \mathcal{S}_4$.) Then, R*3 requires a first semicolon between $\pi_0 = 1$ and $\pi_1 = 0$, a second semicolon between $\pi_1 = 0$ and $\pi_2 = 3$, ..., and an $(N-1)$th semicolon (the maximal number allowed) between $\pi_{N-2} = N - 3$ and $\pi_{N-1} = N - 1$. Hence we are left with the increasing subsequence $\pi_{N-1}, \pi_N = N - 1, N$ in the last block ($\sigma_{N-1} = +$), what violates R*1.

3. Finally, let $N \geq 3$ be odd and $\sigma_0 = \sigma_{N-1} = -$.

(i) Let us prove that all ordinal $(N + 1)$-patterns are allowed for $\Sigma_{(-,+,...,+,-)}$. Given $\pi \in \mathcal{S}_{N+1}$, there are three possibilities: (a) $N = \pi_0$, (b) $N = \pi_n$ with $1 \leq n \leq N - 1$, or (c) $N = \pi_N$. In the first case, π admits the allowed decomposition

$$\pi = \langle N, \pi_1; \pi_2; \ldots; \pi_k; \ldots; \pi_N \rangle .$$

In the second case, π admits the decomposition

$$\pi = \langle \pi_0; \pi_1; \ldots; \pi_{n-1}; N, \pi_{n+1}; \ldots; \pi_N \rangle$$

both if $\sigma_n = +$ or $\sigma_n = -$. In the third case, π admits the decomposition

$$\pi = \langle \pi_0; \pi_1; \ldots; \pi_k; \ldots; \pi_{N-1}, N \rangle .$$

(ii) A forbidden pattern of length $L = N + 2$ can be constructed attending to the blocks with negative sign. Let first $N = 5 \bmod 4$, so that the central sign of σ is $\sigma_{(N-1)/2} = -$. Take the increasing subsequence $\pi_0, \pi_1 = 0, 1$ for $\sigma_0 = -$,

$$\pi = \langle 0, 1, \ldots \rangle ,$$

the decreasing subsequence $\pi_N, \pi_{N+1} = 3, 2$ for $\sigma_{N-1} = -$,

$$\pi = \langle 0, 1, \ldots, 3, 2 \rangle ,$$

the increasing subsequence $\pi_2, \pi_3 = 4, 5$ for $\sigma_2 = -$,

$$\pi = \langle 0, 1, 4, 5, \ldots, 3, 2 \rangle ,$$

the decreasing subsequence $\pi_{N-2}, \pi_{N-1} = 7, 6$ for $\sigma_{N-3} = -$,

$$\pi = \langle 0, 1, 4, 5, \ldots, 7, 6, 3, 2 \rangle ,$$

and so on until arriving at the central block, $\sigma_{(N-1)/2} = -$, for which we take $\pi_{(N-1)/2}, \pi_{(N+1)/2}, \pi_{(N+3)/2} = N - 1, N + 1, N$,

$$\pi = \langle 0, 1, 4, 5, \ldots, N-1, N+1, N, \ldots, 7, 6, 3, 2 \rangle \in \mathcal{S}_{N+2}.$$

(For $N = 5$, $\pi = \langle 0, 1, 4, 6, 5, 3, 2 \rangle \in \mathcal{S}_7$.) Then, R*3 requires a first semicolon between $\pi_0 = 0$ and $\pi_1 = 1$, a second semicolon between $\pi_1 = 1$ and $\pi_2 = 4, \ldots$, an $((N+1)/2)$th semicolon between $\pi_{(N-1)/2} = N-1$ and $\pi_{(N+1)/2} = N+1$ or between $\pi_{(N+1)/2} = N+1$ and $\pi_{(N+3)/2} = N$ (since the central subsequence $N-1, N+1, N$ is not spiraling), $\ldots$, and an $(N-1)$th semicolon (the maximal number allowed) between $\pi_{N-1} = 6$ and $\pi_N = 3$. But the sequence $\pi_N, \pi_{N+1} = 3, 2$ in the last block ($\sigma_{N-1} = -$) violates R*3 because $\pi_N + 1 = 4$ is not on the right of $\pi_{N+1} + 1 = 3$.

In the case $N = 3 \bmod 4$, the central sign of σ is $\sigma_{(N-1)/2} = +$. The construction of a forbidden pattern of length $L = N+2$ follows the same assignment of entry pairs as before for σ_0, σ_{N-1}, σ_2, $\ldots$, $\sigma_{(N-3)/2}$, but takes $\pi_{(N+1)/2}, \pi_{(N+3)/2}, \pi_{(N+5)/2} = N+1, N, N-1$ for $\sigma_{(N+1)/2} = -$:

$$\pi = \langle 0, 1, 4, 5, \ldots, N-2, N+1, N, N-1, \ldots, 7, 6, 3, 2 \rangle \in \mathcal{S}_{N+2}.$$

(For $N = 3$, $\pi = \langle 0, 1, 4, 3, 2 \rangle \in \mathcal{S}_5$.) Then, R*3 requires a first semicolon between $\pi_0 = 0$ and $\pi_1 = 1$, a second semicolon between $\pi_1 = 1$ and $\pi_2 = 4$, $\ldots$, an $((N+1)/2)$th semicolon between $\pi_{(N-1)/2} = N-2$ and $\pi_{(N+1)/2} = N+1$ or between $\pi_{(N+1)/2} = N+1$ and $\pi_{(N+3)/2} = N$ (since the subsequence $\pi_{(N-1)/2}, \pi_{(N+1)/2}, \pi_{(N+3)/2} = N-2, N+1, N$ cannot belong to an s-block with positive sign because $\pi_{(N-1)/2} + 1 = N-1$ is not on the left of $\pi_{(N+3)/2} + 1 = N+1$), $\ldots$, and an $(N-1)$th semicolon (the maximal number allowed) between $\pi_{N-1} = 6$ and $\pi_N = 3$. But the sequence $\pi_N, \pi_{N+1} = 3, 2$ in the last block ($\sigma_{N-1} = -$) violates R*3 because $\pi_N + 1 = 4$ is not on the right of $\pi_{N+1} + 1 = 3$. $\square$

A further signature with general features is $\sigma = (-, -, \ldots, -)$.

Theorem 7 *The shift Σ_σ with $\sigma_0 = \cdots = \sigma_{N-1} = -$, $N \geq 2$, has*

1. *allowed L-patterns for $L \leq N+1$ and*
2. *root forbidden L-patterns for $L \geq N+2$.*

Since $\sigma = (-, \ldots, -) = \sigma_{\mathrm{mirrored}}$, the number of root forbidden patterns for Σ_σ will be even (Corollary 3).

Proof 1. We need to consider only the case $L = N+1$, since all L-patterns with $L \leq N$ are trivially allowed. Given $\pi \in \mathcal{S}_{N+1}$, there are three possibilities: (i) $N = \pi_0$, (ii) $N = \pi_n$ with $1 \leq n \leq N-1$, or (iii) $N = \pi_N$. The decompositions (i)

$$\pi = \langle N, \pi_1; \pi_2; \ldots; \pi_k; \ldots; \pi_N \rangle,$$

(ii)

$$\pi = \langle \pi_0; \pi_1; \ldots; \pi_{n-1}; N, \pi_{n+1}; \ldots; \pi_N \rangle \quad \text{or} \quad \langle \pi_0; \pi_1; \ldots; \pi_{n-1}, N; \pi_{n+1}; \ldots; \pi_N \rangle$$

(since $\pi_{n-1}, \pi_n, \pi_{n+1} = \pi_{n-1}, N, \pi_{n+1}$ does not form a spiraling subsequence), and (iii)

$$\pi = \langle \pi_0; \pi_1; \ldots; \pi_k; \ldots; \pi_{N-1}, N \rangle,$$

show that any $\pi \in \mathcal{S}_{N+1}$ is allowed for Σ_σ, $\sigma_0 = \cdots = \sigma_{N-1} = -$.

2. Consider

$$\pi = \langle 0, 1, 2, \ldots, N - 1, N, N + 1 \rangle \in \mathcal{S}_{N+2}.$$

Then R*2 requires a first semicolon between 0 and 1, a second semicolon between 1 and 2, and an $(N - 1)$th semicolon (the maximal number allowed) between $N - 2$ and $N - 1$. This leads to a last block $\pi_{N-1}, \pi_N, \pi_{N+1} = N - 1, N, N + 1$, which is not a spiraling subsequence. Hence π is forbidden.

The assumption that π is not a root forbidden pattern leads to the fact that π is outgrowth of the forbidden pattern

$$\langle 0, 1, 2, \ldots, N - 1, N \rangle \in \mathcal{S}_{N+1},$$

whether π belongs to group I or II (3.12). But clearly this pattern admits the decomposition

$$\langle 0; 1; 2; \ldots; N - 1, N \rangle,$$

with $N - 1$ semicolons (the maximal number allowed). This contradiction shows that π is not an outgrowth forbidden pattern. Needless to say (Corollary 3),

$$\pi_{\text{mirrored}} = \langle N + 1, N, N - 1, \ldots, 2, 1, 0 \rangle$$

is also a root forbidden pattern.   □

To conclude this chapter, we consider briefly the existence of *root* forbidden patterns for the signed shifts on $N \geq 3$ symbols. For $\sigma = (+, \ldots, +)$ and $\sigma = (-, \ldots, -)$ we know that there exist root forbidden patterns for every $L \geq N+2$ (Theorems 2 and 7, respectively). The structure of the forbidden ordinal patterns depends, of course, on the signature of the signed shift envisaged, thus the construction of root forbidden patterns can only be done, in general, on a case-by-case basis.

To illustrate this point, consider the signed shifts (with mixed signs) on three symbols. Because of the relation between the allowed/forbidden patterns for Σ_σ and $\Sigma_{\sigma_{\text{mirrored}}}$, only the following four cases are really distinct:

$$\text{Case a: } \sigma = (+, +, -), \quad \text{Case b: } \sigma = (+, -, +),$$
$$\text{Case c: } \sigma = (+, -, -), \quad \text{Case d: } \sigma = (-, +, -).$$

These four cases were studied in [17]. There it is proven that all the signed shifts (a)–(d) have root forbidden L-patterns for $L \geq 5$. Furthermore, $\Sigma_{(+,-,+)}$ has two (root) forbidden 4-patterns, $\Sigma_{(+,-,-)}$ has one (root) forbidden 4-pattern, while $\Sigma_{(+,+,-)}$, $\Sigma_{(-,+,-)}$ have no forbidden 4-patterns. Of course, the same holds for any map order isomorphic to those signed shifts, in particular for the corresponding signed sawtooth maps.

Exercise 10 Check the following statements on root forbidden patterns for Σ_σ in the four cases a–d.

(a) The patterns

$$\pi = \langle 0, L-1, 2, 3, \ldots, L-2, 1 \rangle \in \mathcal{S}_L,$$

$L \geq 5$, are root forbidden patterns for $\Sigma_{(+,+,-)}$.
(b) The patterns

$$\pi = \langle L-2, 0, L-4, \ldots, 3, 1, 2, 4, \ldots, L-3, L-1 \rangle \in \mathcal{S}_L$$

if $L \geq 5$ is odd and

$$\pi = \langle L-1, L-3, \ldots, 3, 1, 2, 4, \ldots, L-4, 0, L-2 \rangle \in \mathcal{S}_L$$

if $L \geq 6$ is even, together with their corresponding mirrored patterns, are root forbidden patterns for $\Sigma_{(+,-,+)}$. (If $L = 5$, then $\pi = \langle 3, 0, 1, 2, 4 \rangle$; if $L = 6$, then $\pi = \langle 5, 3, 1, 2, 0, 4 \rangle$.)
(c) The patterns

$$\pi = \langle 2, 1, 0, 3, 4 \rangle \in \mathcal{S}_5,$$
$$\pi = \langle L-3, \ldots, 4, 2, 1, 0, 3, 5, \ldots, L-4, L-2, L-1 \rangle \in \mathcal{S}_L$$

for $L \geq 7$ odd, and

$$\pi = \langle L-1, L-2, L-4, \ldots, 4, 2, 1, 0, 3, 5, \ldots, L-3 \rangle \in \mathcal{S}_L$$

for $L \geq 6$ even, are root forbidden patterns for $\Sigma_{(+,-,-)}$. Although $\sigma = (+,-,-) \neq \sigma_{mirrored} = (-,-,+)$, the mirrored patterns of these patterns are also root forbidden patterns for $\Sigma_{(+,-,-)}$.
(d) The patterns

$$\pi = \langle 0, 1, 4, 3, 2 \rangle \in \mathcal{S}_5,$$
$$\pi = \langle 0, 1, L-1, L-2, \ldots, 3, 2, 4, \ldots, L-3 \rangle \in \mathcal{S}_L$$

if $L \geq 7$ is odd and

$$\pi = \langle 0, 1, L-1, L-2, \ldots, 4, 2, 3, \ldots, L-3 \rangle \in \mathcal{S}_L$$

if $L \geq 6$ is even, together with the corresponding mirrored patterns, are root forbidden patterns for $\Sigma_{(-,+,-)}$.

Exercise 11 Using signed sawtooth maps with alternating signature, construct a *continuous* map whose orbits realize all possible ordinal patterns (*hint*: the construction is similar to Fig. 4.2).

Chapter 6
Metric Permutation Entropy

The word "entropy" was coined by the German physicist R. Clausius (1822–1888), who introduced it in thermodynamics in 1865 to measure the amount of energy in a system that cannot produce work. The fact that the entropy of an isolated system never decreases constitutes the second law of thermodynamics and clearly shows the central role of entropy in many-particle physics. The direction of time is then explained as a consequence of the increase of entropy in all irreversible processes. Later on the concept of entropy was given a microscopic interpretation in the foundational works of L. Boltzmann (1844–1906) on gas kinetics and statistical mechanics [184]. The celebrated Boltzmann's equation reads in the usual physical notation

$$S = k_B \ln \Omega, \tag{6.1}$$

where here S is the entropy of the thermodynamical system, k_B is a physical constant (called Boltzmann's constant, $k_B = 1.3806504(24) \times 10^{-23}$ J/K) and Ω is the number of microscopic states consistent with the macroscopic constraints. In this realm, the entropy is a measure of the microscopic *disorder* of the system, the entropy being higher the more disordered the system.

In 1948 the word entropy came to the fore in the new context of information theory, coding theory, and cryptography through the seminal papers of C.E. Shannon[1] (1916–2001) [186]. This time, entropy measures the average *uncertainty* about the outcome of a random variable. More generally, the *entropy rate* measures the uncertainty per symbol (time unit, channel use, etc.) of a stationary stochastic process, eventually modeling an information source. Instead of associating entropy with uncertainty, one can alternatively speak of the average information gained by performing a random experiment. Entropy plays a paramount role in all information-related fields, being at the heart of the fundamental results.

---

[1] According to [64] "When Shannon had invented his quantity and consulted von Neumann on what to call it, von Neumann replied: 'Call it entropy. It is already in use under that name and besides, it will give you a great edge in debates because nobody knows what entropy is anyway.' "

J.M. Amigó, *Permutation Complexity in Dynamical Systems,*
Springer Series in Synergetics, DOI 10.1007/978-3-642-04084-9_6,
© Springer-Verlag Berlin Heidelberg 2010

Shannon's ideas, properly transformed, were incorporated by A.N. Kolmogorov (1903–1987) into ergodic theory in 1958 [126] to measure the *randomness* of deterministic dynamical systems. Kolmogorov's proposal was improved a short time later by Sinai [189]. The result became the most important invariant in the theory of discrete and continuous dynamical systems.

Since then the concept of entropy has evolved along different ways: Rényi entropy, topological entropy, sequence entropy, Tsallis entropy, directional entropy, permutation entropy, epsilon–tau entropy, etc. The basics of Shannon entropy, metric (Kolmogorov–Sinai or measure-theoretical) entropy, and topological entropy are systematized in Annex B.

Permutation entropy, both in the metric version (this chapter) and in the topological version (next chapter), was introduced by Bandt, Keller, and Pompe in [29] (see [28] as well). The main ingredient of permutation entropy is the ordinal patterns we studied in Chap. 3. As we shall see below, the definition of the metric permutation entropy of an information source is formally the same as Shannon's entropy, except for the fact that now probabilities refer not to length-L blocks of symbols but to the length-L ordinal patterns realized by them (assuming, of course, that those symbols can be ordered).

On defining the metric permutation entropy of maps, we depart from [29] to follow basically Kolmogorov's strategy: coarse-grain the state space with a partition, apply the definition of (in our case, permutation) entropy to the resulting symbolic dynamics, and then refine successively the original partition into the partition into separate points. Moreover, the partitions used may be taken to be product, uniform partitions, making possible the numerical estimation of metric permutation entropy under rather general conditions. Most importantly, we shall show that metric permutation entropy converges to the conventional metric entropy for ergodic self-maps of n-dimensional intervals.

6.1 The Metric Permutation Entropy of a Finite-State Process

Let $\mathbf{X} = \{X_n\}_{n \in \mathbb{N}_0}$ be a random process with finite state space S (see Annex A.3). We take without restriction $S = \{1, 2, \ldots, |S|\}$. As noted in Example 2, the relation between length-L words and length-L ordinal patterns is in general many-to-one. This is due to the fact that ordinal patterns do not take into account the sizes of the elements being compared, but only their relative order. The same happens with the *ranks* or *rank variables*, which are the outputs of a random process $\mathbf{R} = \{R_n\}_{n \in \mathbb{N}_0}$ subsidiary of $\mathbf{X}$, defined as follows:

$$R_n = |\{X_i, 0 \le i \le n : X_i \le X_n\}| = \sum_{i=0}^{n} \delta(X_i \le X_n),$$

where as usual the δ-function of a proposition is 1 if it holds and 0 otherwise. By definition, R_n is a *discrete* random variable with range $\{1, \ldots, n+1\}$, and the sequence

$\mathbf{R} = \{R_n\}_{n \in \mathbb{N}_0}$ builds a discrete-time, non-stationary stochastic process. The point about introducing rank variables is that the relation between length-L ordinal patterns $\pi(x_n^{n+L-1})$ and length-L ranks $r_n^{n+L-1} = r_n, r_{n+1}, \ldots, r_{n+L-1}$ is one-to-one. The many-to-one relation between X_0^{L-1} and R_0^{L-1} will be written as

$$R_0^{L-1} = \mathrm{rank}(X_0^{L-1}). \tag{6.2}$$

Ranks are specially useful in proofs.

Example 10 If, as in Example 2, $S = \{a, b, c\}$ with $a < b < c$ and $x_0^2 = c, a, a$, then $r_0^2 = 1, 1, 2$. All other words defining the same ordinal pattern $\pi(x_0^2) = \langle 1, 2, 0 \rangle$ define also the same rank variables:

$$r_0^2 = 1, 1, 2 = \mathrm{rank}(c, b, b) = \mathrm{rank}(c, a, b) = \mathrm{rank}(b, a, a).$$

Having defined the sibling concepts of ordinal patterns and rank variables of finite-alphabet sequences, we can proceed now very much the same way as we did when defining Shannon's entropy (rate) of stochastic processes or information sources in Sect. 1.1.1 (see also Annex B.1), this time though bookkeeping ordinal patterns instead of symbol blocks.

In this spirit, the *metric permutation entropy* of a stochastic process $\mathbf{X} = \{X_n\}_{n \in \mathbb{N}_0}$ is defined as

$$h^*(\mathbf{X}) = \lim_{L \to \infty} h^*(X_0^{L-1}), \tag{6.3}$$

provided the limit exists, where

$$h^*(X_0^{L-1}) = -\frac{1}{L} \sum_{x_0, \ldots, x_{L-1}} p(\pi(x_0^{L-1})) \log p(\pi(x_0^{L-1}))$$

is the *metric permutation entropy of order $L \geq 2$* of $\mathbf{X}$. Here $p(\pi(x_0^{L-1}))$ is the probability for the length-L block $x_0^{L-1} = x_0, \ldots, x_{L-1}$ to be of type $\pi(x_0^{L-1}) \in S_L$. Alternatively,

$$h^*(X_0^{L-1}) = -\frac{1}{L} \sum_{r_0, \ldots, r_{L-1}} p(r_0^{L-1}) \log p(r_0^{L-1}) = h(R_0^{L-1}), \tag{6.4}$$

where $p(r_0^{L-1})$ is the probability for the block x_0^{L-1} to define the rank vector $r_0^{L-1} = r_0, \ldots, r_{L-1}$ (remember that the relation between $\pi(X_0^{L-1})$ and $R_0^{L-1} = \mathrm{rank}(X_0^{L-1})$ is one-to-one). In both cases,

$$h^*(\mathbf{X}) = h(\pi(\mathbf{X})) = h(\mathbf{R}),$$

where $h(\cdot)$ denotes the Shannon entropy of the corresponding stochastic process.

In case that the random process $\mathbf{X}$ is stationary, there is still a third way to look at its metric entropy permutation. If $(S^{\mathbb{N}_0}, \mathcal{B}_\Pi(S), m, \Sigma)$ is the sequence space model of $\mathbf{X}$ (see Annex A.3), then the non-empty cylinder sets

$$C_\pi = \{(x_n) \in S^{\mathbb{N}_0} : x_0^{L-1} \text{ is of type } \pi \in \mathcal{S}_L\}$$

build a partition of $(S^{\mathbb{N}_0}, \mathcal{B}_\Pi(S), m)$ with $m(C_\pi) = \Pr\{\pi(X_0^{L-1}) = \pi\} = \Pr\{R_0^{L-1} = r_0^{L-1}\}$, where $R_0^{L-1} = \operatorname{rank}(X_0^{L-1})$, and $1 \le r_k \le k+1$ for $k = 0, \dots, L-1$. Therefore

$$h^*(X_0^{L-1}) = -\frac{1}{L} \sum_{\pi \in \mathcal{S}_L} m(C_\pi) \log m(C_\pi). \tag{6.5}$$

As a result, the permutation entropy is sensitive to the measures of non-trivial order relationships observed in a word, as the Shannon entropy is sensitive to the measures of the different word values themselves.

When stationarity is important, as in (6.5), we call $\mathbf{X}$ an information source or just a source.

In the next lemma we use the conditional entropy of a random variable Y given another random variable X, $H(Y|X)$, which is the expected value of the entropies of the conditional distributions averaged over the conditioning variable X (see Annex B, (B.5)).

Lemma 7 *Given an ergodic source* $\mathbf{X} = \{X_n\}_{n \in \mathbb{N}_0}$, *the equality*

$$\lim_{k \to \infty} H(R_k^{k+l} | X_0^{k-1}) = \lim_{k \to \infty} H(X_k^{k+l} | X_0^{k-1})$$

holds for all $l \ge 0$.

That is, given a sufficiently long tail of previously observed symbols, the later ranks can be predicted virtually as well as the symbols themselves. Heuristically, this is because the rank of a late variable is sensitive effectively to the cumulative distribution function of the source, approximated by the normalized sum of X_0^{k-1}. In turn, this means that the information contained in R_k is the same as the information in X_k.

Proof Consider $R_k = \sum_{i=0}^k \delta(X_i \le X_k)$. For $a \in S = \{1, \dots, |S|\}$ define the *sample frequency* of the letter a in the word x_0^k, $k \ge 0$, to be

$$\vartheta_k(a) = \frac{1}{k+1} \sum_{i=0}^k \delta(X_i = a).$$

With the help of $\vartheta_k(a)$ we may express R_k in terms of X_i, $0 \le i \le k$, namely,

$$R_k(X_k) = (k+1) \sum_{a=1}^{X_k} \vartheta_k(a),$$

where we assume the outcomes $X_0, \ldots, X_k$ to be known. Then, the identity

$$\Pr\{R_k = y\} = \sum_{q=1}^{|S|} \Pr\{X_k = q\}\delta\left(R_k(q) = y\right) \tag{6.6}$$

gives us the probability for observing some R_k with value $y \in \{1, \ldots, k+1\}$ by means of $\Pr\{X_k = q\}$, $1 \le q \le |S|$. Since, given X_0^{k-1} ($k \ge 1$), R_k is a deterministic function of the random variable X_k, i.e., $\Pr\{R_k = y|X_k = q\} = \delta(R_k(q) = y)$, (6.6) can be seen as an application of the law of total probability.

Without loss of generality, we may first rearrange the sum in (6.6) to consider only those symbol values q with non-zero $\Pr\{X_k = q\}$, summing to $N \le |S|$. Expand the sum,

$$\begin{aligned}
\Pr\{R_k = y\} =\ & \Pr\{X_k = 1\}\delta\left[y = (k+1)\vartheta_k(1)\right] \\
& + \Pr\{X_k = 2\}\delta\left[y = (k+1)(\vartheta_k(1) + \vartheta_k(2))\right] \\
& + \cdots + \Pr\{X_k = N\}\delta\left[y = (k+1)(\vartheta_k(1) + \cdots + \vartheta_k(N))\right].
\end{aligned}$$

Suppose all the relevant sample frequencies $\vartheta_k(1), \ldots, \vartheta_k(N)$ are greater than zero. This means that for any y, only a single one of the δ-functions can be non-zero, and hence we have a one-to-one transformation taking non-zero elements from the distribution $\Pr\{X_k\}$ without change into some bin for $\Pr\{R_k\}$. Since entropy is invariant to a renaming of the bins, and the remaining zero probability bins add nothing to the entropy, we conclude that, if $\vartheta_k(a) > 0$ for all a where the true probability $\Pr\{X_k = a\} > 0$ (i.e., $a = 1, \ldots, N$ after a hypothetical rearrangement), then $H(R_k|X_0^{k-1}) = H(X_k|X_0^{k-1})$ for $k \ge 1$. Because of the assumed ergodicity, we can make the probability that $\vartheta_k(a) = 0$ when $\Pr\{X_k = a\} > 0$ to be arbitrarily small by taking k to be sufficiently large, and the claim follows for $l = 0$.

This construction can be extended without change to words X_k^{k+l} of arbitrary length $l + 1 \ge 1$ via

$$\begin{aligned}
& \Pr\{R_k^{k+l} = y_0 \ldots y_l\} \\
& = \sum_{q_0,\ldots,q_l=1}^{N} \Pr\{X_k^{k+l} = q_0 \ldots q_l\}\delta(R_k(q_0) = y_0)\ldots\delta(R_{k+l}(q_l) = y_l).
\end{aligned}$$

Observe that if $\vartheta_k(a) > 0$ for $1 \le a \le N$, then the same happens with $\vartheta_{k+1}(a), \ldots, \vartheta_{k+l}(a)$ and $H(R_k^{k+l}|X_0^{k-1}) = H(X_k^{k+l}|X_0^{k-1})$ follows. Again, ergodicity guarantees that there exist realizations of X_0^{k+l} with sufficiently large k, whose sample frequencies fulfill the said condition. $\square$

Example 11 As way of illustration, suppose that $X_n = 0, 1$ are independent random variables with probability $\Pr\{X_n = 0\} = \Pr\{X_n = 1\} = \frac{1}{2}$. Given $x_0^{k-1} =$

$x_0 \ldots x_{k-1} \in \{0,1\}^k$, set $N_0 = \left| \{i : x_i = 0 \text{ in } x_0^{k-1}\} \right|$, $0 \le N_0 \le k$. Consider the case $l = 1$ in Lemma 1. There are two possibilities:

(i) $0 \le N_0 \le k-1$. Then

$$
\begin{array}{lll}
x_k^{k+1} = 0,0 & \Rightarrow & r_k^{k+1} = N_0 + 1, N_0 + 2, \\
x_k^{k+1} = 0,1 & \Rightarrow & r_k^{k+1} = N_0 + 1, k + 2, \\
x_k^{k+1} = 1,0 & \Rightarrow & r_k^{k+1} = k + 1, N_0 + 1, \\
x_k^{k+1} = 1,1 & \Rightarrow & r_k^{k+1} = k + 1, k + 2.
\end{array}
$$

Each of these events has the joint probability

$$
\Pr\{N_0 = v, R_k^{k+1} = r_k^{k+1}\} = \frac{\binom{k}{v}}{2^k} \cdot \frac{1}{4} = \frac{1}{2^{k+2}} \binom{k}{v}
$$

and conditional probability

$$
\Pr\{R_k^{k+1} = r_k^{k+1} | N_0 = v\} = \frac{1}{4},
$$

where $0 \le v \le k-1$ and $r_k^{k+1} = (v+1, v+2)$, $(v+1, k+2)$, $(k+1, v+1)$, or $(k+1, k+2)$.

(ii) $N_0 = k$. Then

$$
\begin{array}{ll}
x_k^{k+1} = 0,0 \ \& \ x_k^{k+1} = 0,1 \ \& \ x_k^{k+1} = 1,1 & \Rightarrow \quad r_k^{k+1} = k+1, k+2, \\
x_k^{k+1} = 1,0 & \Rightarrow \quad r_k^{k+1} = k+1, k+1.
\end{array}
$$

These events have the joint probabilities

$$
\Pr\left\{N_0 = k, R_k^{k+1} = (k+1, k+2)\right\} = \frac{1}{2^k} \cdot \frac{1}{4} \cdot 3 = \frac{3}{2^{k+2}},
$$

$$
\Pr\left\{N_0 = k, R_k^{k+1} = (k+1, k+1)\right\} = \frac{1}{2^k} \cdot \frac{1}{4} = \frac{1}{2^{k+2}}
$$

and conditional probabilities

$$
\Pr\left\{R_k^{k+1} = (k+1, k+2) | N_0 = k\right\} = \frac{3}{4},
$$

$$
\Pr\left\{R_k^{k+1} = (k+1, k+1) | N_0 = k\right\} = \frac{1}{4}.
$$

From Annex (B.5) and (i)–(ii), we get

$$H(R_k^{k+1}|X_0^{k-1}) = -4 \times \sum_{v=0}^{k-1} \frac{1}{2^{k+2}}\binom{k}{v}\log\frac{1}{4} - \frac{3}{2^{k+2}}\log\frac{3}{4} - \frac{1}{2^{k+2}}\log\frac{1}{4}$$

$$= 4 \times \frac{2}{2^{k+2}}(2^k - 1) + \frac{8}{2^{k+2}} - \frac{3}{2^{k+2}}\log 3$$

$$= 2\left(1 - \frac{3}{2^{k+3}}\log 3\right).$$

On the other hand, since the random variables X_n are independent,

$$H(X_k^{k+1}|X_0^{k-1}) = H(X_k^{k+1}) = 2.$$

It follows that $H(R_k^{k+1}|X_0^{k-1})$ and $H(X_k^{k+1}|X_0^{k-1})$ coincide in the limit $k \to \infty$, as guaranteed by Lemma 7.

With Lemma 7 in hand, we turn to the main result.

Theorem 8 *For a finite-alphabet ergodic source* **X**, *the permutation entropy exists and equals the metric entropy:* $h^*(\mathbf{X}) = h(\mathbf{X})$.

Proof We prove inequalities in both directions.

(a) $\limsup_{L\to\infty} h^*(X_0^{L-1}) \leq h(\mathbf{X})$. Given X_0^{L-1}, the corresponding rank variables are uniquely determined via $R_0^{L-1} = \text{rank}(X_0^{L-1})$. By [59, Chap. 2, Exercise 5], $H(\varphi(Z)) \leq H(Z)$ for any discrete random variable Z and function φ, so $H(R_0^{L-1}) \leq H(X_0^{L-1})$ and thus (see (6.4)),

$$\limsup_{L\to\infty} h^*(X_0^{L-1}) = \limsup_{L\to\infty} h(R_0^{L-1}) \leq \limsup_{L\to\infty} h(X_0^{L-1}) = h(\mathbf{X}).$$

(b) $\liminf_{L\to\infty} h^*(X_0^{L-1}) \geq h(\mathbf{X})$. There are several ways to prove this inequality. Consider, for instance,

$$\liminf_{L\to\infty} h^*(X_0^{L-1})$$

$$= \liminf_{L\to\infty} \frac{1}{L} H(R_0^{L-1})$$

$$= \liminf_{L\to\infty} \frac{1}{L}\left(\left[H(R_{L-1}|R_0^{L-2}) + \cdots + H(R_{L^*+1}|R_0^{L^*})\right] + H(R_0^{L^*})\right)$$

for any $L^* < L - 1$, where we have applied the chain rule for entropy (B.9). As $R_1^k = \text{rank}(X_1^k)$ we apply the data processing inequality $H(Y|\varphi(Z)) \geq H(Y|Z)$ [59] to all elements of the first term on the right-hand side:

$$\liminf_{L\to\infty} h(X_0^{L-1})$$

$$\geq \liminf_{L\to\infty} \frac{1}{L}\left(\left[H(R_{L-1}|X_0^{L-2}) + \cdots + H(R_{L^*+1}|X_0^{L^*})\right] + H(R_0^{L^*})\right).$$

By Lemma 7 with $l = 0$, for any $\varepsilon > 0$ there exists some L^* such that

$$\left| H(X_L | X_0^{L-1}) - H(R_L | X_0^{L-1}) \right| < \varepsilon$$

for $L > L^*$, so

$$\liminf_{L \to \infty} h(X_0^{L-1})$$

$$> \liminf_{L \to \infty} \left(\frac{1}{L} \left[H(X_{L-1} | X_0^{L-2}) + \cdots + H(X_1 | X_0) + H(X_0) \right] \right.$$

$$\left. + \frac{1}{L} \left[H(R_0^{L^*}) - H(X_0^{L^*}) \right] - \left(\frac{L - L^* - 1}{L} \right) \varepsilon \right)$$

$$= h(\mathbf{X}) - \varepsilon,$$

since $H(X_0^{L^*}) = H(X_0) + H(X_1 | X_0) + \cdots + H(X_{L^*} | X_0^{L^*-1})$ (B.9).
The existence of the limit and equality follows from (a) and (b).   □

Observe in the proof of Theorem 8 that the ergodicity hypothesis was used only in part (b) via Lemma 7, while part (a) is completely general. We highlight this particular result in the following corollary for further reference.

Corollary 4 *For finite-alphabet sources* **X**,

$$\limsup_{L \to \infty} h^*(X_0^{L-1}) \le h(\mathbf{X})$$

holds.

In order to deal further with the general, nonergodic case, we appeal to the theorem on ergodic decompositions [114]: if Ω is a compact metrizable space and $T:(\Omega, \mathcal{B}, \mu) \to (\Omega, \mathcal{B}, \mu)$ is a continuous transformation, then there is a partition of Ω into T-invariant subsets Ω_w, each equipped with a sigma-algebra $\mathcal{B}_w$ and a probability measure μ_w, such that T acts ergodically on each probability space $(\Omega_w, \mathcal{B}_w, \mu_w)$, the indexing set being another probability space $(W, \mathcal{F}, \nu)$. Furthermore,

$$\mu(E) = \int_W \int_E d\mu_w d\nu(w) = \int_W \mu_w(E) d\nu(w) \quad (E \in \mathcal{B}).$$

The family $\{\mu_w : w \in W\}$ is called the *ergodic decomposition* of μ.

If Σ is the shift on the (compact, metric) sequence space $(S^{\mathbb{N}_0}, \mathcal{B}_\Pi(S), m)$, the indexing set can be taken to be $S^{\mathbb{N}_0}$, i.e.,

$$m(C) = \int_{S^{\mathbb{N}_0}} \int_C dm_s dm(s) = \int_{S^{\mathbb{N}_0}} m_s(C) dm(s) \quad (C \in \mathcal{B}_\Pi(S)), \tag{6.7}$$

where $m_{\Sigma(s)} = m_s$ [89]. This result shows that any source which is not ergodic can be represented as a mixture of ergodic subsources. The next lemma states that such a decomposition holds also for the entropy.

Lemma 8 (Ergodic Decomposition of the Entropy) [89] *Let* $(S^{\mathbb{N}_0}, \mathcal{B}_{\Pi}(S), m, \Sigma)$ *be the sequence space model of a stationary finite-alphabet random process* $\mathbf{X} = \{X_n\}_{n \in \mathbb{N}_0}$. *Let* $\{m_s{:}s \in S^{\mathbb{N}_0}\}$ *be the ergodic decomposition of* m. *If* $h_{m_s}(\mathbf{X})$ *is* m-integrable, then

$$h(\mathbf{X}) = \int_{S^{\mathbb{N}_0}} h_{m_s}(\mathbf{X}) dm(s).$$ (6.8)

Theorem 9 *Under the assumptions of Lemma 8,*

$$\liminf_{L \to \infty} h^*(X_0^{L-1}) \geq h(\mathbf{X})$$ (6.9)

for any finite-alphabet source $\mathbf{X}$.

Proof Fix $L \geq 2$. From (6.5) and (6.7),

$$h^*(X_0^{L-1}) = -\frac{1}{L} \sum_{\pi \in S_L} \left(\int_{S^{\mathbb{N}_0}} m_s(C_\pi) dm(s) \right) \log \left(\int_{S^{\mathbb{N}_0}} m_s(C_\pi) dm(s) \right)$$

$$\geq -\frac{1}{L} \sum_{\pi \in S_L} \left(\int_{S^{\mathbb{N}_0}} m_s(C_\pi) \log m_s(C_\pi) dm(s) \right)$$ (6.10)

$$= \int_{S^{\mathbb{N}_0}} \left(-\frac{1}{L} \sum_{\pi \in S_L} m_s(C_\pi) \log m_s(C_\pi) \right) dm(s)$$

$$= \int_{S^{\mathbb{N}_0}} h^*_{m_s}(X_0^{L-1}) dm(s),$$

where in (6.10) we have used Jensen's inequality,

$$\Phi \left(\int_{S^{\mathbb{N}}} f d\mu \right) \leq \int_{S^{\mathbb{N}}} \Phi \circ f d\mu,$$

with $\Phi(t) = t \log t$ convex in $[0, \infty)$ and $f(s) = m_s(C_\pi) \geq 0$.

Therefore,

$$\liminf_{L\to\infty} h^*(X_0^{L-1}) \geq \liminf_{L\to\infty} \int_{S^{\mathbb{N}_0}} h^*_{m_s}(X_0^{L-1}) dm(s)$$

$$\geq \int_{S^{\mathbb{N}_0}} \left(\liminf_{L\to\infty} h^*_{m_s}(X_0^{L-1}) \right) dm(s) \qquad (6.11)$$

$$= \int_{S^{\mathbb{N}_0}} h^*_{m_s}(\mathbf{X}) dm(s),$$

where we have applied Fatou's lemma in (6.11) to the sequence of positive and (by hypothesis) m-measurable functions $h^*_{m_s}(X_0^{L-1})$. Observe that $h^*_{m_s}(\mathbf{X})$ exists for all $s \in S^{\mathbb{N}_0}$ (and is m-integrable as a function of s) since $h^*_{m_s}(\mathbf{X}) = h_{m_s}(\mathbf{X})$ by Theorem 8 ($\mathbf{X}$ is ergodic with respect to m_s). Therefore,

$$\lim_{L\to\infty} \inf h^*(X_0^{L-1}) \geq \int_{S^{\mathbb{N}_0}} h_{m_s}(\mathbf{X}) dm(s) = h(\mathbf{X})$$

by (6.8).   □

Corollary 4 and Theorem 9 yield the following result.

Corollary 5 *Under the assumptions of Lemma 8, $h^*(\mathbf{X}) = h(\mathbf{X})$ holds for any finite-alphabet source* $\mathbf{X}$.

6.2 Permutation Metric Entropy of Maps

In this section we shall use the previous results on finite-alphabet stochastic processes to show that the equality between permutation and metric entropies holds also for ergodic self-maps on domains homeomorphic to q-dimensional compact intervals.

We say that a set $D \subset \mathbb{R}^q$ is a (q-dimensional) *simple domain* if it is homeomorphic to a q-dimensional compact interval (hence D is compact). In particular, one-dimensional simple domains are close intervals. As a subset of $\mathbb{R}^q$, D is also ordered. Let D be a q-dimensional simple domain and $f : D \to D$ a μ-preserving map, with μ being a probability measure on $(D, \mathcal{B} \cap D)$ and $\mathcal{B}$ being the Borel sigma-algebra of $\mathbb{R}^q$. In order to define the permutation entropy of f, consider a q-dimensional compact interval $I \supset D$ and product partitions

$$\iota = \prod_{k=1}^{q} \{I_{1,k}, \ldots, I_{N_k,k}\} \qquad (6.12)$$

of I into $|\iota| = N_1 \cdots N_q$ subintervals of lengths $\Delta_{j,k}$, $1 \leq j \leq N_k$, in each coordinate k. As for the norm of ι (see (1.13)), the perhaps most popular are the *Euclidean norm*,

$$\|\iota\| = \max_{j_1,\dots,j_q} \left(\sum_{k=1}^{q} \Delta_{j_k,k}^2 \right)^{1/2} =: \|\iota\|_2 \qquad (6.13)$$

(i.e., $\|\iota\|_2$ is the longest diagonal of the bins $I_{j_1,1} \times \cdots \times I_{j_q,q} \in \iota$) and the *supremum norm*,

$$\|\iota\| = \max_{j,k} \Delta_{j,k} =: \|\iota\|_\infty . \qquad (6.14)$$

For definiteness, the intervals are lexicographically ordered in each dimension, that is, points in $I_{j,k}$ are smaller than points in $I_{j+1,k}$ and, for the multiple dimensions, $I_{j,k} < I_{j,k+1}$, so there is an order relation between all the N partition elements, and we can enumerate them with a single index $i \in \{1, \dots, |\iota|\}$:

$$\iota = \{I_i : 1 \le i \le |\iota|\}, \quad I_i < I_{i+1}$$

(i.e., points in I_i are smaller than points in I_{i+1}).

Below we shall consider refinements of product and general partitions. As usual we write $\alpha \le \beta$ to mean that the partition β is a *refinement* of the partition α (of $(D, \mathcal{B} \cap D)$ or of any other measurable space for that matter), meaning that the elements of α are unions of the elements of β. By an *increasing sequence of partitions* we mean therefore a sequence of partitions, $(\alpha_n)_{n \in \mathbb{N}}$, such that $\alpha_n \le \alpha_{n+1}$ for all n. If, as in the present case, the state space is a product space, then by a *product refinement* of partition (6.12) we mean any product partition of I obtained by subdividing some or all of the intervals $\{I_{1,k}, \dots, I_{N_k,k}\}$, $1 \le k \le q$.

Furthermore, let κ be the partition of D defined as

$$\kappa = \iota \cap D = \{I_i \cap D \ne \emptyset : 1 \le i \le |\iota|\} = \{K_j : 1 \le j \le |\kappa|\}.$$

In words, κ consists of all subintervals $I_i \in \iota$ contained in the interior of D, together with the overlaps with D of those I_i that intersect the boundary of D. Partitions κ of the form $\kappa = \iota \cap D$, where ι is a product partition and D a simple domain, will be called *quasi-product partitions*; if, moreover, ι is a box (i.e., uniform) partition, κ will be called a *quasi-box partition*. For simplicity, we set $\|\kappa\| = \|\iota\|$.

Next let $\mathbf{X}^\kappa = \{X_n^\kappa\}_{n \in \mathbb{N}_0}$ be the symbolic dynamics associated with $f:D \to D$ with respect to the partition κ:

$$X_n^\kappa(x) = j \quad \text{if} \quad f^n(x) \in K_j, \, n = 0, 1, \dots .$$

Hence $\mathbf{X}^\kappa$ is a stationary, $|\kappa|$-state random process on $(D, \mathcal{B} \cap D, \mu)$ with alphabet $S^\kappa = \{1, \dots, |\kappa|\}$.

Example 12 If $I = [0, 1]$ and $\kappa = \{K_j : 1 \le j \le 10^k\}$, with $K_j = [(j-1)10^{-k}, j10^{-k})$ for $1 \le j \le 10^k - 1$ and $K_{10^k} = [1 - 10^{-k}, 1]$, then $\mathbf{X}^\kappa$ can be written as follows: $X_n^\kappa(x) = \lfloor f^n(x) \cdot 10^k \rfloor + 1$ for $0 \le x < 1$ and $X_n^\kappa(1) = 10^k$.

According to (B.16) (with $\alpha = \kappa$), the entropy of the symbolic dynamics $\mathbf{X}^\kappa$ equals the metric entropy of f with respect to κ:

$$h_\mu(f, \kappa) = h_\mu(\mathbf{X}^\kappa). \tag{6.15}$$

If we take now an increasing sequence of product refinements $\kappa \equiv \kappa_0 \leq \kappa_1 \leq \cdots$ such that $\|\kappa_n\| \to 0$, then we deduce from Theorem 25 that $h_\mu(f) = \lim_{n\to\infty} h_\mu(\mathbf{X}^{\kappa_n})$. This suggests to define the metric permutation of f as $h_\mu^*(f) = \lim_{n\to\infty} h_\mu(\mathbf{X}^{\kappa_n})$. The fact that the limit $n \to \infty$ proceeds by successive refinements of κ_0 and the way product partitions are being numbered guarantees that the order relations are preserved. This means, in particular, that if $X_k^{\kappa_n}(x) = i < j = X_{k+1}^{\kappa_n}(x)$ $(1 \leq i, j \leq |\kappa_n|)$, then $X_k^{\kappa_{n+1}}(x) = i' < j' = X_{k+1}^{\kappa_{n+1}}(x)$ $(1 \leq i', j' \leq |\kappa_{n+1}|)$ for all $x \in D$ and $k \in \mathbb{N}_0$. Thus $h_\mu^*(f)$ has a good chance to exist.

Definition 3 Given a measure-preserving dynamical system $(D, \mathcal{B} \cap D, \mu, f)$, and a lexicographically ordered, quasi-product partition κ_0 of $(D, \mathcal{B} \cap D, \mu)$, the metric permutation entropy of f with respect to the measure μ is defined by

$$h_\mu^*(f) = \lim_{n\to\infty} h_\mu^*(\mathbf{X}^{\kappa_n}) \tag{6.16}$$

(provided the limit exists), where $(\kappa_n)_{n\in\mathbb{N}}$ is a sequence of successive product refinements of κ_0 such that $\|\kappa_n\| \to 0$ and $\mathbf{X}^{\kappa_n}$ is the symbolic dynamics of f with respect to κ_n.

It is plain that this definition is independent from the auxiliary interval $I \supset D$ used to construct κ_0 and also independent from the particular collection of product refinements κ_n used, as long as $\|\kappa_n\| \to 0$. This being the case, we may take quasi-box partitions in (6.16).

One practical reason for using product partitions is that they make numerical calculations much easier. But most importantly, we claim that $\lim_{\|\alpha_n\|\to 0} h_\mu^*(\mathbf{X}^{\alpha_n})$ does not depend on the particular increasing sequence $(\alpha_n)_{n\in\mathbb{N}_0}$ of successive refinements of a general finite partition α_0 of $(D, \mathcal{B} \cap D, \mu)$, as long as (i) they converge to the point partition of D, $\epsilon = \{\{x\} : x \in D\}$, and (ii) the numbering of the elements of $\alpha_1, \alpha_2, \ldots$ preserves the order relations through the process of refinement. Condition (i) requires that α_n consists of connected sets for all n and $\lim_{n\to\infty} \|A\| = 0$ for all $A \in \alpha_n$. Condition (ii) means that if $A_i, A_j \in \alpha_n$ and $i < j$, then $i' < j'$ whenever $A_i \supset A_{i'}' \in \alpha_{n+1}$ and $A_j \supset A_{j'}' \in \alpha_{n+1}$ (this is automatically satisfied by the lexicographically ordered, product refinements ι_n).

Lemma 9 *Let $(D, \mathcal{B} \cap D, \mu, f)$ be a measure-preserving dynamical system, α_0 a finite partition of $(D, \mathcal{B} \cap D, \mu)$, and $(\alpha_n)_{n\in\mathbb{N}}$ a sequence of successive refinements of α_0 preserving the order relations and converging to the point partition. Then*

$$h_\mu^*(f) = \lim_{n\to\infty} h_\mu^*(\mathbf{X}^{\alpha_n}),$$

where $\mathbf{X}^{\alpha_n}$ is the symbolic dynamics of f with respect to the partition α_n.

Proof Roughly speaking, the increasing sequences $\cdots \leq \kappa_n \leq \kappa_{n+1} \leq \cdots$ and $\cdots \leq \alpha_n \leq \alpha_{n+1} \leq \cdots$ are equivalent in the sense that, given κ_n there is a partition α_m with $\|\alpha_m\| \lesssim \|\kappa_n\|$ which can resolve the orbits of f with the same precision as κ_n does—and reciprocally. Of course, the ordinal patterns of length $L = 2, 3, \ldots$ of a given orbit will be, in general, different, depending on the partitions used. Nevertheless, there will be a one-to-one relation between the ordinal L-patterns realized by $\mathbf{X}^{\alpha_n}$ and $\mathbf{X}^{\kappa_n}$ in the limit $n \to \infty$, and the same holds for the corresponding probabilities. Therefore,

$$\lim_{n\to\infty} h_\mu^*(\mathbf{X}^{\alpha_n}) = \lim_{n\to\infty} h_\mu^*(\mathbf{X}^{\kappa_n}) = h_\mu^*(f).$$

$\square$

The partitions $\mathcal{P}_L$, Eq. (3.5) build a sequence of successive refinements, but they do not preserve in general the order relations because their elements eventually decompose into different components. For the same reason, they cannot converge in general to the partition of D into separate points, ϵ, nor are their norms otherwise expected to vanish as $L \to \infty$.

Having shown that the metric permutation entropy does not depend on the partitions used in its calculation (with the provisos stated in Lemma 9), we turn to the main result of this chapter.

Theorem 10 *Let $f:D \to D$ be ergodic with respect to the measure μ, and suppose that $h_\mu^*(f)$ exists. Then $h_\mu^*(f) = h_\mu(f)$.*

Proof Let κ_0 be a quasi-box partition of $(D, \mathcal{B} \cap D, \mu)$ and $(\kappa_n)_{n\in\mathbb{N}}$ a sequence of successive product refinements of κ_0. Then,

$$h_\mu(f, \kappa_n) = h_\mu(\mathbf{X}^{\kappa_n})$$

by (6.15), where $\mathbf{X}^{\kappa_n} = \{X_k^{\kappa_n}\}_{k\in\mathbb{N}_0}$ is the symbolic dynamics of f with respect to the partition κ_n. Furthermore, $h_\mu(\mathbf{X}^{\kappa_n}) = h_\mu^*(\mathbf{X}^{\kappa_n})$ by Theorem 8, since $\mathbf{X}^\kappa$ is ergodic with respect to the measure μ if f is ergodic with respect to μ. Putting together, we have so far

$$h_\mu^*(f) = \lim_{n\to\infty} h_\mu^*(\mathbf{X}^{\kappa_n}) = \lim_{n\to\infty} h_\mu(\mathbf{X}^{\kappa_n}) = \lim_{n\to\infty} h_\mu(f, \kappa_n).$$

From Theorem 25 (Annex B) it follows then

$$\lim_{n\to\infty} h_\mu(f, \kappa_n) = h_\mu(f)$$

and we are done.

$\square$

If instead of Theorem 8, we use Corollary 5 in the previous proof for every process $\mathbf{X}^\kappa$, we conclude also $h_\mu^*(f) = h_\mu(f)$ for μ-preserving maps. This requires the technical assumption that $h_{m_s}(\mathbf{X}^\kappa)$ is m-integrable, where $\{m_s : s \in S^{\mathbb{N}_0}\}$, $S =$

$\{1, \ldots, |\kappa|\}$, is the ergodic decomposition of m, and m the shift-invariant measure of the sequence space model $(S^{\mathbb{N}_0}, \mathcal{B}_\Pi(S), m, \Sigma)$ of $\mathbf{X}^\kappa$—and this for every partition κ.

Theorem 11 *Let $f: D \to D$ be μ-preserving, and suppose that $h_\mu^*(f) = \lim_{n \to \infty} h_\mu^*$ $(\mathbf{X}^{\kappa_n})$ exists. Under the assumptions of Lemma 8 for each $\mathbf{X}^{\kappa_n}$, the equality $h_\mu^*(f) = h_\mu(f)$ holds.*

6.3 On the Definition of Metric Permutation Entropy for Maps

The original definition of permutation entropy by Bandt, Keller, and Pompe [29] was presented in Sect. 1.2. Recall that it involves closed *one-dimensional* intervals I, maps $f: I \to I$, and sets of the form

$$P_\pi = \left\{ x \in I : f^{\pi_0}(x) < f^{\pi_1}(x) < \cdots < f^{\pi_{L-1}}(x) \right\},$$

where $\pi = \langle \pi_0, \ldots, \pi_{L-1} \rangle \in \mathcal{S}_L, L \geq 2$. Recall once again that

$$\mathcal{P}_L = \{ P_\pi \neq \emptyset : \pi \in \mathcal{S}_L \}.$$

In most situations of interest, $\mathcal{P}_L$ will be a partition of $(I, \mathcal{B} \cap I, \mu)$, where $\mathcal{B}$ is the Borel sigma-algebra of $\mathbb{R}$ and μ is an f-invariant measure. This is going to be our setting throughout this section.

Bandt, Keller, and Pompe define then the metric permutation entropy of order L as[2]

$$h_\mu^{*\mathrm{BKP}}(f, L) = -\frac{1}{L-1} \sum_{\pi \in \mathcal{S}_L} \mu(P_\pi) \log \mu(P_\pi) \tag{6.17}$$

and the permutation entropy of f to be

$$h_\mu^{*\mathrm{BKP}}(f) = \lim_{L \to \infty} h_\mu^{*\mathrm{BKP}}(f, L), \tag{6.18}$$

provided the limit exists.

As compared to conventional entropy, $h_\mu^{*\mathrm{BKP}}(f)$ has at least one remarkable feature: it involves only one infinite limit over the length of the word, while $h_\mu(f)$ involves additionally a second infinite process, namely, a supremum over partitions—unless a generating partition is known. This fact can be rephrased by saying that the sequence $\mathcal{P}_L$ builds a "generator" for $h_\mu^{*\mathrm{BKP}}$.

Let us highlight at this point the main result concerning $h_\mu^{*\mathrm{BKP}}(f)$:

Theorem 12 [29] *If $f: I \to I$ is piecewise monotone, then $h_\mu^{*\mathrm{BKP}}(f) = h_\mu(f)$.*

---

[2] Bandt, Keller, and Pompe chose the factor $1/(L-1)$ instead of $1/L$ (see (1.30)) because $\pi(x_0^0)$ contributes nothing to the entropy. Of course, either choice yields the same limit when $L \to \infty$.

Example 13 For the symmetric tent map (1.17), the elements of $\mathcal{P}_2$ are

$$P_{\langle 0,1\rangle} = (0, \tfrac{2}{3}), \qquad P_{\langle 1,0\rangle} = (\tfrac{2}{3}, 1);$$

the elements of $\mathcal{P}_3$ are

$$P_{\langle 0,1,2\rangle} = (0, \tfrac{1}{3}), \qquad P_{\langle 0,2,1\rangle} = (\tfrac{1}{3}, \tfrac{2}{5}), \qquad P_{\langle 2,0,1\rangle} = (\tfrac{2}{5}, \tfrac{2}{3}),$$

$$P_{\langle 1,0,2\rangle} = (\tfrac{2}{3}, \tfrac{4}{5}), \qquad P_{\langle 1,2,0\rangle} = (\tfrac{4}{5}, 1);$$

and the elements of $\mathcal{P}_4$ are

$$P_{\langle 0,1,2,3\rangle} = (0, \tfrac{1}{6}), \quad P_{\langle 0,1,3,2\rangle} = (\tfrac{1}{6}, \tfrac{1}{5}), \quad P_{\langle 0,3,1,2\rangle} = (\tfrac{1}{5}, \tfrac{2}{9}) \cup (\tfrac{2}{7}, \tfrac{1}{3}),$$

$$P_{\langle 3,0,1,2\rangle} = (\tfrac{2}{9}, \tfrac{2}{7}), \quad P_{\langle 0,2,1,3\rangle} = (\tfrac{1}{3}, \tfrac{2}{5}), \quad P_{\langle 2,0,3,1\rangle} = (\tfrac{2}{5}, \tfrac{4}{9}) \cup (\tfrac{4}{7}, \tfrac{3}{5}),$$

$$P_{\langle 2,3,0,1\rangle} = (\tfrac{4}{9}, \tfrac{4}{7}), \quad P_{\langle 2,0,1,3\rangle} = (\tfrac{3}{5}, \tfrac{2}{3}), \quad P_{\langle 3,1,0,2\rangle} = (\tfrac{2}{3}, \tfrac{4}{5}),$$

$$P_{\langle 1,3,2,0\rangle} = (\tfrac{4}{5}, \tfrac{5}{6}), \quad P_{\langle 1,2,0,3\rangle} = (\tfrac{6}{7}, \tfrac{8}{9}), \quad P_{\langle 1,2,3,0\rangle} = (\tfrac{5}{6}, \tfrac{6}{7}) \cup (\tfrac{8}{9}, 1).$$

See Fig. 6.1 and compare with Fig. 1.7; owing to the order isomorphy of the symmetric tent map and the logistic map, there is a one-to-one relation between their admissible ordinal L-patterns. Computation of the metric permutation entropies of orders 2, 3, and 4 of the symmetric tent map Λ (the invariant measure μ is here the Lebesgue measure) yields the following results:

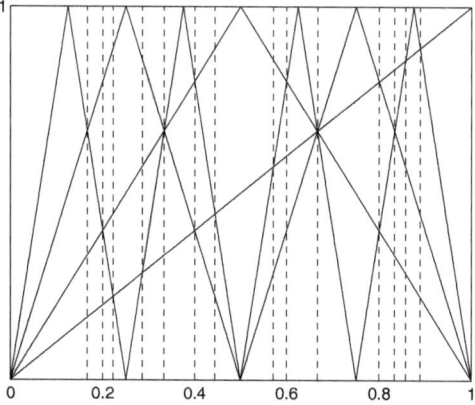

Fig. 6.1 Graphs of the identity, Λ, Λ^2, and Λ^3. The *vertical, dashed lines* separate different P_π, $\pi \in \mathcal{S}_4$

$$h_\mu^{*\text{BKP}}(\Lambda, 2) = \tfrac{2}{3} \log \tfrac{3}{2} + \tfrac{1}{3} \log 3 = 0.9183 \text{ bit/symbol},$$

$$h_\mu^{*\text{BKP}}(\Lambda, 3) = 1.0746 \text{ bit/symbol},$$

$$h_\mu^{*\text{BKP}}(\Lambda, 4) = 1.1807 \text{ bit/symbol}.$$

By Theorem 12,

$$h_\mu^{*\text{BKP}}(\Lambda) = h_\mu(\Lambda) = \log 2 = 1 \text{ bit/symbol}.$$

But in the case of general maps, it seems that only inequality (6.19) below (formally similar to (6.9)) can be proved. Comparing such one-dimensional results with the dimensional generality of Theorem 10, we may conclude that the definition (6.16) of permutation entropy offers some advantages.

Note that the central distinction, which makes formulation (6.16) easier and more natural, is that (6.16) takes the limit of infinite long conditioning ($L \to \infty$) first and the discretization limit ($\|\kappa_n\| \to 0$) last, similar to Kolmogorov–Sinai entropy, and as opposed to (6.18), where an explicit discretization is not taken. Thus we have two limits to take (while $h_\mu^{*\text{BKP}}(f)$ involves only one limit), but the second, $\|\kappa_n\| \to 0$, is harmless and, in principle, can be numerically approximated. We conjecture that for "non-pathological" dynamical systems of the sort one might observe in nature, the two formulations are equivalent, but there are likely to be some non-trivial technicalities involved in a rigorous analysis. More on this, in the next chapter.

Transformations with an infinite number of monotonicity segments are not unusual in ergodic theory.

Example 14 The *Gauss transformation, $f.[0, 1) \twoheadrightarrow [0, 1)$* with

$$f(x) = \begin{cases} 0 & \text{if } x = 0 \\ \frac{1}{x} \pmod 1 & \text{if } x \neq 0 \end{cases},$$

is an ergodic map [52, Chap. 5] with infinitely many monotonicity segments, see Fig. 6.2.

The next theorem shows that, in general, $h_\mu^{*\text{BKP}}(f)$ can only be expected to be an upper bound of $h_\mu(f)$.

Theorem 13 [29] *If $f{:}I \to I$ is a μ-preserving map with $h_\mu(f) < \infty$, then*

$$\liminf_{L \to \infty} h_\mu^{*\text{BKP}}(f, L) \geq h_\mu(f). \tag{6.19}$$

It follows $h_\mu^{*\text{BKP}}(f) \geq h_\mu(f)$, provided $h_\mu^{*\text{BKP}}(f)$ exists.

Proof Let $\iota = \{I_j, 1 \leq j \leq |\iota|\}$ be a partition of $(I, \mathcal{B} \cap I, \mu)$, where $I_j \subset I$ are intervals. This being the case, let $c_1 < c_2 < \cdots < c_{|\iota|-1}$ be the points that subdivide the interval $I = [a, b]$ into the $|\iota|$ intervals I_j of the partition ι. We consider a fixed

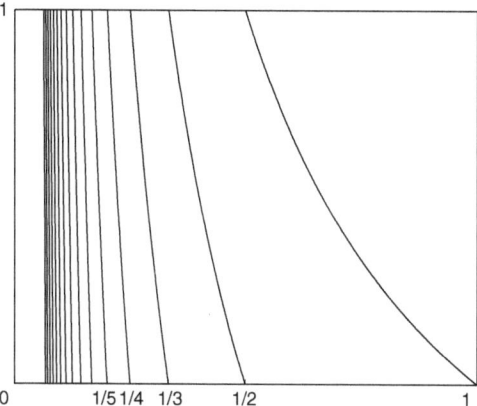

Fig. 6.2 Some monotony intervals of the Gauss transformation

$P_\pi \in \mathcal{P}_L$ and show that it can intersect at most $(L+1)^{|\iota|-1}$ sets of the partition $\iota_0^{L-1} := \vee_{i=0}^{L-1} f^{-i}(I_{j_i})$ with $I_{j_0}, \dots, I_{j_{L-1}} \in \iota$. For $x \in P_\pi$, let $\Delta_L[x]$ denote the set in ι_0^{L-1} that contains x. Thus, $\Delta_L[x]$ can be written as $I_{j_0} \cap f^{-1}(I_{j_1}) \cap \cdots \cap f^{-(L-1)}(I_{j_{L-1}})$ with $I_{j_0}, \dots, I_{j_{L-1}} \in \iota$, so that it can be specified by the n-tupel $j[x] = (j_0, \dots, j_{L-1}) \in \{1, \dots, |\iota|\}^L$.

Now, π is given by inequalities $x_{k_1} < \cdots < x_{k_L}$ with $\{k_1, \dots, k_L\} = \{0, \dots, L-1\}$ and $x_k = f^k(x)$. For each $x \in P_\pi$ we can extend these inequalities so that they give the common order of the c_r and the x_{k_l}, where $1 \le r \le |\iota| - 1$ and $1 \le l \le L$. It follows that there are at most $(L+1)^{|\iota|-1}$ possible extended orders since each c_r has $L+1$ possible bins to go among the x_{k_l}. Moreover, when we know the common order of the c_r and x_{k_l}, then $j[x]$ is uniquely determined (since $c_{j-1} < x_k < c_j$ implies $x_k \in I_j$ and thus $x \in f^{-k}(I_j)$, with $1 \le j \le |\iota|$, $c_0 = a$, and $c_{|\iota|} = b$).

Each $P_\pi \in \mathcal{P}_L$ is then the union of at most $(L+1)^{|\iota|-1}$ sets $V_k \in \iota_0^{L-1} \vee \mathcal{P}_L$ with total measure $\mu(P_\pi)$. Hence,

$$- \sum_{k=1}^{(L+1)^{|\iota|-1}} \mu(V_k) \log \mu(V_k)$$

$$\le - \sum_{k=1}^{(L+1)^{|\iota|-1}} \frac{\mu(P_\pi)}{(L+1)^{|\iota|-1}} \log \frac{\mu(P_\pi)}{(L+1)^{|\iota|-1}}$$

$$= -\mu(P_\pi) \log \mu(P_\pi) + (|\iota| - 1)\mu(P_\pi) \log (L+1)$$

and summing over all $\pi \in \mathcal{S}_L$,

$$H_\mu(\iota_0^{L-1}) \le H_\mu(\iota_0^{L-1} \vee \mathcal{P}_L) \le H_\mu(\mathcal{P}_L) + (|\iota| - 1) \log (L+1). \qquad (6.20)$$

It follows that

$$\frac{1}{L-1}H_\mu(\mathcal{P}_L) \geq \frac{1}{L-1}\left[H_\mu(\iota_0^{L-1}) - (|\iota| - 1)\log(L+1)\right]$$

and

$$\liminf_{L\to\infty}\frac{1}{L-1}H_\mu(\mathcal{P}_L) \geq \liminf_{L\to\infty}\frac{1}{L-1}H_\mu(\iota_0^{L-1}), \qquad (6.21)$$

since $\frac{1}{L-1}\log(L+1) \to 0$ as $L \to \infty$.

On the other hand, the sequence $\frac{1}{L-1}H_\mu(\iota_0^{L-1})$ converges to $h_\mu(f,\iota)$ when $L \to \infty$, hence

$$\liminf_{L\to\infty} h_\mu^{*\mathrm{BKP}}(f,L) = \liminf_{L\to\infty}\frac{1}{L-1}H_\mu(\mathcal{P}_L) \geq h_\mu(f,\iota),$$

for any partition ι. Finally,

$$\liminf_{L\to\infty} h_\mu^{*\mathrm{BKP}}(f,L) \geq \sup_\iota h_\mu(f,\iota) = h_\mu(f).$$

$\square$

6.4 Numerical Issues

Our way to the metric permutation entropy of maps was paved by partitions of the state space and the corresponding symbolic dynamics, very much the same way as it happens with the Kolmogorov–Sinai entropy. Therefore, calculating the metric permutation entropy of maps and information sources turns out to be essentially the same task, except for the fact that in the first case this calculation has, in principle, to be repeated with ever finer partitions. In practice, one estimates the true value of the permutation entropy by taking a "sufficiently" fine partition once and for all. This corresponds, by the way, to the numerical practice, as we shall presently explain. If, furthermore, the map (and hence the ensuing source) is ergodic, then it suffices to consider one or a small sample of coarse-grained orbits.

As a by-product of the previous results on metric permutation entropy, the practitioner of time-series analysis will find an alternative way to envision or, eventually, numerically estimate the Kolmogorov–Sinai entropy of real sources. It is worth reminding (see Chap. 1) that the entropy of information sources can be measured by a variety of techniques that go beyond counting word statistics and comprise different definitions of "complexities" such as, for example, counting the patterns along a digital (or digitalized) data sequence [137, 211, 6]. Bandt and Pompe refer in [28] to the permutation entropy of time series as complexity. That the entropy can also be computed by counting ordinal patterns shows once again that it is a so general concept that can be captured with different and seemingly blunt approaches.

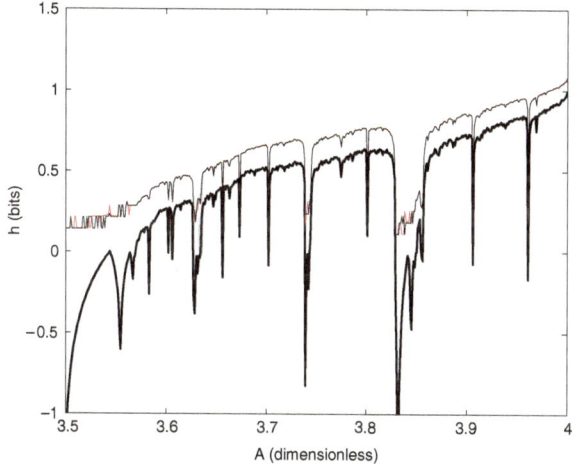

Fig. 6.3 Lyapunov exponent (*black thick line*) of the logistic map g_A, $3.5 \leq A \leq 4$, and metric permutation entropy (rate) estimates $\hat{h} = h^*(X_0^{13})$ in bits/symbol for $N = 10^6$ length time series from the map (*black thin lines*). The metric permutation entropy estimate tracks changes in the Lyapunov exponent well, with a nearly constant bias. Periodic orbits give a finite permutation entropy, but the rate estimate would tend to zero given a sufficiently long word

We demonstrate numerical results on time series $x_{n+1} = g_A(x_n)$ from the logistic map $g_A(x) = Ax(1 - x)$, where $0 \leq A \leq 4$ and $0 \leq x \leq 1$. Figure 6.3 shows an estimate of the metric permutation entropy on noise-free data as a function of A, comparing the Lyapunov exponent $L_\mu(g_A)$ (computed from the orbit knowing the equation of motion) to the metric permutation entropy of g_A for $3.5 \leq A \leq 4$. To be precise, we are estimating $h_\mu^*(\mathbf{X})$ with $\mathbf{X}$ discretized from the logistic map iterated at the discretization of double-precision numerical representation, i.e., $\mathbf{X}$ is the output of a standard numerical iteration and μ is the natural invariant measure with density $d\mu/dx = \frac{1}{\pi\sqrt{x(1-x)}}$. The entropy estimator of the block ranks was the plug-in estimator (substituting observed frequencies for probabilities) plus the classical bias correction, first order in $1/N$, N being here the number of samples (which can be taken, for instance, from sliding windows of fixed length L along the orbit/orbits considered) [167]. Let us remind that

$$h_\mu(g_4) = L_\mu(g_4) = \int_0^1 \log\left|g_4'(x)\right| d\mu(x) = \log 2.$$

Thus, in practice the BKP approach (Sect. 6.3) and our approach (Sect. 6.2) boil down to the same recipe: generate orbits and count ordinal patterns in sliding windows of increasing sizes; for more details, see Chap. 9. The most intriguing characteristic of order relations is that they define, on their own, partitions $\mathcal{P}_L$ for the mapping from continuous values (as the discretization level $\|\kappa_n\|$ goes to zero) to a lower precision symbolic representation which has the natural structure for entropy. When estimating entropy from the discrete information source induced from a *fixed* discretization, the entropy of the symbol stream will not generally equal the Kolmogorov–Sinai entropy unless a generating partition is used, and that can be difficult to find, especially for observed data alone, although some recent works show progress in this direction (e.g., [40] and references therein). The "magic" in using ordinal patterns is that the self-defined partitions $\mathcal{P}_L$ give the Kolmogorov–Sinai entropy, at least asymptotically. Permutation entropy may offer a significant opportunity to advance analytical computations of entropies for various dynamical systems, where generating partitions might be too difficult to find rigorously.

It turns out that using metric permutation entropy to accurately estimate the Kolmogorov–Sinai entropy is more difficult than using it as a very rapid and easy-to-compute *relative* quantification of entropy or complexity which can be computed without requiring a fixed partition (see, e.g., [45]). The key issue in using permutation entropy for empirical data analysis as an entropy estimator is the same as with standard Shannon entropy estimation: balancing the tension between larger word lengths L, to capture more dependencies, and the loss of sufficient sampling for good statistics in the ever larger discrete space. Extracting permutation entropies is rapid and easy—but taking the limits is not at all simple numerically. The finite L performance and convergence rate and bias of any specific computational method are major issues when it comes to accurately estimating the entropy of a source from observed data. It is now appreciated that numerically estimating the Shannon block entropy from finite data and, especially, the asymptotic entropy can be surprisingly tricky [195, 127, 6, 121, 122]. The theoretical definitions of entropy do not necessarily lead to good statistical methods, and superior alternatives have been developed over the many years since Shannon. We believe that some of these ideas may similarly be applicable to the permutation entropy situation, either in terms of using some of the superior entropy estimation methods for block entropies or developing algorithms based on more sophisticated data compression principles to extract the entropy itself.

Also important for practical time-series analysis is the usual situation where observations of a predominantly deterministic source is contaminated with a small level of observational noise. Here, we recommend that the user *fix* some discretization level $\|\kappa_n\|$ characteristic of the noise and evaluate the permutation entropies via entropies of rank words evaluated from the discretized observables.

In regard to vector-valued sources, we used (without restriction) lexicographic ordering in the theoretical part because of definiteness and simplicity. For analyzing chaotic observed data, however, it may be acceptable to still use but one scalar projection subject to the traditional caveats of time-delay embedology. We would expect that for appropriately mixing sources and generic observation functions, the Kolmogorov–Sinai entropy estimated through that scalar still equals the true value, and likewise so might permutation entropy. We have found that numerically this appears to work in practice. Moreover, the lexicographic ordering will effectively reduce to this case anyway except for the few cases where the symbols on the dominant coordinate match, which will be less frequent as L increases. More on this in Chaps. 7 and 9.

Chapter 7
Topological Permutation Entropy

Permutation entropy, as conventional entropy, comes in the metric version (Chap. 6) and in the topological version (this chapter). Topological permutation entropy was also introduced by Bandt et al. [29], together with metric permutation entropy. Let us stress once more that the concept of metric permutation entropy of a map introduced in the last chapter differs from the original one, the difference consisting basically in the order of an iterated limit (first the length of the orbit, then the precision of the measurement, as in the definition of the Kolmogorov–Sinai entropy). This technical change made possible to generalize one of the main results of [29], namely, the equality of metric entropy and metric permutation entropy for piecewise monotone maps on one-dimensional intervals to higher dimensions at the expense of requiring ergodicity (Theorem 10).

In this chapter we will apply the same approach to topological entropy with the parallel result that the equality of topological entropy and topological permutation entropy for piecewise monotone maps on one-dimensional intervals (the other main result of [29]) can also be generalized to higher dimensions, this time requiring the map to be expansive (Theorem 15). The possibility of going higher dimensional is an advantage of the definitions of metric and topological permutation entropies used in this book.

7.1 Topological Permutation Entropy of Sources

Let $\mathbf{X} = \{X_n\}_{n \in \mathbb{N}_0}$ be an information source with finite alphabet S. We define the *topological entropy of order L of $\mathbf{X}$* as

$$h_{\text{top}}(X_0^{L-1}) = \frac{1}{L} \log N(\mathbf{X}, L), \tag{7.1}$$

where X_0^{L-1} is shorthand for the block of random variables $X_0, \ldots, X_{L-1}$ and $N(\mathbf{X}, L)$ is the number of sequences (words, blocks, etc.) of length L, $x_0^{L-1} = x_0, \ldots, x_{L-1}$, that $\mathbf{X}$ can output. Put in a different way, $N(\mathbf{X}, L)$ is the number of words of length L, built by consecutive letters, that are *allowed* or *admissible* in

J.M. Amigó, *Permutation Complexity in Dynamical Systems*,
Springer Series in Synergetics, DOI 10.1007/978-3-642-04084-9_7,
© Springer-Verlag Berlin Heidelberg 2010

the messages of $\mathbf{X}$ (since $\mathbf{X}$ is stationary, we may restrict ourselves to an initial segment). The *topological entropy* of $\mathbf{X}$ is then defined as

$$h_{\text{top}}(\mathbf{X}) = \lim_{L\to\infty} h_{\text{top}}(X_0^{L-1}), \tag{7.2}$$

provided the limit exists. In an information-theoretical framework, $h_{\text{top}}(\mathbf{X})$ is called the *capacity* of $\mathbf{X}$ [186]. If, furthermore,

$$h_\mu(X_0^{L-1}) = -\frac{1}{L} \sum_{x_0,\dots,x_{L-1}\in S} p(x_0,\dots,x_{L-1}) \log p(x_0,\dots,x_{L-1}) \tag{7.3}$$

is the *Shannon* (or *metric*) *entropy of order L* of $\mathbf{X}$, then clearly $h_\mu(X_0^{L-1}) \le h_{\text{top}}(X_0^{L-1})$ (for any logarithm base > 1). Therefore

$$h_\mu(\mathbf{X}) = \lim_{L\to\infty} h_\mu(X_0^{L-1}) \le h_{\text{top}}(\mathbf{X}), \tag{7.4}$$

where $h_\mu(\mathbf{X})$ is the *Shannon* (or *metric*) *entropy* of $\mathbf{X}$. Also

$$h_\mu(\mathbf{X}) = h_{\text{top}}(\mathbf{X}) \Leftrightarrow p(x_0,\dots,x_{L-1}) = \frac{1}{N(\mathbf{X},L)} \quad \forall L \ge 1.$$

Suppose now that the alphabet S of the source $\mathbf{X}$ is endowed with a total ordering $\le$, so that one can also define the corresponding *permutation entropies* of order L via the ordinal patterns realized by the words of finite lengths $L > 2$. Then the topological *permutation* entropy of an information source is defined analogous to the topological entropy, using *rank variables*.

Thus, the *topological permutation entropy* of $\mathbf{X}$, $h_{\text{top}}^*(\mathbf{X})$, is defined as

$$h_{\text{top}}^*(\mathbf{X}) = \lim_{L\to\infty} h_{\text{top}}^*(X_0^{L-1}), \tag{7.5}$$

provided the limit exists, with

$$h_{\text{top}}^*(X_0^{L-1}) \equiv h_{\text{top}}(R_0^{L-1}) = \frac{1}{L} \log N(\mathbf{R}, L). \tag{7.6}$$

Analogous to (7.1), $N(\mathbf{R}, L)$ stands for the number of allowed words of length L of the process $\mathbf{R} = \{R_n\}_{n\in\mathbb{N}_0}$ (see Sect. 6.1). Note that

$$N(\mathbf{R}, L) \le N(\mathbf{X}, L), \tag{7.7}$$

since several finite symbol sequences may produce the same sequence of rank variables (i.e., $x_0^{L-1} \mapsto r_0^{L-1} = \text{rank}(x_0^{L-1})$ is many-to-one).

As in (7.4), the metric permutation entropy,

$$h_\mu^*(\mathbf{X}) = -\lim_{L\to\infty} \frac{1}{L} \sum p(r_0,\ldots,r_{L-1}) \log p(r_0,\ldots,r_{L-1}),$$

is upper bounded by the topological permutation entropy,

$$h_\mu^*(\mathbf{X}) \le h_{\text{top}}^*(\mathbf{X}) \tag{7.8}$$

and, moreover,

$$h_\mu^*(\mathbf{X}) = h_{\text{top}}^*(\mathbf{X}) \Leftrightarrow p(r_0,\ldots,r_{L-1}) = \frac{1}{N(\mathbf{R},L)} \quad \forall L \ge 2.$$

From these definitions and (7.7), it follows that

$$h_{\text{top}}^*(\mathbf{X}) \le h_{\text{top}}(\mathbf{X}). \tag{7.9}$$

Therefore, the topological permutation entropy is always a lower bound of the topological entropy for information sources.

Remark 2 The topological permutation entropy of sources can also be introduced using ordinal patterns instead of rank variables:

$$h_{\text{top}}^*(X_0^{L-1}) = \frac{1}{L} \log N^*(\mathbf{X},L), \tag{7.10}$$

where $N^*(\mathbf{X},L)$ is the number of admissible ordinal L-patterns in the messages produced by $\mathbf{X}$.

7.2 Constrained Sequences

Let $N(\mathbf{X},L)$ be as before the number of allowed sequences of length L of a source $\mathbf{X}$ with finite alphabet. If all possible sequences of length L are allowed, i.e., $N(\mathbf{X},L) = |S|^L$, then

$$h_{\text{top}}(\mathbf{X}) = \lim_{L\to\infty} \frac{1}{L} \log |S|^L = \log |S|.$$

To calculate $h_{\text{top}}^*(\mathbf{X})$ for an unconstrained source $\mathbf{X}$, we assume for simplicity a binary alphabet. Remember from Example 11, that, given the length-L word x_0^{L-1}, $L \ge 1$, then

$$\begin{aligned} x_L = 0 &\Rightarrow r_L = N_0 + 1, \\ x_L = 1 &\Rightarrow r_L = L + 1, \end{aligned}$$

where N_0 is the number of 0's in x_0^{L-1} (remember also that $1 \leq r_L \leq L + 1$). How many distinct ranks of length $L + 1$, r_0^L, can produce a word x_0^L?

The case $r_L = 1$ is only possible if $x_0 = x_1 = \cdots = x_{L-1} = 1$ (i.e., $N_0 = 0$) and $x_L = 0$.

The case $r_L = 2$ requires $N_0 = 1$ and $x_L = 0$. If $x_i = 0, 0 \leq i \leq L - 1$, (otherwise 1), then

$$r_0^L = 1, 2, \ldots, i, 1, i+2, \ldots, L, 2.$$

This case contributes $L = \binom{L}{1}$ distinct rank blocks of length $L + 1$.

The case $r_L = 3$ requires $N_0 = 2$ and $x_L = 0$. If $x_i = x_j = 0, 0 \leq i < j \leq L - 1$, (otherwise 1), then

$$r_0^L = 1, 2, \ldots, i, 1, i+2, \ldots, j, 2, j+2, \ldots, L, 3.$$

This case contributes $\binom{L}{2}$ distinct rank blocks of length $L + 1$.

Proceeding further in this way, we conclude that the case $r_L = k$, $1 \leq k \leq L$ contributes $\binom{L}{k-1}$ distinct rank blocks of length $L + 1$.

Finally, the case $r_L = L + 1$ requires $N_0 = L$ and $x_L = 0$, or $0 \leq N_0 \leq L$ and $x_L = 1$. There are $1 + 2^L$ such cases. Therefore, for $L \geq 1$,

$$N(\mathbf{R}, L + 1) = 1 + \binom{L}{1} + \cdots + \binom{L}{L-1} + 1 + 2^L = 2^{L+1}$$

and

$$h_{\text{top}}^*(\mathbf{X}) = \lim_{L \to \infty} \frac{1}{L+1} \log N(\mathbf{R}, L + 1) = \lim_{L \to \infty} \frac{1}{L+1} \log 2^{L+1}$$
$$= \log |S|.$$

In general, the information source $\mathbf{X}$ has forbidden words. In this case, one speaks also of *constrained sequences* or *constrained sources* [186]. Constrained sequences are very important in information theory, where the constrains are imposed by technological feasibility or convenience. For example, to ensure proper synchronization in magnetic recording, it is often necessary to limit the length of runs of 0's between two 1's when reading and recording bits. Also to reduce intersymbol interference, it may be required at least one 0 between any two 1's [59].

Alternatively, a constrained source can be defined as the set of sequences generated by walks on a labeled, oriented graph G. Formally, an oriented graph G is an ordered pair of sets, $G = (V, E)$, where E is a subset of *ordered* pairs of V. The elements of V are called vertices, and will be denoted as i, j, etc.; the elements $(i, j) \in E$ are called (oriented or directed) edges, with initial vertex i and terminal vertex j, and will denoted by e_{ij}. Without restriction we take $V = \{1, 2, \ldots, |V|\}$. The vertices i of the graph represent "states" and the directed edges e_{ij} show the state transitions allowed to the system. The system outputs the letter attached to each oriented edge when performing the corresponding transition. Depending on how the transition probabilities p_{ij} are defined, we have different kinds of stochastic processes: Markovian, finite type, etc.

Example 15 [59] Suppose that in the example mentioned above, borrowed from magnetic recording, we are required to have at least one 0 and at most two 0's between any pair of 1's in a sequence. The forbidden words are 11 and any word of the form 10...01 containing more than two 0's. Show that the set of constrained sequences is the same as the set of allowed paths on the state diagram in Fig. 7.1.

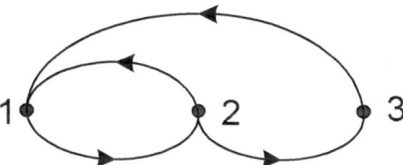

Fig. 7.1 Allowed paths between the nodes 1, 2, and 3

Given an oriented graph G, the *connection matrix* of G is a $|V| \times |V|$ matrix A_G whose entries $(A_G)_{i,j}$, $1 \le i,j \le |V|$, are defined as follows:

$$(A_G)_{i,j} = \begin{cases} 1 & \text{if } (j,i) \in E, \\ 0 & \text{otherwise.} \end{cases}$$

A *path P* of length l is a graph of the form

$$V(P) = \{i_0, i_1, \ldots, i_l\}, \quad E(P) = \{e_{i_0 i_1}, e_{i_1 i_2}, \ldots, e_{i_{l-1} i_l}\}.$$

An oriented graph is irreducible if, given any two vertices, there exists a path from the first vertex to the second. If $N_i(L)$ is the number of valid paths of lengths L ending at node (or state) i and $\mathbf{N}(L)$ is the column vector

$$\mathbf{N}(L) = (N_1(L), N_2(L), \ldots, N_{|V|}(L))^\top,$$

where the upper index $\top$ stands for "transposed," then

$$\mathbf{N}(L) = A_G \mathbf{N}(L-1),$$

and by induction,

$$\mathbf{N}(L) = A_G^{L-1} \mathbf{N}(1),$$

where the entries in A_G^L correspond to paths in G of length L.

By the *Perron–Frobenius theorem* [202] for non-negative matrices, there is an eigenvalue $\lambda \ge 0$ such that no other eigenvalue of A_G has absolute value greater than λ. Corresponding to λ there is a non-negative left (row) eigenvector $\mathbf{u} = (u_1, \ldots, u_{|V|})$ and a non-negative right (column) eigenvector $\mathbf{v} = (u_1, \ldots, u_{|V|})^\top$. Moreover, if A_G is *irreducible* (i.e., for any pair i, j there is some $n > 0$ such that $(A_G^n)_{i,j} > 0$), then $\lambda > 0$ (in fact, $\min_i \sum_{j=1}^{|V|} (A_G)_{i,j} \le \lambda \le \max_i \sum_{j=1}^{|V|} (A_G)_{i,j}$), λ is a simple eigenvalue, and the corresponding eigenvectors are strictly positive (i.e., $u_i > 0$, $v_i > 0$ for all i).

The connection matrix A_G is *irreducible* and *aperiodic* if there exists $n \ge 1$ such that $(A_G^n)_{i,j} > 0$ for all i, j. In this case [202],

$$\lim_{n \to \infty} \frac{1}{\lambda^n} (A_G^n)_{i,j} = u_j v_i = (\mathbf{v} \otimes \mathbf{u})_{i,j},$$

where $\mathbf{v} \otimes \mathbf{u}$ denotes the tensor product of the vectors $\mathbf{v}$ and $\mathbf{u}$. This means that the matrices A_G^n and $\lambda^n (\mathbf{v} \otimes \mathbf{u})$ have the same limit when $n \to \infty$.

Lastly,

$$\lim_{L \to \infty} \frac{1}{L} \log N_i(L)$$

$$= \lim_{L \to \infty} \frac{1}{L} \log \sum_{j=1}^{|V|} (A_G^{L-1})_{i,j} N_j(1)$$

$$= \lim_{L \to \infty} \frac{1}{L} \log \lambda^{L-1} \sum_{j=1}^{|V|} (\mathbf{v} \otimes \mathbf{u})_{i,j} N_j(1)$$

$$= \lim_{L \to \infty} \frac{1}{L} \log \lambda^{L-1} + \lim_{L \to \infty} \frac{1}{L} \log \sum_{j=1}^{|V|} (\mathbf{v} \otimes \mathbf{u})_{i,j} N_j(1)$$

$$= \log \lambda.$$

This shows that the number of allowed sequences of length L grows as λ^L for large L and provides sufficient conditions for the limit $h_{\text{top}}(\mathbf{X})$ to exist.

Proposition 7 [186] *If $\mathbf{X}$ is a constrained source such that the connection matrix A_G of its oriented graph is irreducible and aperiodic, then*

$$h_{top}(\mathbf{X}) = \log \rho(A_G),$$

where $\rho(A_G)$ is the spectral radius of the matrix A_G,

$$\rho(A_G) = \max\{|\lambda| : \lambda \text{ is an eigenvalue of } A_G\}.$$

7.3 Topological Permutation Entropy of Maps

Once more let D be a simple domain of $\mathbb{R}^q$ endowed with the Borel sigma-algebra $\mathcal{B}$, and let f be a map from D to itself. Furthermore, consider a quasi-box partition

$$\kappa_0 = \{K_j : 1 \leq i \leq |\kappa_0|\}, \quad K_j < K_{j+1},$$

of D and an increasing sequence $(\kappa_n)_{n \in \mathbb{N}}$ of refinements of κ_0 with $\|\kappa_n\| \to 0$ (see Sect. 6.2).

Analogous to the definition of the metric permutation entropy of f with respect to an f-invariant measure μ on $(D, \mathcal{B} \cap D)$ (6.16),

$$h_\mu^*(f) = \lim_{n \to \infty} h_\mu^*(\mathbf{X}^{\kappa_n}),$$

where $\mathbf{X}^{\kappa_n}$ is the symbolic dynamics of f with respect to the partition κ_n, we define now the *topological permutation entropy of f* as

$$h_{\text{top}}^*(f) = \lim_{n \to \infty} h_{\text{top}}^*(\mathbf{X}^{\kappa_n}). \tag{7.11}$$

Note that limit (7.11) exists or diverges to $+\infty$, since $h_{\text{top}}^*(\mathbf{X}^{\kappa_n})$ is non-decreasing with ever finer partitions κ_n. Moreover, as shown in the proof of Lemma 9, this limit does not depend on the particular initial partition α_0 and its successive refinements α_n as long as $(\alpha_n)_{n \in \mathbb{N}}$ converges to the partition of D into separated points, and the order relations are preserved when going from α_n to α_{n+1}. This implies the following result.

Theorem 14 *Let D_1, D_2 be two simple domains of $\mathbb{R}^q$, and suppose that the maps $f_i : D_i \to D_i$, $i = 1, 2$, are order isomorphic by means of a homeomorphism $\phi : D_1 \to D_2$. If the topological permutation entropy exists for one of the maps, then it also exists for the other map, and in this case*

$$h_{top}^*(f_1) = h_{top}^*(f_2).$$

Proof Let κ be a quasi-box partition of D_1. Then $\phi(\kappa)$ is a partition of D_2 which, furthermore, generates an increasing sequence of partitions preserving the order relations and converging to the partition of D_2 into separate points as $\|\kappa\| \to 0$.

Let $\mathbf{X}^\kappa$ be the symbolic dynamics of f_1 with respect to the partition $\kappa = \{K_j : 1 \leq j \leq |\kappa|\}$ and $\mathbf{Y}^{\phi(\kappa)}$ be the symbolic dynamics of f_2 with respect to the partition $\phi(\kappa) = \{\phi(K_j) : 1 \leq j \leq |\kappa|\}$. Then

$$
\begin{aligned}
X_n^\kappa(x) = j &\Leftrightarrow f_1^n(x) \in K_j \\
&\Leftrightarrow \phi^{-1} \circ f_2^n \circ \phi(x) \in K_j \\
&\Leftrightarrow f_2^n \circ \phi(x) \in \phi(K_j) \\
&\Leftrightarrow Y^{\phi(\kappa)}(\phi(x)) = j.
\end{aligned}
$$

It follows that $\mathbf{X}^\kappa$ and $\mathbf{Y}^{\phi(\kappa)}$ have the same admissible ordinal patterns of any length, hence

$$h^*_{\text{top}}(f_1) = \lim_{\|\kappa\| \to 0} h^*_{\text{top}}(\mathbf{X}^\kappa) = \lim_{\|\phi(\kappa)\| \to 0} h^*_{\text{top}}(\mathbf{Y}^{\phi(\kappa)}) = h^*_{\text{top}}(f_2). \ \square$$

Note for further reference that (7.8) implies

$$h^*_\mu(f) \le h^*_{\text{top}}(f). \tag{7.12}$$

Therefore, the topological permutation entropy is always an upper bound of the topological entropy for maps, as it happens with the conventional metric and topological entropies.

Since the (conventional) topological entropy is usually defined for continuous maps (see Sect. B.3.1), we shall assume continuity in the following propositions. In dimension 1, continuity may be replaced by piecewise monotonicity.

Lemma 10 *Let $f : D \to D$ be a continuous map. Then*

$$h_{\text{top}}(f) \le h^*_{\text{top}}(f). \tag{7.13}$$

Proof From Theorem 10, $h_\mu(f) = h^*_\mu(f)$ holds for all $\mu \in E(D,f)$, the set of f-invariant, ergodic measures on $(D, \mathcal{B} \cap D)$. Thus, in virtue of the variational principle (B.27),

$$h_{\text{top}}(f) = \sup_{\mu \in E(D,f)} h^*_\mu(f) \le h^*_{\text{top}}(f), \tag{7.14}$$

where the last inequality follows from (7.12). $\square$

Observe from (7.14) that if a variational principle like (B.27) would also hold for the metric and topological permutation entropies, that is,

$$\sup_{\mu \in E(D,f)} h^*_\mu(f) = h^*_{\text{top}}(f), \tag{7.15}$$

then $h_{\text{top}}(f) = h^*_{\text{top}}(f)$ would follow.

Proposition 8 *Let $f : D \to D$ be a continuous map. Then the variational principle (7.15) holds if and only if $h_{\text{top}}(f) = h^*_{\text{top}}(f)$.*

Another equivalent condition for the variational principle (7.15) to hold follows from the inequality (7.9) applied to the sources $\mathbf{X}^{\kappa_n}$ in (7.11):

$$h^*_{\text{top}}(\mathbf{X}^{\kappa_n}) \le h_{\text{top}}(\mathbf{X}^{\kappa_n}).$$

Letting $n \to \infty$, we conclude

$$h^*_{\text{top}}(f) \leq \lim_{n \to \infty} h_{\text{top}}(\mathbf{X}^{\kappa_n}), \tag{7.16}$$

provided $\lim_{n \to \infty} h^*_{\text{top}}(\mathbf{X}^{\kappa_n})$ converges.

Proposition 9 *Let $f : D \to D$ be a continuous map. Then the variational principle (7.15) holds if and only if $\lim_{n \to \infty} h_{top}(\mathbf{X}^{\kappa_n}) = h_{top}(f)$.*

Proof If $\lim_{n \to \infty} h_{\text{top}}(\mathbf{X}^{\kappa_n}) = h_{\text{top}}(f)$, then (7.16) implies $h^*_{\text{top}}(f) \leq h_{\text{top}}(f)$. On the other hand, $h_{\text{top}}(f) \leq h^*_{\text{top}}(f)$ holds true in general (Lemma 10). Apply now Proposition 8. □

7.4 Relation Between Topological Entropy and Topological Permutation Entropy

One of the main interests of $h^*_{\text{top}}(f)$ is that, under some assumptions on f, it coincides with $h_{\text{top}}(f)$, the topological entropy of f, thus eventually providing an estimator of it.

Lemma 11 *Let $D \subset \mathbb{R}^q$, $q \geq 2$, be a simple domain and $f : D \to D$ a positively expansive map. Then*

$$\lim_{n \to \infty} h_{\text{top}}(\mathbf{X}^{\kappa_n}) = h_{\text{top}}(f), \tag{7.17}$$

where $(\kappa_n)_{n \in \mathbb{N}}$ is an increasing sequence of quasi-box partitions of D and $\mathbf{X}^{\kappa_n}$ is the symbolic dynamics of f with respect to κ_n.

Intuitively speaking, a self-map is positively expansive if every pair of sufficiently close points eventually separate by a finite distance under iteration of the map. Expansive and positively expansive maps are defined in Sect. B.3.1, Definition 26. Typical examples of positively expansive maps are the one- and two-sided shifts. The condition $q \geq 2$ recalls that one-dimensional closed intervals do not admit expansive maps. To establish a connection between $h_{\text{top}}(\mathbf{X}^{\kappa_n})$ and $h_{\text{top}}(f)$, we will use (n, ε)-separated sets (Definition 23).

Proof For definiteness we will take the metric d in $\mathbb{R}^q$ to be the Euclidean distance (any other equivalent distance would do as well). Let $A \subset D$ be (n, ε)-separated with respect to f, i.e., $x, y \in A$, $x \neq y$, implies $d_n(x, y) > \varepsilon$, where

$$d_n(x, y) = \max_{0 \leq i \leq n-1} d(f^i(x), f^i(y)).$$

Lay on D a quasi-box partition $\kappa = \{K_j : 1 \leq j \leq |\kappa|\}$ such that

$$\|\kappa\| < \varepsilon,$$

so as points lying at a distance greater than ε belong necessarily to different bins of κ. Then,

$$d_n(x,y) > \varepsilon \quad \Leftrightarrow \quad d(f^i(x), f^i(y)) > \varepsilon \quad \text{for some } 0 \le i \le n-1$$
$$\Rightarrow \quad (X^\kappa)_0^{n-1}(x) \ne (X^\kappa)_0^{n-1}(y).$$

Thus, every point $x \in A \cap K_{j_0}$, $1 \le j_0 \le |\kappa|$, generates a different sequence $(X^\kappa)_0^{n-1}(x) = j_0, \ldots$ of length n. Of course, there can be points $x' \in K_{j_0}$, $x' \notin A$, such that $(X^\kappa)_0^{n-1}(x') = j_0, \ldots \ne (X^\kappa)_0^{n-1}(x)$ for all $x \in A \cap K_{j_0}$, but the number of such points will vanish when $n \to \infty$ if $\varepsilon \le \delta$, δ being an expansiveness constant for f (see Definition 26). In this limit (and $\varepsilon \le \delta$) we also have $A \cap K_j \ne \emptyset$ for $\forall j$, $1 \le j \le |\kappa|$, hence there is a one-to-one relation between points in A and outputs $(x^\kappa)_0^\infty$ of X^κ. If, as in Definition 23, $s_n(\varepsilon, D)$ denotes the largest cardinality of any (n, ε)-separated subset of D with respect to f and $N(X^\kappa, n)$ denotes the number of distinct symbolic sequences of length n, it follows that

$$\limsup_{n \to \infty} \frac{1}{n} \log N(X^\kappa, n) = \limsup_{n \to \infty} \frac{1}{n} \log s_n(\varepsilon, D),$$

for $\varepsilon \le \delta$, and thus (see (7.11) and (B.25))

$$\lim_{\|\kappa\| \to 0} h_{\text{top}}(X^\kappa) = \lim_{\|\kappa\| \to 0} \limsup_{n \to \infty} \frac{1}{n} \log N(X^\kappa, n)$$
$$= \lim_{\varepsilon \to 0} \limsup_{n \to \infty} \log s_n(\varepsilon, I)$$
$$= h_{\text{top}}(f). \qquad \square$$

Theorem 15 *Let D be a q-dimensional simple domain, $q \ge 2$, and $f : D \to D$ a positively expansive map. Then*

$$h_{\text{top}}^*(f) = h_{\text{top}}(f) \tag{7.18}$$

and

$$\sup_{\mu \in E(D,f)} h_\mu^*(f) = h_{\text{top}}^*(f).$$

Proof Apply Lemma 11 and Propositions 8 and 9. $\square$

From the proof of Lemma 11, it should be clear where the need for expansiveness comes from: it can otherwise happen that points x of the (n, ε)-separated subset $A \subset D$ have neighboring points x'_ε that shadow their trajectories at arbitrarily close distance (hence $x'_\varepsilon \notin A$) but define symbolic sequences $X^\kappa(x'_\varepsilon) \ne X^\kappa(x)$. This will be certainly the case when, for instance, x belongs to the stable manifold of a hyperbolic fixed point $p \in D$ or, more generally, whenever the state space have lower dimensional manifolds whose points are not sensitive to initial conditions. The good news for the practitioner is that, since such local manifolds have Lebesgue measure zero, at least for sufficiently smooth dynamics, equality (7.18) will hold in numerical calculations for smooth maps with sensitivity to initial conditions almost everywhere (with respect to the Lebesgue measure). The bad news is that expansive maps are difficult to approximate numerically: small errors in computations (like those due to round-off) get magnified upon iteration.

From Theorems 14 and 27 (Sect. B.3) it follows:

Corollary 6 *Let D_1, D_2 be simple domains of $\mathbb{R}^q$, and $f_i : D_i \rightarrow D_i$, $i = 1, 2$, positively expansive maps. Suppose that $\phi : D_1 \rightarrow D_2$ is a homeomorphism such that $\phi \circ f_1 = f_2 \circ \phi$. Then*

$$h_{top}^*(f_1) = h_{top}^*(f_2).$$

Thus topological conjugacy is a sufficient condition for two positively expansive self-maps of simple domains to have the same topological permutation entropy.

Let us remark at this point that the original definition of the topological permutation entropy of a self-map f of a closed one-dimensional interval I, given by Bandt, Keller, and Pompe in [29], is

$$h_{top}^{*BKP}(f) = \lim_{n \rightarrow \infty} h_{top}^{*BKP}(f, L), \tag{7.19}$$

where

$$h_{top}^{*BKP}(f, L) = \frac{1}{L-1} \log |\mathcal{P}_L| \tag{7.20}$$

is the topological permutation entropy of f of order L, and remember from (3.4) and (3.5),

$$|\mathcal{P}_L| = |\{P_\pi \neq \varnothing : \pi \in \mathcal{S}_L\}| \tag{7.21}$$

gives the number of ordinal patterns realized by the orbits of length L, $(f^n(x))_{n=0}^{L-1}$ with $x \in I$. The following result holds.

Theorem 16 [29] *If I is a closed one-dimensional interval and $f : I \rightarrow I$ is piecewise monotone, then $h_{top}^{*BKP}(f) = h_{top}(f)$, where $h_{top}(f)$ is the topological entropy of f.*

On the other hand, Misiurewicz proved that this result is not true if the map is not piecewise monotone [157]. His counterexample is a continuous map with infinite monotonicity segments that has zero topological entropy but positive topological permutation entropy. He also shows in [157] that for piecewise monotone interval maps, the topological entropy can be computed by counting the permutations exhibited by the periodic orbits.

Example 16 For the symmetric tent map Λ, the partitions $\mathcal{P}_2$, $\mathcal{P}_3$, and $\mathcal{P}_4$ have cardinalities 2, 5, and 12 (Example 13), respectively. Hence, the topological permutation entropies of orders 2, 3, and 4 are the following:

$$h_{top}^{*BKP}(\Lambda, 2) = \log |\mathcal{P}_2| = \log 2 = 1 \; bit/symbol,$$

$$h_{top}^{*BKP}(\Lambda, 3) = \frac{1}{2} \log |\mathcal{P}_3| = \frac{1}{2} \log 5 = 1.1610 \; bit/symbol,$$

$$h_{top}^{*BKP}(\Lambda, 4) = \frac{1}{3} \log |\mathcal{P}_4| = \frac{1}{3} \log 12 = 1.1950 \; bit/symbol.$$

By Theorem 16,

$$h_{top}^{*BKP}(\Lambda) = h_{top}(\Lambda) = \log 2 = 1 \; bit/symbol.$$

To conclude, it was pointed out in Sect. 3.4.1 that order-isomorphic maps have the same admissible and forbidden ordinal patterns of any length. This fact together with Theorem 16 lead to the following results.

Corollary 7 *Let I_1, I_2 be two closed intervals of $\mathbb{R}$, and suppose that the maps $f_i : I_i \to I_i$, $i = 1, 2$, are order isomorphic. Then,*

(1) $h_{top}^{*BKP}(f_1) = h_{top}^{*BKP}(f_2)$, provided one of them exists.
(2) Furthermore, if f_1 and f_2 are piecewise monotone, then $h_{top}(f_1) = h_{top}(f_2)$.

7.5 Estimating Topological Entropy

Estimation of topological entropies from naive numerical simulation of long orbits is notoriously difficult. Metric entropy by itself can be quite tricky and difficult, requiring very long data sets for increasing L, but topological entropy is worse yet, because it weighs each pattern equally. This means that patterns which are exceptionally infrequent on the natural measure of the attractor can still have a significant influence on the result. Attempting to estimate the same quantities using empirical occurrences of ordinal patterns is even more difficult, requiring more data than would a good, low-alphabet generating partition for ordinary symbolic dynamics.

For the present purpose, we consider a continuous system in greater than one dimension, with a chaotic attractor, and whose topological entropy can be found by independent rigorous means. The *Lozi map*,

$$x_{i+1} = y_i,$$
$$y_{i+1} = 1 + bx_i - a|y_i|,$$

with parameters $a, b \in \mathbb{R}$, $b \neq 0$, satisfies all these criteria. A mathematical proof for the existence of an attractor for the Lozi map was given by Misiurewicz [156]. In particular, $a = 6/5, b = -2/15$ yield a low-entropy chaotic attractor (roughly 0.3 bits/iteration) and for those parameters, the topological entropy has been bounded rigorously with computer-assisted analytical computations [102, 178], and we use their results.

We found that the best numerical procedure was to look at the "outgrowth ratio" of ordinal patterns of a given length L. The outgrowth ratio for some pattern of length L is the cardinality of the set of distinct ordinal patterns of length $L + 1$ which have the given length-L pattern as a prefix. More concretely, we find vectors of length $L + 1$ from an orbit of the map. The ordinal pattern on the first L points is the prefix pattern. Regardless of the dynamics, there can be at most $L + 1$ ordinal patterns of length $L + 1$ conditioned on the length-L ordinal pattern, since the single new element belongs to the alphabet $\{1, \ldots, L + 1\}$.

Indeed, according to definitions (7.11), (7.5), and (7.6), the topological permutation entropy $h_{\text{top}}^*(f)$ is the scaling rate of the logarithm of the number of patterns with L of the "coarse-grained" dynamics $\mathbf{X} \equiv \mathbf{X}^\kappa$ for κ sufficiently fine, i.e.,

$$\log N(\mathbf{R}, L) \approx L h_{\text{top}}^*(\mathbf{X}),$$

(R_0^{L-1} are the rank variables defined by X_0^{L-1}), so

$$\log \frac{N(\mathbf{R}, L + 1)}{N(\mathbf{R}, L)} \approx h_{\text{top}}^*(\mathbf{X}).$$

Therefore, a reasonable estimator for $h_{\text{top}}^*(f)$ is the logarithm of the outgrowth ratio averaged uniformly over all extant prefix patterns. This value, for sufficiently large L and sufficiently large simulation sets, ought to be $h_{\text{top}}^*(f)$ on average. Note that independent white noise would give an estimate of $\log(L + 1)$, i.e., not converging with L.

Figure 7.2 shows the numerical result of estimating $h_{\text{top}}^*(f)$ on long orbits of the Lozi map with $a = 6/5, b = -2/15$, using two specific instantiations of the outgrowth method. The dotted lines are the bounds on the true topological entropy.

The first strategy involves computing $N_1 = 50 \times 10^6$ ordinal patterns of length $L + 1$ and their length L prefix. For every element in the prefix set we accumulate the number of distinct elements in the conditioning set and average the logarithm of the number of distinct occurrences over the observed length-L ordinal patterns—as long as each of those ordinal patterns had at least two successors. This method will typically have a bias downward for large L on account of undersampling the space.

The second strategy starts by computing $N_2 = 10^6$ ordinal patterns of length $L+1$ from orbits of the map. The set of distinct order-L prefixes forms the "conditioning" set. The N_2 length $L + 1$ ordinal patterns from these are accumulated, and then the map is iterated and ordinal patterns computed, until there have been $(K-1)N_2$ more observations of length-L ordinal patterns which were in the prefix, so that there are $KN_2 = N_1$ with $K = 50$ observations, all of whose order L prefixes are in the conditioning set. Then similarly the logarithm of the outgrowth ratio is estimated over the conditioning set for all conditioning patterns with at least two observations. This method has positive and negative biases due to finiteness of observations. First, because of finite K there is a downward bias, as the number of observed outgrowths is a strict lower bound on the number of allowed outgrowths in the dynamical

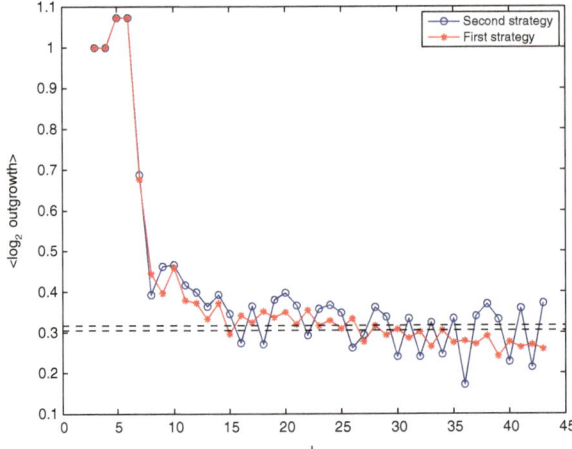

Fig. 7.2 Logarithmic outgrowth ratios for the Lozi map vs L. The *dotted lines* represent rigorous bounds on the topological entropy rate, computed by computer-assisted analytical methods. The outgrowth ratio approximates the topological permutation entropy rate, is practical for computing, and can scale to significant L

system. There is a more subtle upward bias, which changes with L as well. It is because the ordinal patterns which were selected as conditioning states came from an ergodic sample on the natural measure which does not sample the support uniformly. More frequently occurring patterns are more likely to occur in the conditioning set—and we have observed heuristically that in chaotic systems the outgrowth ratio tends to be roughly correlated in the same direction as the frequency of the conditioning pattern. The measure on the allowable patterns does vary very widely hence it can take very long simulations to find more of the allowable conditioning patterns even though their total number is far smaller than the number of samples from the map. This effect is also present in the first method as well, but appears to be dominated by the downward bias.

7.6 Existence of Forbidden Ordinal Patterns

We turn to the study of forbidden patterns for self-maps of q-dimensional simple domains and, more specifically, to the issue of finding sufficient conditions that guarantee forbidden patterns of any length. The existence of one-dimensional interval maps with no forbidden patterns (Fig. 4.2) shows that this question is pertinent.

Let D be a q-dimensional simple domain and $f : D \to D$ a map with $h^*_{\text{top}}(f) < \infty$. According to the definition of $h^*_{\text{top}}(f)$, (7.11), given $\varepsilon > 0$ arbitrarily small there exists a quasi-box partition κ_0 of D such that

$$\left| h^*_{\text{top}}(f) - h^*_{\text{top}}(\mathbf{X}^\kappa) \right| < \frac{\varepsilon}{2}$$

whenever the quasi-box partition κ is a refinement of κ_0. Furthermore, according to the definition of $h^*_{\text{top}}(\mathbf{X}^\kappa)$ ((7.5) and (7.10) with $\mathbf{X}^\kappa$ instead of $\mathbf{X}$), there exists a length L_0 such that

$$\left| h^*_{\text{top}}(\mathbf{X}^\kappa) - \frac{1}{L} \log N^*(\mathbf{X}^\kappa, L) \right| < \frac{\varepsilon}{2}$$

whenever $L \geq L_0$, where $N^*(\mathbf{X}^\kappa, L)$ is the number of *admissible* ordinal L-patterns of the symbolic dynamics $\mathbf{X}^\kappa$ with respect to κ. Therefore, with κ sufficiently fine and L sufficiently large we have

$$\left| h^*_{\text{top}}(f) - \frac{1}{L} \log N^*(\mathbf{X}^\kappa, L) \right| < \varepsilon,$$

hence,

$$N^*(\mathbf{X}^\kappa, L) = e^{L h^*_{\text{top}}(f)} + \mathcal{O}_L(\varepsilon), \tag{7.22}$$

where the term $\mathcal{O}_L(\varepsilon)$ depends also on L, as indicated by the subindex, and $\mathcal{O}_L(\varepsilon) \to 0$ when $\varepsilon \to 0$ (or $\|\kappa\| \to 0$).

On the other hand, we already know that the number of *possible* ordinal L-patterns, $|\mathcal{S}_L| = L!$, grows superexponentially with L, (3.8). We conclude from (7.22) that the symbolic dynamics $\mathbf{X}^\kappa$ has forbidden patterns whenever $h^*_{\text{top}}(f)$ exists and is finite. Then, the same must happen with maps, since their dynamic can be approximated by symbolic dynamics.

Theorem 17 *Let $D \subset \mathbb{R}^q$ be a simple domain and $f : D \to D$ a map. Then*

$$\lim_{\|\kappa\| \to 0} N^*(\mathbf{X}^\kappa, L) = |\mathcal{P}_L|,$$

where we use the notation $|\mathcal{P}_L|$ as in (7.21) for the number of admissible ordinal L-patterns for f.

Proof We claim that the admissible L-patterns for f will coincide with the admissible L-patterns for the corresponding symbolic dynamics $\mathbf{X}^\kappa$ with respect to a quasi-box partition $\kappa = \{K_j\}_{1 \leq j \leq |\kappa|}$ in the limit $\|\kappa\| \to 0$. Indeed, if $x \in D$ is of type $\pi \in \mathcal{S}_L$, the only way that the length-L word $x, f(x), \ldots, f^{L-1}(x)$ does not define π when observed with the precision set by κ is that at least two letters, say $f^{i_1}(x)$ and $f^{i_2}(x)$, $0 \leq i_1 < i_2 \leq L - 1$, fall in the same bin $K_{j_0} \in \kappa$, since then we cannot discern the order relation between both letters. But this will not happen when κ is so fine that $x, f(x), \ldots, f^{L-1}(x)$ fall in different bins. We conclude that the number of such discrepancies will diminish as the partition κ gets finer, and finally vanish in the limit $\|\kappa\| \to 0$. $\qquad \square$

Theorem 17 and (7.22) imply the following result.

Corollary 8 *The number of allowed L-patterns of self-maps f of q-dimensional simple domains grows asymptotically with L as*

$$|\mathcal{P}_L| \sim e^{Lh_{top}^*(f)}, \tag{7.23}$$

provided $h_{top}^*(f)$ exists and is finite.

The same conclusion follows directly from (7.19) when $h_{top}^*(f)$ is replaced by $h_{top}^{*BKP}(f)$ in (7.23) for one-dimensional interval maps. Since calculating $h_{top}^*(f)$ requires in practice the calculation of the growth rate with L of the allowed L-patterns for f, we use Theorems 15 and 16 to provide more natural conditions for (7.23).

Corollary 9 *Let $D \subset \mathbb{R}^q$ be a simple domain and $f : D \to D$ a map with $h_{top}(f) < \infty$. (i) If $q = 1$ and f is piecewise monotone or (ii) $q \geq 2$ and f is positively expansive, then*

$$|\mathcal{P}_L| \sim e^{Lh_{top}(f)}.$$

Corollary 9 provides sufficient conditions for the existence of forbidden ordinal patterns since, as already pointed out in some previous passages, the number of possible ordinal L-patterns grows superexponentially with L: $|\mathcal{S}_L| = L!$. In more quantitative terms, forbidden patterns proliferate in these two cases as (see (3.8))

$$|\{P_\pi = \varnothing : \pi \in \mathcal{S}_L\}| \sim L! - e^{Lh_{top}(f)} = e^{L\ln L}\left(1 - e^{-L(\ln L - h_{top}(f))}\right).$$

It is an open problem to find a more general condition than expansiveness in higher dimensional dynamics for the existence of forbidden pattern. Numerical simulations support the existence of forbidden patterns also for non-expansive multi-dimensional maps (see next section).

Apart from the superexponential scaling law with L, it is quite difficult to make more specific statements on the forbidden patterns for a map like, for instance, the minimal length of its forbidden patterns or the lengths of its root forbidden patterns. One important exception is the shift and signed shift transformations (and all order-isomorphic maps) we studied in Chaps. 4 and 5.

Last but not the least, forbidden patterns, be in one-dimensional dynamics or in higher dimensional dynamics, have the properties discussed in Sect. 3.4.

7.7 Numerical Simulations

We demonstrate numerical evidence for the existence of forbidden ordinal patterns in multi-dimensional maps. Of course, direct simulation of dynamical systems directly yields only *allowed* ordinal patterns. The failure to observe any given ordinal pattern in any finite time series does not mean of course that it is forbid-den (probability zero) but only that its probability is sufficiently low in the natural measure induced by the dynamics that it has not yet been seen.

However, with reasonable L (as effort and memory increases radically with L) and robust computational ability we can infer in many cases, the existence of forbid-den patterns by examining the convergence of allowed patterns with N, the number of data emitted by the source. In particular, we suggest examining the logarith-mic ratio of the cardinality of all L-patterns to the number of observed L-patterns $\log (L!/P_{obs})$ vs $\log N$. If a system has a "core" of forbidden patterns, as with deter-ministic systems, then we expect that this ratio will decline with N and eventually level off with increasing N, assuming the asymptotic behavior can be observed. Here, P_{obs} is the naive, biased-downward, estimator of the unknown $P_{allowed}$, the number of allowed L-patterns.

When N is much larger than $P_{allowed}$, P_{obs} is likely to be a good estimator, assum-ing most patterns have a reasonable probability of occurring. With increasing L, however, this is difficult to achieve practically because of memory limitations, as the identities and counts of each observed patterns (a subset of the allowed patterns) must be retained. The number of allowed patterns increases exponentially with L in deterministic chaos and faster than exponentially with noise, and therefore one must increase N, the number of iterates, substantially to permit a commensurately large number of distinct patterns to be actually observed.

This motivates using a superior statistical estimator of $P_{allowed}$. This equivalent problem has a significant history, motivated especially from the ecology community. Consider a situation where one can observe a finite sample of individual organ-isms, from a presumably large population. What is the estimated number of distinct species, the biodiversity, and how can we estimate this given the individual counts of observed species? (For reviews of approaches to this problem, see [41, 100].) This is analogous to our situation where we can distinguish individual ordinal patterns but each observation is drawn from the natural distribution induced by typical orbits of the dynamical system. For our needs we wish to go reasonably deep into the under-sampled regime and impose few probabilistic priors. We adopt the non-parametric estimator of Chao [49], motivated by comments in the reviews and our experience, as a simple but reasonably effective improvement:

$$P_{Chao} = P_{obs} + \frac{c_1^2}{2c_2^2}, \qquad (7.24)$$

where c_k are the "meta-counts" of observations, i.e., c_1 is the number of distinct ordinal patterns which were observed exactly once in the sample, c_2 the number which were observed exactly twice, etc. In practice this is accomplished by count-ing frequencies of observed patterns through a hash table and in a second phase, counting the frequencies of such frequencies with a similar hash table. Note that if the sample size is particularly small (relative to what is necessary to see a substantial fraction of allowed patterns), P_{Chao} will still be an underestimate. Consider that its maximum value is obtained with $c_1 = N - 1$ and $c_2 = 1$, i.e., one doubleton and all remaining observations being unique (all unique naturally leads to an undefined estimate), and so P_{Chao} is bounded by $(P_{obs}^2 + 1)/2$. Bunge and Fitzpatrick [41] call P_{Chao} to be an "estimated lower bound." We believe that no statistical estimator can perform well in the extremely undersampled regime and there is no substitute for substantial computational effort when L becomes sufficiently large; however, we will see an improvement over the naive estimator.

Our first numerical example is Arnold's *cat map*:

$$x_{i+1} = x_i + y_i \qquad \text{mod } 1,$$
$$y_{i+1} = x_i + 2y_i \qquad \text{mod } 1.$$

As a hyperbolic toral automorphism, this is an expansive transformation [115]. We start with initial conditions drawn uniformly in $[0, 1) \times [0, 1)$ and iterate. Ordinal patterns are computed using order relations on the x-coordinate only; since coincidences in the x-coordinate are unlikely, this amounts in practice to using lexicographic order in $[0, 1) \times [0, 1)$. Figure 7.3 shows the strong numerical evidence for forbidden patterns characteristic of deterministic systems. As a demonstration of the genericity of the results, Fig. 7.4 shows the equivalent except that the observable upon which ordinal patterns were computed is $3x^3 - y$. Results are nearly identical, as one expects.

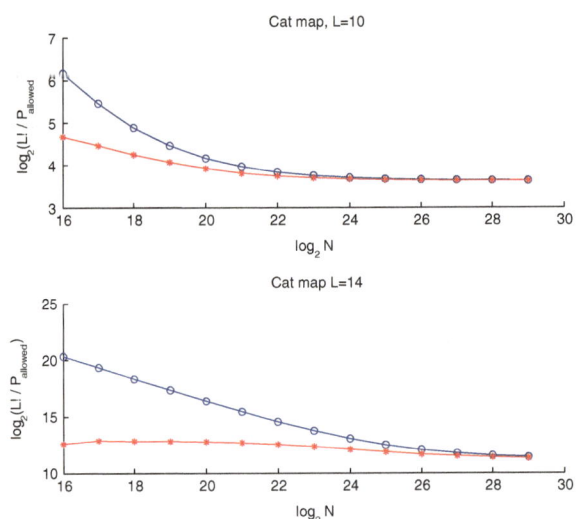

Fig. 7.3 Convergence of estimated forbidden patterns with N, cat map. *Circles* (o) are for $P_{allowed}$ estimated by P_{obs}, *asterisks* (*) have $P_{allowed}$ estimated by P_{Chao}. *Top, $L = 10$, bottom $L = 14$.* Both figures show clear evidence of convergence to a constant, evidence of true forbidden patterns as $N \rightarrow \infty$. In the *lower figure* especially, the improved estimator P_{Chao} "senses" the approach to a convergence earlier than the naive counting estimator. Note the differing scales on the y-axes

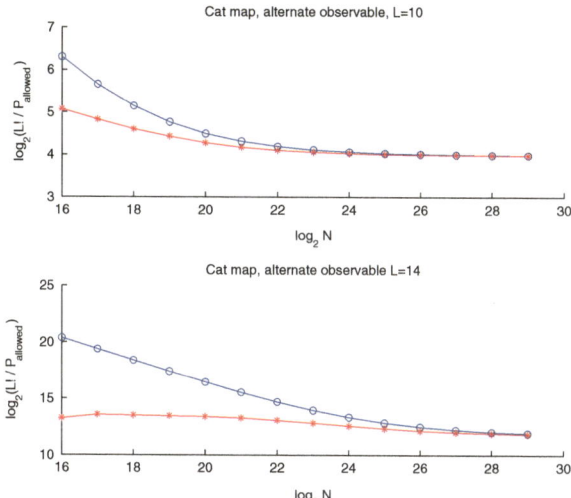

Fig. 7.4 Convergence of estimated forbidden patterns with N, cat map, alternative observable. *Circles* (o) are for $P_{\text{allowed}} = P_{\text{obs}}$, *asterisks* (*) have $P_{\text{allowed}} = P_{\text{Chao}}$. *Top*, $L = 10$, *bottom* $L = 14$. Both figures show clear evidence of convergence to a constant, evidence of true forbidden patterns as $N \to \infty$. In the *lower figure* especially, the improved estimator P_{Chao} "senses" the approach to a convergence earlier than the naive counting estimator. Note the differing scales on the *y*-axes

By comparison, consider Fig. 7.5, generated by an i.i.d. noise source (ordinal patterns are insensitive to changes in distribution). Here, the observed patterns imply convergence to zero forbidden patterns with increasing N. More remarkably, the estimator P_{Chao} senses this long before and predicts zero forbidden patterns with orders of magnitude lower than N, apparently because the assumptions made by the estimator of equiprobable patterns for both observed and unobserved are exactly fulfilled.

As an example of a non-expansive map, we turn to a chaotic system, the *Hénon map*,

$$x_{i+1} = 1 - ax_i^2 + by_i,$$
$$y_{i+1} = x_i,$$

with $a = 1.4$, $b = 0.3$, observable being the x-coordinate. This map is not uniformly hyperbolic (it has two fixed points, one attractive and one repellent), more characteristic of real dynamics seen in nature. (The Hénon map is non-expansive for "almost all" values of the parameter a [154].) In Fig. 7.6, we see convergence to a finite core of forbidden patterns with larger N. Note that the performance of P_{Chao} is still improved over the naive estimator but it is not as good as with noise, because with real dynamics there is a wide variation in the probability of the various allowed patterns, and so larger N feels the "tail" of the distribution of rare patterns. By comparison consider Fig. 7.7, which shows results from the same dynamics but

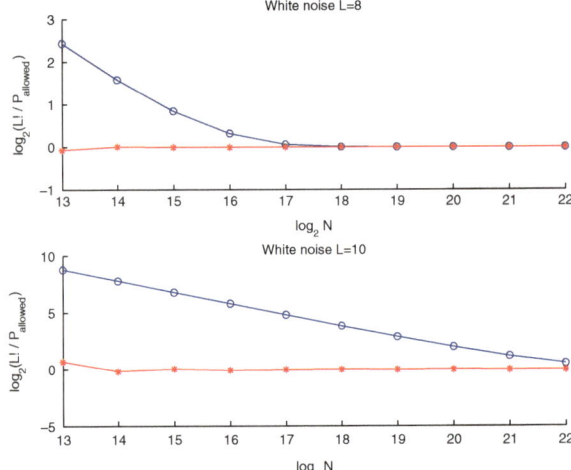

Fig. 7.5 Convergence of estimated forbidden patterns with N, i.i.d. noise. *Circles* (o) are for $P_{allowed} = P_{obs}$, *asterisks* (*) have $P_{allowed} = P_{Chao}$. *Top, L = 8, bottom L = 10*. P_{obs} shows convergence to zero forbidden patterns; P_{Chao} estimates zero forbidden patterns well before convergence of naive estimator

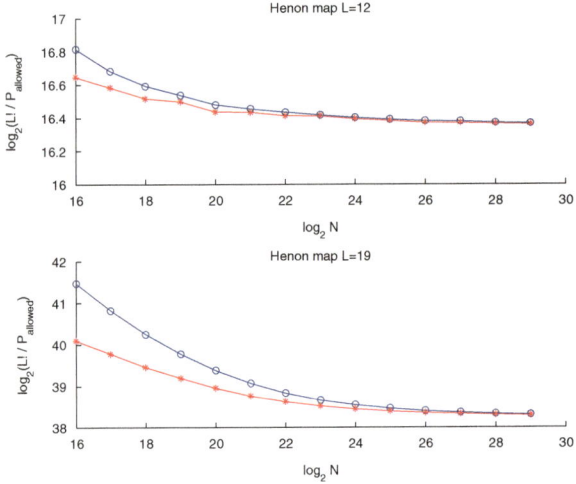

Fig. 7.6 Convergence of estimated forbidden patterns with N, Hénon map. *Circles* (o) are for $P_{allowed}$ estimated by P_{obs}, *asterisks* (*) have $P_{allowed}$ estimated by P_{Chao}. *Top, L = 12, bottom L = 19*. Both naive and improved estimators show convergence to a finite number

each observable was contaminated with uniform i.i.d. noise $\eta \in [0, 0.2)$. This time, increasing N clearly shows increasing allowed/decreasing forbidden patterns, proportional to N as expected with noise. As a matter of fact, arbitrarily small noise will eventually lead to noise-like scaling, but the size of the word necessary to

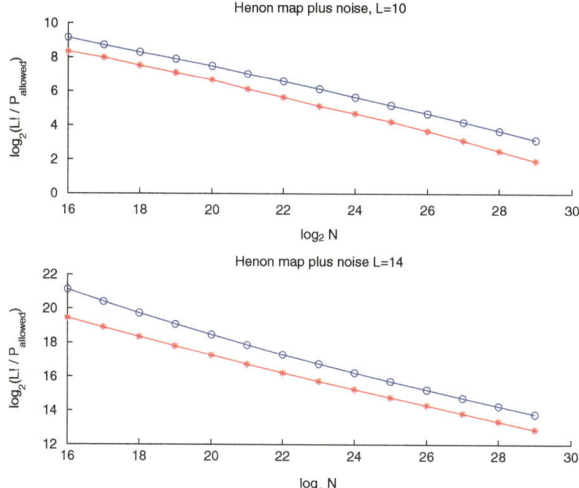

Fig. 7.7 Lack of convergence of estimated forbidden patterns with N, Hénon map with additive i.i.d. noise. *Circles* (o) are for $P_{\text{allowed}} = P_{\text{obs}}$, *asterisks* (*) have $P_{\text{allowed}} = P_{\text{Chao}}$. *Top, $L = 12$, bottom $L = 14$*. Both naive and improved estimators show continued increase in allowed patterns (decrease in forbidden patterns) with increasing N

see this (and consequently the size of the data set necessary to see the effect) will grow astronomically. If the noise support is bounded (or, we conjecture, thin-tailed), then fairly small noise levels will not be visible in the ordinal patterns if they are substantially smaller than typical sizes of $x_i - x_j$ for $1 \leq j \leq L$, and hence will not change the patterns. The behavior with N clearly distinguishes low-dimensional dynamics from noise.

As a philosophical point it is true that the "noise" generator in a computer software is but a deterministic dynamical system on its own, but in practice it has an extremely long period and virtually no correlation, and hence if one wanted to see ordinal pattern scaling different from true noise, one would need exceptionally long L and impractically astronomical memory requirements. We use a validated high-quality random number generator [148] from the Boost C++ library.

Chapter 8
Discrete Entropy

From a mathematical point of view, entropy made its first appearance in continuous-time dynamical systems (more exactly, in Hamiltonian flows), and from there it was extended to quantum mechanics by von Neumann, to information theory by Shannon, and to discrete-time dynamical systems by Kolmogorov and Sinai. In all these cases we observe that (i) if the state space is discrete and/or finite (like in quantum mechanics and finite-alphabet information sources), then the evolution is random and (ii) if the evolution is deterministic (like in continuous- and discrete-time dynamical systems), then the state space is infinite. Still today one speaks of random dynamical systems in the first case and of deterministic dynamical systems in the second case. But not all dynamical systems of interest fall under one of the previous categories. An important example of a deterministic physical system where both state space and dynamics are discrete is a digital computer; this entails that any dynamical trajectory in computer becomes eventually periodic—a well-known effect in the theory and practice of pseudo-random number generation. Dynamical systems with discrete and even with a finite number of states have been considered by a number of authors, in particular in the development of *discrete chaos* [125]—an attempt to formalize the idea that maps on finite sets may have different diffusion and mixing properties. From this perspective, it seems desirable to export some concepts and tools from the general theory to this new setting. This is the rationale behind, e.g., the discrete Lyapunov exponent [124, 125].

The topic of this chapter is precisely the extension of entropy to maps on finite sets—a concept we call *discrete entropy*. When going from the conventional framework to considering maps F on sets S with cardinality $|S| < \infty$ (and, eventually, an atomic measure), one main difficulty arises at the very beginning: the entropy of F with respect to any partition of S vanishes, rendering null entropy. It is not clear how to modify the concept of entropy while still gauging the "randomness" generated by F on S in the limit $|S| \to \infty$. Thanks to its combinatorial nature, permutation entropy lends itself especially well to the methods of discrete mathematics, allowing in fact to define a *discrete entropy* concept (Sect. 8.1). The definition of discrete entropy can then be justified by showing that, for a large class of maps, the discrete entropy converges to the measure-theoretic entropy in the "infinite" limit (Sect. 8.2). More precisely, let $f: I^d \to I^d$ be a d-dimensional interval map, which is ergodic with respect to a measure μ, and let F_M be a permutation on M elements

J.M. Amigó, *Permutation Complexity in Dynamical Systems*,
Springer Series in Synergetics, DOI 10.1007/978-3-642-04084-9_8,
© Springer-Verlag Berlin Heidelberg 2010

obtained from f via discretization and orbit truncation (see Sect. 8.2 for details). Then $\lim_{M \to \infty} h_\delta(F_M) = h_\mu(f)$, where $h_\delta(F_M)$ is the *discrete entropy* of F_M, and $h_\mu(f)$ is the metric entropy of f with respect to μ. An alternative approach using *topological* entropy is also possible and will be discussed in Sect. 8.3.

Apart from their role as entropy-like tools of discrete chaos, metric and topological discrete entropy can be viewed also as estimators of the corresponding "continuous" counterpart, thanks to the infinite limit mentioned above. This more practical side of discrete entropy is somewhat hampered by the fact that discrete entropy requires large amounts of data to converge—albeit a property shared with most of the entropy estimators.

8.1 Discrete Entropy

Let $A = \{a_1, \ldots, a_L\}$ be a finite set endowed with a linear ordering $\leq$, $F{:}A \to A$ a bijection, and $\pi = \langle \pi_0, \ldots, \pi_{n-1} \rangle \in S_n$, $2 \leq n \leq L$. Define

$$Q_\pi(n) = \left\{ a \in A{:}F^{\pi_0}(a) < \cdots < F^{\pi_{n-1}}(a) \right\} \tag{8.1}$$

and

$$q_\pi(n) = \frac{|Q_\pi(n)|}{\sum_{\tau \in S_n} |Q_\tau(n)|} \tag{8.2}$$

if $\sum_{\tau \in S_n} |Q_\tau(n)| \neq 0$ (in which case, $\sum_{\pi \in S_n} q_\pi(n) = 1$) and $q_\pi(n) = 0$ otherwise. We shall drop the argument n of Q_π and q_π when it is clear from the context that $\pi \in S_n$. We say that $a \in A$ defines the ordinal pattern $\pi \in S_n$ if $s \in Q_\pi$.

Without restriction we take $A = \{0, \ldots, L-1\}$ with the natural order inherited from $\mathbb{N}_0$. Then we write the permutation F as in (1.22),

$$F = [F(0), F(1), \ldots, F(L-1)].$$

On the other hand, F can also be written as a product of cycles. As in Sect. 2.4, we denote by $(i_1, i_2, \ldots, i_n)$ the *cycle* $i_1 \mapsto F(i_1) = i_2 \mapsto \cdots \mapsto F(i_{n-1}) = i_n \mapsto F(i_n) = i_1$ of length n. If $F = (i_1, \ldots, i_L)$ (i.e., a cycle of maximal length), we say that F is *irreducible*, otherwise it is *reducible*.

In view of (6.17), we introduce the following concept.

Definition 4 The discrete entropy of F of order $n \geq 2$, is

$$h_\delta^{(n)}(F) = -\frac{1}{n-1} \sum_{\pi \in S_n} q_\pi \log q_\pi.$$

The subscript δ stands for "discrete" but also for "Dirac measure" on A. Observe that (i) $a \in Q_\pi(n)$ as long as $n \leq \mathrm{Per}(a)$, the period of a and (ii) alternatively, one can take the truncated orbits $orb(a) = \{a, F(a), \ldots, F^{p-1}(a)\}$, with $p = Per(a)$, and normalize the count of a's defining the ordinal pattern $\pi \in S_n$ for $n = 2, \ldots, p_{max} = \max_{a \in A}\{Per(a)\}$.

So to speak, $h_\delta^{(n)}(F)$ senses the mixing properties of F in the short run $1 \leq n \leq p_{max}$—before periodicity sets in on the whole "phase space" A. This is the timescale we are interested in. This explains also why we do not allow for repetition of symbols and use strict order in (8.1) instead of ranks. In the infinite limit $L \to \infty$ (Sect. 8.2) it makes no difference, but for finite L we want to switch off periodicities.

Let us tackle this discretization of entropy by considering first some special cases.

Case 1. If $F = (i_1, i_2, \ldots, i_L)$, then each $a \in A$ defines permutations $\pi \in S_n$ for $2 \leq n \leq L$, the corresponding sets $Q_\pi(n)$ building thus partitions of A, and we can write down a whole hierarchy of entropies of orders $2, \ldots, L$. In particular, $q_\pi(L) = 1/L$ if $Q_\pi \neq \emptyset$, so

$$h_\delta^{(L)}(F) = \frac{1}{L-1} \log L.$$

As a result, $h_\delta^{(L)}(F)$ and possibly other entropies of lower orders cannot discriminate two permutations of the same maximal length L.

Example 17 For the right shift modulo $L - 1$, defined as $\theta_L(i) = i + 1$ for $i = 0, 1, \ldots, L - 2$ and $\theta_L(L - 1) = 0$, i.e.,

$$\theta_L = (0, 1, \ldots, L - 1), \tag{8.3}$$

we get

$$h_\delta^{(n)}(\theta_L) = \frac{L - n + 1}{L(n-1)} \log \frac{L}{L - n + 1} + \frac{1}{L} \log L \tag{8.4}$$

for $2 \leq n \leq L$. In particular, for $L = 4$ we have

$$h_\delta^{(2)}(\theta_4) = 0.811, \quad h_\delta^{(3)}(\theta_4) = 0.750, \quad h_\delta^{(4)}(\theta_4) = 0.667,$$

in bit/symbol.

Case 2. On the opposite (non-trivial) end, let $F = (i_1, i_2)(j_1, j_2) \cdots (k_1, k_2)$ with, say, $i_1 < i_2, j_1 < j_2, \ldots, k_1 < k_2$. In this case, every $a \in A$ defines only one ordinal pattern of order 2, the symbols $i_1, \ldots, k_1$ belonging to $Q_{\langle 0,1 \rangle}$ and the symbols $i_2, \ldots, k_2$ to $Q_{\langle 1,0 \rangle}$. Hence, $q_{\langle 0,1 \rangle} = q_{\langle 1,0 \rangle} = 1/2$ and $h_\delta^{(2)}(F) = 1$; entropies of higher order are not defined ($Q_\pi(n) = \emptyset$ for $n \geq 3$).

In general, $F = (i_1, \ldots, i_{p_1})(j_1, \ldots, j_{p_2}) \ldots (k_1, \ldots, k_{p_r})$ with $1 \leq p_1, \ldots, p_r \leq L$ ($p_i = 1$ for the fixed points), $p_1 + \cdots + p_r = L$ and $p_{max} := \max\{p_1, \ldots, p_r\} \geq 2$ (otherwise, F is the identity). If the symbol $a \in A$ appears in a cycle of length p, then a defines ordinal patterns of order $2, 3, \ldots, p$. Hence, F has entropies of order $2, 3, \ldots, p_{max}$, although from some order on (depending on F), both the number and cardinality of the sets $Q_\pi(n)$ will decrease with n, rendering their contribution less and less significant.

Let us mention in passing that the normalized expected maximum cycle length of a random permutation of L symbols tends to $0.62432\ldots$ as $L \to \infty$, a result first observed experimentally by Golomb [86] and proved by Shepp and Lloyd [187]. So, we expect on average

$$h_\delta^{p\max}(F) \approx \frac{1}{0.6L - 1} \log 0.6L.$$

Remark 3 By definition, the discrete entropy of order n and, thus, the discrete entropy do not sense the presence of fixed points. For example, $\theta_L = (0, 1, \ldots, L-1)$ (Example 17) and $F_{L+1} = (0, 1, \ldots, L-1)(L)$ or $F_{L+2} = (0, 1, \ldots, L-1)(L)(L+1)$ have the same entropies (8.4).

Example 18 Given a permutation F_L of $\{0, 1, .., L-1\}$, we call

$$\lambda_{F_L} = \frac{1}{L-1} \sum_{i=0}^{L-2} \log |F_L(i+1) - F_L(i)|,$$

the *discrete Lyapunov exponent* [125] of F_L. If $L = 2l$, it can be proved [13, Thm. II.2] that $\lambda_{F_{2l}}$ is maximal for the permutation

$$\Gamma_{2l} = [l, 0, l+1, 1, l+2, 2, \ldots, 2l-1, l-1], \tag{8.5}$$

in which case

$$\lambda_{F_{2l}} \le \lambda_{\Gamma_{2l}} = \frac{l}{2l-1} \ln l + \frac{l-1}{2l-1} \ln(l+1).$$

For $l = 2$ we get

$$h_\delta^{(2)}(\Gamma_4) = 1, \quad h_\delta^{(3)}(\Gamma_4) = 1, \quad h_\delta^{(4)}(\Gamma_4) = 0.667,$$

in bit/symbol. Comparison with Example 17 shows that $h_\delta^{(n)}(\theta_4) \le h_\delta^{(n)}(\Gamma_4)$ for $n = 2, 3, 4$. In particular, the smaller orders $n = 2, 3$ show that Γ_4 is more "random" than θ_4.

The possibly simplest way to encapsulate in a single number the information contained in the whole hierarchy $h_\delta^{(2)}(F), \ldots, h_\delta^{(n_{\max})}(F)$, $n_{\max} = \max\{n : h_\delta^{(n)}(F) \ne 0\}$, without having to dissect F into cycles, is taking the arithmetic mean of it.

Definition 5 We call

$$h_\delta(F) = \frac{1}{n_{\max} - 1} \sum_{n=2}^{n_{\max}} h^{(n)}(F) \tag{8.6}$$

the discrete entropy (or just the entropy) of F.

Hence, $h_\delta(F)$ takes into account both high and, most importantly, low and middle orders on an equal footing. Indeed, although the number of summands in $h_\delta^{(n)}(F)$ grows as $n!$, the sum of the non-zero terms (before getting multiplied by $1/(n-1)$) actually scales linearly in n, rendering the different $h_\delta^{(n)}(F)$ of comparable magnitudes. Moreover, if we let formally $n_{max} \to \infty$ (the limit of ordered sets with arbitrary cardinality), we recover the usual definition of entropy, $h_\delta(F) = \lim_{n \to \infty} h_\delta^{(n)}(F)$, since a convergent sequence and the arithmetic mean of their successive terms (*Césaro mean*) have the same limit.

Example 19 In cryptography, any substitution on n-bit blocks is called an $n \times n$ S-box (for "substitution box"). The cryptographic security of S-boxes can be analyzed with a variety of tools. Consider, for instance, the 4×4 S-boxes defined by the permutations

$$F_1 = [15, 12, 2, 1, 9, 7, 10, 4, 6, 8, 5, 11, 0, 3, 13, 14]$$
$$= (0, 15, 14, 13, 3, 1, 12)(2)(4, 9, 8, 6, 10, 5, 7)(11)$$

and

$$F_2 = [8, 2, 4, 13, 7, 14, 11, 1, 9, 15, 6, 3, 5, 0, 10, 12]$$
$$= (0, 8, 9, 15, 12, 5, 14, 10, 6, 11, 3, 13)(1, 2, 4, 7).$$

The action of the corresponding S-box on the binary block $b_1 b_2 b_3 b_4$ is identified with the action of F_1 or F_2 on the number $b_1 2^3 + b_2 2^2 + b_3 2^1 + b_4 \in \mathbb{Z}_{16}$. F_1 and F_2 share some standard properties of secure S-boxes, like being 0/1 balanced, nonlinear, and fulfilling the maximum entropy criterion [172]. But from the discrete entropy point of view, they are quite different. The discrete entropies of F_1 in bit/symbol are

$$h_\delta^{(2)}(F_1) = 0.99, \quad h_\delta^{(3)}(F_1) = 1.04, \quad h_\delta^{(4)}(F_1) = 0.96,$$
$$h_\delta^{(5)}(F_1) = 0.84, \quad h_\delta^{(6)}(F_1) = 0.70, \quad h_\delta^{(7)}(F_1) = 0.58,$$

and $h_\delta(F_1) = 0.85$. The discrete entropies of F_2 in bits/symbol are

$$h_\delta^{(2)}(F_2) = 0.99, \quad h_\delta^{(3)}(F_2) = 1.08, \quad h_\delta^{(4)}(F_2) = 1.17,$$
$$h_\delta^{(n)}(F_2) = 3.59/(n-1) \quad \text{for } n = 5, \ldots, 12$$

and $h_\delta(F_2) = 0.68$. Thus F_1, with a more even cycle decomposition than F_2, has a higher discrete entropy. Whether discrete entropy is useful for S-box design is an open problem in discrete chaos [125].

Exercise 12 A primitive root for a modulo m is a cyclic generator of $\mathbb{Z}_m^*$, the multiplicative group built by the residues modulo m coprime to m. Prove that the permutation Γ_{2l}, (8.5), is irreducible if and only if $2l + 1$ is a prime with primitive root 2 (i.e., $\mathbb{Z}_{2l+1}^*$ is cyclic and generated by 2). The primes under 100 with primitive root 2 are the following [2, Table 24.8]:

$$3, 5, 11, 13, 19, 29, 37, 53, 59, 61, 67, \text{ and } 83.$$

(Hint: consider the permutation $\widetilde{\Gamma}_{2l}:\{1, 2, \ldots, 2l\} \to \{1, 2, \ldots, 2l\}$ defined as $\widetilde{\Gamma}_{2l}(i) = \Gamma_{2l}(i-1) + 1$ and show that

$$orb(1) = \{2^k \mod (2l+1) : 0 \leq k \leq 2l - 1\}$$

under the permutation $(\widetilde{\Gamma}_{2l})^{-1}$).

8.2 The Infinite Limit

Next we want to establish a more quantitative link between "continuous" and discrete entropies. The transition from the former to the latter proceeds over the discretization and truncation of orbits.

For simplicity, we will consider an *ergodic* map f on the unit interval $I = [0, 1]$ endowed with the Borel sigma-algebra, preserving a measure μ. Without loss of generality, let $\iota = \{I_i : 0 \leq i \leq 10^k - 1\}$, with $I_i = [i10^{-k}, (i+1)10^{-k})$ for $0 \leq i \leq 10^k - 2$ and $I_{10^k-1} = [1 - 10^{-k}, 1]$, be a box partition of I with norm $\|\iota\| = 10^{-k}$. Therefore, the *alphabet* of the ensuing ergodic symbolic dynamic $\mathbf{X}^\iota$ of f with respect to the partition ι is $S = \{0, 1, \ldots, 10^k - 1\}$. Furthermore, let $\{x_j = f^j(x_0) : j \geq 0\}$ be a generic trajectory. Given x_0 and $\|\iota\|$, there is a maximal $M \leq |S| = 10^k$ such that all points in the initial segment $\{x_j = f^j(x_0) : 0 \leq j \leq M - 1\}$ fall in different bins I_i of the partition ι, hence $S_i^\iota(x_0) \neq S_j^\iota(x_0)$ for all $0 \leq i, j \leq M - 1$ and $i \neq j$. This allows us to define a permutation (actually, a cycle) F_M on $S_M = \{0, 1, \ldots, M - 1\}$ in the following way. First, arrange the symbols $s_n = X_n^\iota(x_0) \in S$, $0 \leq n \leq M - 1$, according to their sizes,

$$s_{n_0} < s_{n_1} < \cdots < s_{n_{M-1}}. \tag{8.7}$$

Then define

$$F_M(i) = j \Leftrightarrow \begin{cases} \text{(i) } n_i \neq M - 1 \text{ and } s_{n_i+1} = s_{n_j} \text{ or} \\ \text{(ii) } n_i = M - 1 \text{ and } n = M - 1. \end{cases}$$

By construction, F_M is order isomorphic (Definition 1) to the permutation $\widetilde{F}_M : S \to S$ defined as

$$\widetilde{F}_M(s_n) = \begin{cases} s_{n+1} & \text{for } n = 0, \ldots, M - 2, \\ s_0 & \text{for } n = M - 1. \end{cases}$$

Note that $\widetilde{F}_M$ is a coarse-grained version of f, conveniently "short circuited" at the last orbit point by sending it back to the first one. Let $\phi:(S, <) \to (S_M, <)$ be the order isomorphism $s_{n_i} \mapsto i$ (so, $\widetilde{F}_M = \phi^{-1} \circ F_M \circ \phi$). In particular, if $s_i = s_{n_{k(i)}}$ then s_i and $n_{k(i)}$ define the same ordinal patterns of lengths $l = 2, \ldots, M$ under $\widetilde{f}_M$ and

F_M, respectively. With $\|\iota\| \to 0$, it follows that $x_i \in I$, $s_i \in S$, and $n_{k(i)} \in S_M$ define the same ordinal patterns $\pi \in S_l$ for arbitrarily long l.

On the other hand, since the map $f{:}I \to I$ is ergodic with respect to the measure μ, its entropy and permutation entropy can be determined from a typical trajectory, i.e., except for a set of initial conditions of measure zero. To be specific, let (i) $S^{\mathbb{N}_0}$ be the sample path space of the ergodic process $\mathbf{X}^\iota$, (ii) m_ι the measure induced by μ on $S^{\mathbb{N}_0}$

$$m = \mu \circ \Phi^{-1}$$

(see (B.22)), where $\Phi{:}I \to S^{\mathbb{N}_0}$ is the coding map (1.6) with respect to the partition ι, and (iii) Σ the shift transformation (1.8). Furthermore, for $L \leq M$ and $\pi \in S_L$ set

$$P_\pi = \{s_0^\infty \in S^{\mathbb{N}_0}{:}s_{\pi_0} < s_{\pi_1} < \cdots < s_{\pi_{L-1}}\} \in \mathcal{P}_L$$

(notation as in (3.4) and (3.5)), and

$$Q_\pi = \{i \in S_M{:}F_M^{\pi_0}(i) < \cdots < F_M^{\pi_{L-1}}(i)\}$$

(notation as in (8.1)). Observe that in virtue of the order isomorphy ϕ between the permutations $\tilde{F}_M$ and F_M, $s_0^\infty \in P_\pi$ if and only if $\phi(s_0) \in Q_\pi$.

More generally, consider the shift Σ on the sequences $s_0^\infty \in S^{\mathbb{N}_0}$. Then $\Sigma^n(s_0^\infty) = (s_n, s_{n+1}, \ldots) \in P_\pi$, $0 \leq n \leq M - L$, if and only if $\phi(s_n) \in Q_\pi$, $\pi \in S_L$. Apply now the ergodic theorem (Theorem 21) to the dynamical system $(S^{\mathbb{N}_0}, \mathcal{B}_\Pi(S), m_\iota, \Sigma)$ to conclude that, for any $\varepsilon_1 > 0$, there exists a uniform partition ι_0 of I such that

$$\left| m_\iota(P_\pi) - \frac{|\{\Sigma^n(s_0^\infty) \in P_\pi{:}0 \leq n \leq M - L\}|}{M - L + 1} \right| < \varepsilon_1$$

for all $\pi \in S_L$ and almost all $s_0^\infty \in S^{\mathbb{N}_0}$, if $\|\iota\| \leq \|\iota_0\|$ (and, consequently, $M \geq M_0$). The greater the window size L, the greater the sample size $M - L + 1$ (hence, the greater M) we need to estimate $m_\iota(P_\pi)$ with the same precision. Furthermore,

$$\left| q_\pi(L) - \frac{|\{\Sigma^n(s_0^\infty) \in P_\pi{:}0 \leq n \leq M - L\}|}{M - L + 1} \right| < \varepsilon_2, \tag{8.8}$$

where, similar to (8.2),

$$q_\pi(L) = \frac{|Q_\pi(L)|}{\sum_{\pi \in S_L} |Q_\pi(L)|} = \frac{|Q_\pi(L)|}{M} \tag{8.9}$$

(since F_M is a cycle of length M), and the error $\varepsilon_2 = O(1/(M - L))$ stems from the different denominators in (8.8) and (8.9), and also from the last $L - 1$ points $s_{M-L+1}, \ldots, s_{M-1}$, whose size-$L$ windows stretch outside the orbit segment s_0^{M-1}. All in all,

$$\left| h^*((X^{\iota})_0^{L-1}) - h_{\delta}^{(L)}(F_M) \right|$$

$$= \frac{1}{L-1} \left| \sum_{\pi \in \mathcal{S}_L} m_{\iota}(P_{\pi}) \log m_{\iota}(P_{\pi}) - \sum_{\pi \in \mathcal{S}_L} q_{\pi} \log q_{\pi} \right|$$

$$\leq \frac{1}{L-1} \left(\varepsilon_1 |\mathcal{P}_L| \log |\mathcal{P}_L| + O\left(\frac{1}{M-L}\right) |\mathcal{P}_L| \log |\mathcal{P}_L| \right)$$

$$+ \text{ terms of higher order in } M \text{ and } L,$$

i.e., the permutation entropy of order $L \ll M$ of the process $\mathbf{X}^{\iota}$ coincides approximately with the discrete entropy of order L of F_M, the permutation of $\{0, 1, \ldots, M-1\}$ obtained from f in the way explained before.

The first term of the error,

$$e_1 = \frac{\varepsilon_1}{L} |\mathcal{P}_L| \log |\mathcal{P}_L|,$$

can be made arbitrarily small by taking M sufficiently large. In fact, since $|\mathcal{P}_L| = O(L^{L+1/2}e^{-L}) = O(e^{L(\ln L - 1) + (1/2)\ln L})$ (in general, a rough estimate), it suffices to take (a) $M \geq \max\{M_0, -\ln \varepsilon_1 / \ln L\}$ to derive $e_1 = o(L^{-(M-L)})$. As for the second term,

$$e_2 = \frac{1}{L-1} O\left(\frac{1}{M-L}\right) |\mathcal{P}_L| \log |\mathcal{P}_L|,$$

we need (b) $M - L > O(L^{L+1/2}e^{-L} \ln L)$, i.e., $M - L > O(e^{L(\ln L - 1) + (1/2)\ln L + \ln \ln L})$, to make e_2 vanish when $M, L \to \infty$. Therefore if we set, say, $M = Ce^{L \ln L} =: \vartheta^{-1}(L)$ with $C > 0$ large enough so that (a) is also fulfilled, then

$$\frac{1}{\vartheta(M)-1} \sum_{n=2}^{\vartheta(M)} \left| h^*((X^{\iota})_0^{n-1}) - h_{\delta}^{(n)}(F_M) \right| \leq e(M, \vartheta(M)),$$

where $e(M, L) = e_1 + e_2 + $ terms of higher order in M and L, and $e(M, \vartheta(M)) \to 0$ when $M \to \infty$. Letting now $\|\iota\| \to 0$ (hence $M \to \infty$), we get

$$h_{\mu}^*(f) = \lim_{\|\iota\| \to 0} h^*(\mathbf{X}^{\iota}) = \lim_{M \to \infty} h_{\delta}(F_M),$$

provided $h_{\mu}^*(f)$ exists. Since f is ergodic by assumption, Theorem 10 implies

$$\lim_{M \to \infty} h_{\delta}(F_M) = h_{\mu}(f). \tag{8.10}$$

A final caveat. We have supposed that $\mathcal{O}_f(x_0)$ was generic for μ. In order to avoid that different orbits lead to (8.10) with different (ergodic) measures, we suppose furthermore that f is uniquely ergodic (i.e., f is continuous and it has only one invariant measure, see Sect. A.1).

This proves the one-dimensional version of the following theorem.

Theorem 18 *Let $I \subset \mathbb{R}$ be a closed interval and $f:I \to I$ a uniquely ergodic map. Furthermore, let F_M be the permutation of $\{0, 1, \ldots, N - 1\}$ obtained from f in the way explained above. Then $\lim_{M \to \infty} h_\delta(F_M) = h_\mu(f)$, where μ is the only f-invariant Borel measure on I and $h_\mu(f)$ is the metric entropy of f with respect to μ.*

The proof of the general case is analogous to the one-dimensional case. Theorem 18 justifies calling h_δ discrete entropy.

Example 20 In the following numerical simulations, we have used $M = 500,000$ and $2 \leq L \leq 9$. Figure 8.1 compares the discrete entropy with the Lyapunov exponent for the one-dimensional quadratic maps

$$f_a(x) = ax(1 - x), \quad 0 \leq x \leq 1, \quad 3.5 \leq a \leq 4.0. \tag{8.11}$$

Figure 8.2 compares the discrete entropy with the largest Lyapunov exponent for the two-dimensional quadratic maps

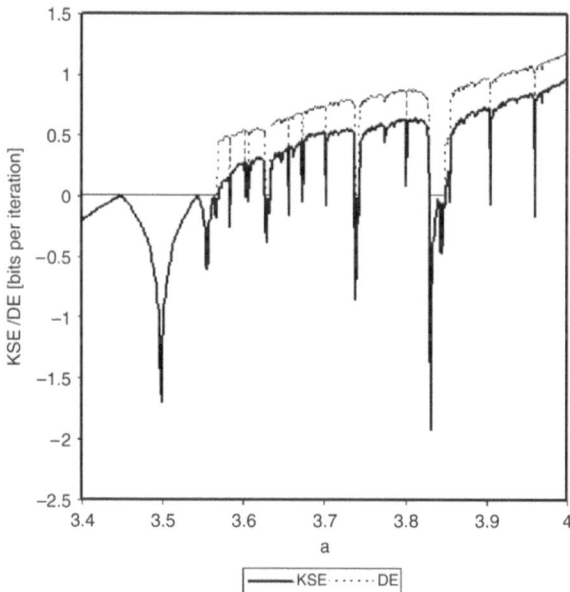

Fig. 8.1 Discrete entropy (*dashed line*) for the one-dimensional quadratic maps (8.11). The discrete entropy tracks the positive part of the Lyapunov exponent (*bold line*) with a uniform bias over the parameter values

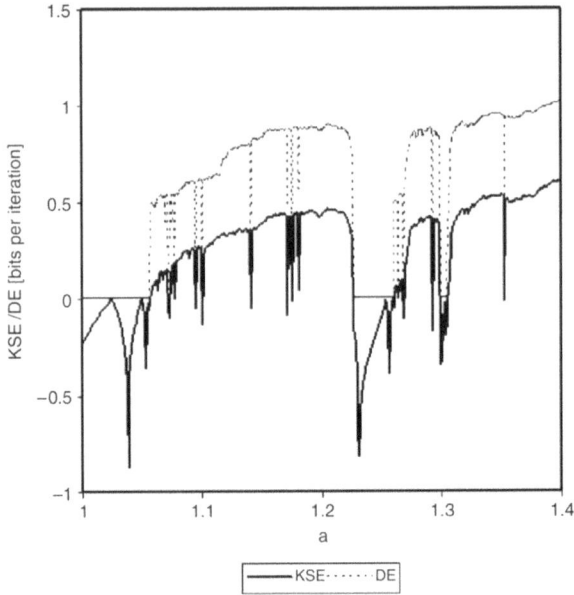

Fig. 8.2 Discrete entropy (*dashed line*) for the two-dimensional quadratic maps (8.12). The discrete entropy traces the positive part of the largest Lyapunov exponent (*bold line*) with a uniform bias over the parameter values

$$f_a(x, y) = (1 - ax^2 + 0.3y, x), \quad 0 \le x, y \le 1, \quad 1 \le a \le 1.4. \tag{8.12}$$

We observe that in both cases, the discrete entropy follows the profile of the positive part of the corresponding Lyapunov exponent with an approximately constant bias, due to the slow (and seemingly uniform) convergence of discrete entropy to its continuous counterpart.

8.3 Discrete Topological Entropy

As a matter of fact, the previous approach to discrete entropy admits some variations, in both concept and implementation. We shall elaborate here only on one of them, based on the topological permutation entropy of a piecewise monotone one-dimensional interval map $f : I \to I$ (Sect. 7.4).

Given a permutation F on $S_M = \{0, 1, \ldots, M - 1\}$, define the partition of S_M

$$\mathcal{Q}_n = \{Q_\pi \ne \emptyset : \pi \in S_n\},$$

with $Q_\pi = \{s \in S_M : F^{\pi_0}(s) < \cdots < F^{\pi_{n-1}}(s)\}$ as in (8.1). Similar to (7.20) and (7.19), we propose the following definition.

Definition 6 We call

$$h_{\delta, \text{top}}^{(n)}(F) = \frac{1}{n - 1} \log |\mathcal{Q}_n|$$

the discrete topological entropy of the permutation F of order n, and

$$h_{\delta,top}(F) = \frac{1}{n_{max} - 1} \sum_{n=2}^{n_{max}} h_{\delta,top}^{(n)}(F)$$

($n_{max} = \max\{n : \mathcal{Q}_n \neq \emptyset\}$) the discrete topological entropyof F.

In order to treat $h_\delta(F)$, (8.6), and $h_{\delta,top}(F)$ on the same footing, one could refer to the first as discrete permutation entropy and write $h_{\delta,per}(F)$ instead.

Let $F_M : S_M \to S_M$ be the permutation obtained from f as explained in Sect. 8.2. The analogue of Theorem 18 for discrete topological entropy holds as well.

Theorem 19 *Let $f{:}I \to I$ be a uniquely ergodic and piecewise monotone map. Then $\lim_{M \to \infty} h_{\delta,top}(F_M) = h_{top}(f)$.*

Proof From the proof of Theorem 18, it follows that $q_\pi \to m_\iota(P_\pi)$ for every $\pi \in S_L$ as $\|\iota\| \to 0$ (or $M \to \infty$). Therefore, $|\mathcal{Q}_n| = |\mathcal{P}_n|$ in that limit, and we get $\lim_{M \to \infty} h_{\delta,top} = h_{top}^*(f) = h_{top}(f)$, the last equality following from Theorem 16. $\square$

From

$$h_{\delta,per}^{(n)}(F) = -\frac{1}{n-1} \sum_{\pi \in \mathcal{S}_n} q_\pi \log q_\pi \leq \frac{1}{n-1} \log |\mathcal{Q}_n| = h_{\delta,top}^{(n)}(F)$$

for $n = 2, 3, \ldots, n_{max}$, we deduce that the discrete topological entropy is an upper bound of the discrete (permutation) entropy—the same as for their "continuous" counterparts.

Example 21 For the permutation

$$F = [3, 5, 1, 7, 0, 6, 2, 4] = (0, 3, 7, 4)(1, 5, 6, 2),$$

we get

$$h_{\delta,top}^{(2)}(F) = \log 2 = 1 \text{ bit/symbol},$$

$$h_{\delta,top}^{(3)}(F) = \frac{1}{2} \log 6 = 1.2925 \text{ bit/symbol},$$

$$h_{\delta,top}^{(4)}(F) = \frac{1}{3} \log 8 = 1 \text{ bit/symbol},$$

so that $h_{\delta,top}(F) = 1.0975$ bit/symbol. As for

$$\Gamma_8 = [4, 0, 5, 1, 6, 2, 7, 3] = (0, 4, 6, 7, 3, 1)(2, 5),$$

permutation (8.5) on $S_8 = \{0, 1, \ldots, 7\}$ with maximal discrete Lyapunov exponent, we find

$$h^{(2)}_{\delta,top}(\Gamma_8) = \log 2 = 1 \text{ bit/symbol},$$

$$h^{(3)}_{\delta,top}(\Gamma_8) = \frac{1}{2} \log 4 = 1 \text{ bit/symbol},$$

$$h^{(4)}_{\delta,top}(\Gamma_8) = \frac{1}{3} \log 6 = 0.8617 \text{ bit/symbol},$$

$$h^{(5)}_{\delta,top}(\Gamma_8) = \frac{1}{4} \log 6 = 0.6462 \text{ bit/symbol},$$

$$h^{(6)}_{\delta,top}(\Gamma_8) = \frac{1}{5} \log 6 = 0.5170 \text{ bit/symbol},$$

and $h_{\delta,top}(\Gamma_8) \approx 0.8050$ bit/symbol. Thus $h_{\delta,top}(\Gamma_8) < h_{\delta,top}(F)$. The same is true for the discrete permutation entropy: $h_{\delta,per}(\Gamma_8) = 0.7968$ bit/symbol $< h_{\delta,per}(F) = 1.0833$ bit/symbol.

Chapter 9
Detection of Determinism

In Chap. 2 we have illustrated the applications of ordinal patterns with four examples. In this chapter we present a further application, this time to the detection of determinism in noisy time series. Following the common usage of the term in applied science, "determinism" is meant here as the opposite to statistical independence, hence it includes colored noise as well. This application hinges on two basic properties of ordinal patterns: existence of forbidden patterns in the orbits of maps (Sects. 1.2, 3.3, and 7.7) and robustness to observational noise (Sects. 3.4.3, and 9.1). We shall actually present two detection methods.

Method I is based on the number of missing ordinal patterns. It proceeds by (i) counting the number of missing ordinal patterns in sliding, overlapping windows of size L along the data sequence, (ii) randomizing the sequence, and (iii) repeating (i) with the randomized sequence. Is the result of step (iii) clearly greater than the result of step (i), so may we conclude that the original noisy sequence has a deterministic component.

Method II is based on the distribution of the visible ordinal patterns. This method proceeds by (i) counting the number of ordinal patterns in sliding, non-overlapping windows of size L along the data sequence and (ii) performing a χ^2 test based on the results of (i), the null hypothesis being that the data are white noise. Hold the null hypothesis, so should all possible ordinal L-patterns be visible and evenly distributed over sufficiently many windows, at variance with what happens in the case of noisy deterministic data. In the latter case, the number of missing ordinal patterns is higher, its decay rate with L is slower, and the distribution of patterns is not necessarily uniform.

Both methods, as other applications of permutation entropy, are conceptually simple and computationally fast for moderate values of L. But not only this: Method II compares favorably to the popular Brock–Dechert–Scheinkman (BDS) independence test when applied to time series projected from the attractors of the Lorenz map and the time-delayed Hénon map. The bottom line is that determinism in noisy multivariate time series can be detected by observing a single component, a possibility that can come in handy in experimental situations.

Noisy univariate and multivariate time series have been intensively studied in the last few decades [1, 112]. Depending on the noise level of the data, one can expect to recover the full deterministic dynamics, to reconstruct the geometry of the noise-free

J.M. Amigó, *Permutation Complexity in Dynamical Systems,*
Springer Series in Synergetics, DOI 10.1007/978-3-642-04084-9_9,
© Springer-Verlag Berlin Heidelberg 2010

signal in some appropriate space, or just to ascertain the existence of an underlying determinism. The ordinal pattern-based methods described in this chapter falls in the third category. As a compensation for such a seemingly modest accomplishment, it has a remarkable success even with very high levels of noise. Besides the BDS method, which is based on the correlation dimension, other detection methods for determinism use the smoothness of the measure along reconstructed trajectories [164], functionals of probabilistic distributions [176], or the Higuchi fractal dimension on Poincaré sections [85].

9.1 Dynamical Robustness Against Observational Noise

Ordinal patterns are robust against small additive perturbations on account of being defined by inequalities. This property was called conditional robustness in Sect. 3.4.3. Yet, this property alone would not explain the persistence of forbidden patterns in the very noisy deterministic sequences that we are going to study in the next section. It turns out that, in deterministic sequences, there is a second mechanism for robustness, also in case of multi-dimensional maps—the dynamics itself. The result is an enhancement of the robustness of ordinal patterns against additive noise, which we call *dynamical robustness*. A simple explanation follows.

In the sequel we deal with a time series of the form

$$\xi_n = f^n(x_0) + w_n = x_n + w_n \tag{9.1}$$

($n \in \mathbb{N}_0$, or in practice $0 \le n \le N - 1$), where f is a self-map of the interval $[a, b] \subset \mathbb{R}$ and w_n are independent and uniformly distributed random variables (i.e., uniform white noise) in the interval $[-\eta, \eta]$. In order that the noise destroys a given allowed or forbidden pattern $\pi = \langle \pi_0, \ldots, \pi_{L-1} \rangle$ of the noise-free sequence $(x_n)_{n \in \mathbb{N}_0}$, it must happen that

$$x_{\pi_i} < x_{\pi_{i+1}}$$

but

$$x_{\pi_i} + w_{\pi_i} > x_{\pi_{i+1}} + w_{\pi_{i+1}}$$

for some $0 \le i \le L - 2$ and $w_{\pi_i}, w_{\pi_{i+1}} \in [-\eta, \eta]$. If η is small, this will be only possible if $x_{\pi_i} \approx x_{\pi_{i+1}}$, i.e., if $f^{\min\{\pi_i, \pi_{i+1}\}}(x_0)$ is an "approximately" periodic point with period $|\pi_i - \pi_{i+1}|$. We conclude that, indeed, the dynamics imposes an extra condition on $x_{\pi_i}, x_{\pi_{i+1}}$ so that a small amplitude perturbation can reverse their order.

To put some numbers on this argument, take $f(x) = 4x(1 - x)$, $0 \le x \le 1$, the logistic map. We know that for $\eta = 0$ this map has one forbidden 3-pattern, namely, $\langle 2, 1, 0 \rangle$ (Fig. 1.6). In other words, there exists no $x \in [0, 1]$ such that $f^2(x) < f(x) < x$. The pattern $\langle 2, 1, 0 \rangle$ can appear in the noisy sequence (9.1) by a single order reversal if the noise changes the order of x_n, x_{n+1} or the order of x_{n+1}, x_{n+2} in the allowed patterns

$$x_{n+2} < x_n < x_{n+1} \quad \text{or} \quad x_{n+1} < x_{n+2} < x_n,$$

respectively. In the first case, this requires $x_n \approx x_{n+1} = f(x_n)$, i.e., x_n must be close to any of the two fixed points of the map: $x = 0$ or $x = \frac{3}{4}$ (see Fig. 1.5). In the second case, the same applies to x_{n+1} and $x_{n+2} = f(x_{n+1})$. Therefore, it suffices to discuss the first case.

Consider the fixed point $x = 0$ and take $x_n = \delta > 0$. Then $x_{n+1} = f'(0)\delta + R\delta^2$, where R can be estimated with the remainder of the Taylor series. Since $\xi_n \in [x_n - \eta, x_n + \eta] =: I_n$, the inequality $\xi_{n+1} < \xi_n$ can be fulfilled only if the intervals I_n and I_{n+1} overlap, i.e., if

$$\delta \le \delta_0(\eta) = \frac{1 - f'(0) + \sqrt{(1 - f'(0))^2 + 8R\eta}}{2R}. \tag{9.2}$$

One can analogously estimate $\delta_+(\eta) > 0$ and $\delta_-(\eta) > 0$ such that if $x_n \in [\frac{3}{4} - \delta_-(\eta), \frac{3}{4} + \delta_+(\eta)]$, then x_n is sufficiently close to $x = \frac{3}{4}$ again in the sense that the inequality $\xi_{n+1} < \xi_n$ can hold for η small.

Thus, the probability $\Pr(\eta)$ for two consecutive orbit points (x_n, x_{n+1} or x_{n+1}, x_{n+2}) to lie sufficiently close to either fixed point so as the pattern $\langle 2, 1, 0 \rangle$ becomes observable in a noisy orbit of the logistic map by means of a single order reversal is

$$\Pr(\eta) = \mu([0, \delta_0(\eta)]) + \mu([\tfrac{3}{4} - \delta_-(\eta), \tfrac{3}{4} + \delta_+(\eta)]),$$

where μ is the natural invariant measure for the logistic map,

$$\mu([c, d]) = \int_c^d \frac{dx}{\pi \sqrt{x(1 - x)}}$$

(see (1.20)). To make the argument even simpler, observe that once two consecutive orbit points in x_n, x_{n+1}, x_{n+2} are close to a fixed point, we may assume that the third one is around as well. In this case, the type of $\xi_n, \xi_{n+1}, \xi_{n+2}$ is going to depend basically on the type of w_n, w_{n+1}, w_{n+2}.

Consider now a string of length N, $\xi_0^{N-1} = \xi_0, \xi_1, \ldots, \xi_{N-1}$, along with the $\lfloor \frac{N}{3} \rfloor$ independent random vectors $\xi_n^{n+2} = \xi_n, \xi_{n+1}, \xi_{n+2}, n = 0, 3, 6, \ldots$. If we pick one of those vectors, the probability $\Pr(\langle 2, 1, 0 \rangle)$ that $\xi_{n+2} < \xi_{n+1} < \xi_n$ holds is then

$$\Pr(\langle 2, 1, 0 \rangle) \approx \Pr(\eta)\Pr\{w_{n+2} < w_{n+1} < w_n\}$$
$$= \Pr(\eta) \cdot \frac{1}{6}.$$

In order to verify these results, the probability P of finding at least once the pattern $\langle 2, 1, 0 \rangle$ in any of the $\lfloor \frac{N}{3} \rfloor$ windows $\xi_{3n}, \xi_{3n+1}, \xi_{3n+2}$ of the noisy time series $(\xi_n)_{n=0}^{N-1}$, (9.1), was calculated numerically. From the reasoning above, this probability should be close to $1 - (1 - \Pr(\eta)/6)^{\lfloor \frac{N}{3} \rfloor}$ for the logistic map contaminated with additive, uniform white noise of small amplitude η, whereas it should be $1 - (1 - 1/6)^{\lfloor \frac{N}{3} \rfloor}$ for uniform white noise only (i.e., $\xi_n = w_n$ in (9.1)). Clearly, the former probability is greater than the latter because $\Pr(\eta)$ is going to be very small. This is confirmed by Fig. 9.1.

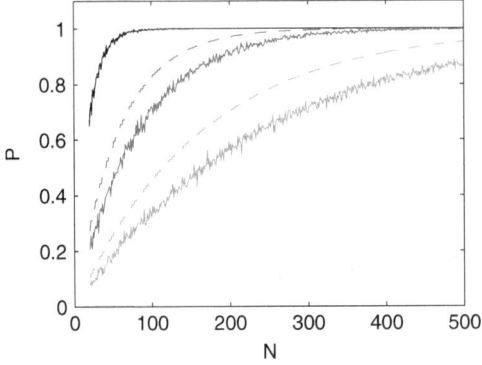

Fig. 9.1 Numerical computation (*continuous line*) and analytical estimation (*dashed*) of the probability P of finding the pattern $\langle 2, 1, 0 \rangle$ in any of the $\lfloor \frac{N}{3} \rfloor$ windows ξ_{3n}, ξ_{3n+1}, ξ_{3n+2} of a time series of length N generated with the logistic map. The noise amplitude is $\eta = 0.0001$ (*light gray*), $\eta = 0.01$ (*gray*), $\eta = 0.1$ (*dark gray*). The *top curve* corresponds to uniform white noise. Clearly the probability P is smaller for a noisy, deterministic time series than for uniform white noise

9.2 Detection of Determinism I: Number of Missing Ordinal Patterns

We already know (Sect. 1.2) that if $(x_n)_{n \in \mathbb{N}_0}$ is a univariate time series generated by a piecewise monotone interval map f, then there exist ordinal patterns which are forbidden for f. The theoretical situation in higher dimensions is less satisfactory in that the existence of forbidden patterns has been proved so far only under the somewhat restrictive condition of expansiveness (Sect. 7.6). There is nevertheless numerical evidence that forbidden ordinal patterns are also a general feature of higher dimensional dynamics. Since, on the other hand, univariate and multivariate random sequences have no forbidden patterns with probability 1, we conclude that the existence of forbidden patterns can be used as a fingerprint of deterministic orbit generation. Here "random sequence" means generated by an unconstrained, stochastic process taking on values in an interval. In summary, the difference between deterministic and random time series is clear-cut from an ordinal-theoretical point of view: the former have forbidden patterns while the latter have not.

However, when it comes to exploit this forbidden pattern-based strategy to detect determinism, two important practical issues arise: finiteness and noise contamination. Finiteness produces *false forbidden patterns*, that is, ordinal patterns which are missing in a finite (segment of a) random sequence without constraints. Noise destroys forbidden patterns; for instance, a forbidden pattern of the "clean" sequence can turn visible because of additive random fluctuations. Let us mention in passing that were not for the observational noise, determinism could be easily ascertained, for example, with graphical methods. It is therefore interesting that ordinal patterns themselves provide the remedy to the two said issues. First of all, the number of false forbidden patterns of a fixed length always decreases with the length of the time series. Second, "true" forbidden patterns (i.e., forbidden patterns for an underlying deterministic dynamics) possess an additional dynamical robustness against additive noise (Sect. 9.1). This translates into a greater number of missing ordinal patterns in a noisy deterministic sequence than in a random one, and also to a slower decay rate with the length of the sequence. We shall shortly present numerical evidence that forbidden patterns persist in very noisy deterministic data—so noisy that the traditional methods [1, 112, 152] fail to uncover the underlying deterministic dynamics. But before coming to this point, let us dwell on some practical issues.

In practice one uses sliding windows of size L to comb a finite sequence $(x_n)_{n=0}^{N}$ for visible ordinal L-patterns. Note that a sequence of length N allows $N - L + 1$ windows of size L, for $2 \leq L \leq N$. Thus, in order to allow every possible ordinal pattern of length L to occur in a time series of length N, the condition $L! \leq N - L + 1$ must hold. Moreover, in cases where undersampling might occur, $N \gg L! + L - 1$ should also hold. As a rule of thumb we chose $(L + 1)! \leq N$ in the numerical simulations below, although $L! \leq N$ would do also in our case (very noisy data). Furthermore, $(x_n)_{n=0}^{N}$ will be initial segments of variable length $N \leq N_{\max} = 8000$, taken from a sequence $(x_n)_{n=0}^{N_{\max}}$. All these constraints leave $L = 4, 5, 6$ as interesting choices for L. In general one takes also moderate values for L, not least because of the sharp increase of the function $L!$.

Under these provisos, suppose now that the ordinal pattern $\pi \in \mathcal{S}_L$ is missing in a finite noise-free time series. Of course, the odds that a *false* forbidden pattern persists in a random or deterministic sequence (or sample of sequences) will decrease exponentially with the number of data (see, e.g., Sect. 9.1). As a result, the number of false forbidden patterns in $(x_n)_{n=0}^{N}$ will decay as N increases up to $N_{\max}$, the number of data at our disposal. Otherwise, if $(x_n)_{n=0}^{N_{\max}}$ is a deterministic *noise-free* time series and π is a forbidden pattern, then π will be missing in $(x_n)_{n=0}^{N}$ for all $N \leq N_{\max}$. In other words, the number of *true* forbidden patterns in $(x_n)_{n=0}^{N}$ does not depend on N.

Consider a fixed initial condition x and suppose that $\pi_{\mathrm{forb}} = \langle \pi_0, \ldots, \pi_{L-1} \rangle$ is a forbidden pattern for f. Suppose furthermore that we switch on a discrete-time random perturbation w_k, $|w_k| \leq w_{\max}$, such that π_{forb} is still missing in the finite sequence $\left(f^k(x) + w_k\right)_{k=0}^{N-1}$ (due to robustness). Observe that the *noisy* time series $\xi_k = f^k(x) + w_k$ can be viewed both as a perturbation of an underlying deterministic dynamics and as a random process correlated with the deterministic dynamics[1] f.

---

[1] Sometimes *colored noise* (i.e., a random process whose variables are statistically dependent) is numerically simulated in this way. For other methods, see, e.g., [113, 83].

If the orbit of x would be infinitely long, then the noisy time series had no missing patterns and π_{forb} would be visible with probability 1. In the finite-length case we are considering, this is in general not the case; rather, there is a threshold $\theta = \theta(N)$ (the greater N, the smaller θ) such that π_{forb} will do appear in $(\xi_k)_{k=0}^{N-1}$ only if $w_{\max} > \theta$. We conclude that amplifying a random perturbation destroys progressively the forbidden patterns of the underlying deterministic dynamic.

In the following we are going to test numerically one of the properties discussed above, namely, the robustness of *true* forbidden patterns against additive random perturbations. In order to estimate the average number $\langle n(L, N) \rangle$ of missing ordinal L-patterns in a finite, noisy sequence of length N,

$$\xi_k = x_k + w_k, \quad 0 \le k \le N - 1,$$

with $x_{k+1} = f(x_k)$ and w_k a random process, we generate 100 samples of length $N_{\max} = 8000$ and normalize the corresponding count of missing patterns of lengths $4 \le L \le 6$. To check the decay of $\langle n(L, N) \rangle$ with N, this parameter is allowed to vary in the range $(L + 1)! \le N \le N_{\max}$. We highlight next a few results obtained with f being the logistic map and w_k being white noise uniformly distributed in the interval $[-w_{\max}, w_{\max}]$, $0 \le w_{\max} \le 1$.

Figure 9.2 shows $\langle n(L, N) \rangle$ when (a) $w_{\max} = 0.25$, (b) $w_{\max} = 0.50$, and (c) $w_{\max} = 1$ and $f^k(x) = 0$ (noise only), respectively. Note the different orders of magnitude of the vertical scales. Needless to say, $\langle n(L, N) \rangle$ decays with increasing N because the greater the N, the more unlikely that an L-pattern is missing in a noisy or random sequence of length N; this is a statistical effect. The important features for us are the magnitude of $\langle n(L, N) \rangle$ and its decay rate with N, since these two properties are tightly related to the forbidden patterns of the underlying deterministic dynamic via robustness: the smaller the $w_{\max}$, the closer we are to the deterministic case, therefore, the more missing ordinal patterns and the slower their decrease with N.

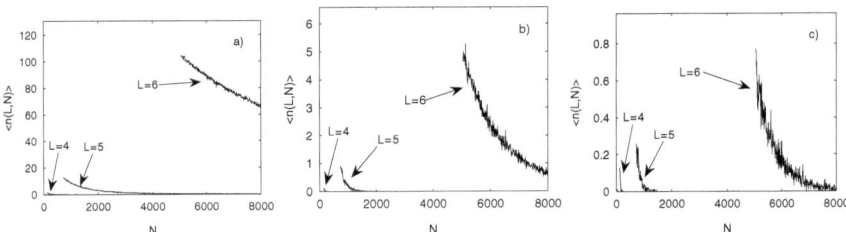

Fig. 9.2 Average number of missing ordinal patterns of length L found in a time series of length N, $\langle n(L, N) \rangle$, for noisy series of the logistic map with $w_{\max} = 0.25$ (**a**), $w_{\max} = 0.5$ (**b**), and for a series of uniformly distributed noise (**c**)

Figure 9.3 depicts ξ_{k+1} vs ξ_k in the previous cases (a) and (b). The higher order of magnitude of, e.g., $\langle n(6, N) \rangle$ in Fig. 9.2(b) as compared to Fig. 9.2(c) signalizes an underlying deterministic law, in spite of the fact that Fig. 9.3(b) hardly gives any clue about this.

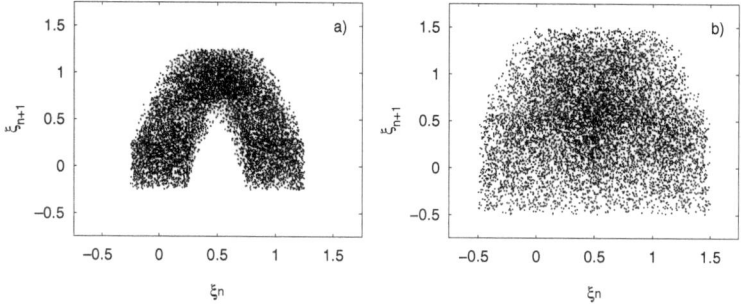

Fig. 9.3 Return map for noisy time series from the logistic map with $w_{max} = 0.25$ (**a**) and $w_{max} = 0.5$ (**b**). In the latter case, the high noise level does not allow to recognize the underlying deterministic dynamics. However, the number of missing ordinal patterns is sensibly higher than in the purely random case

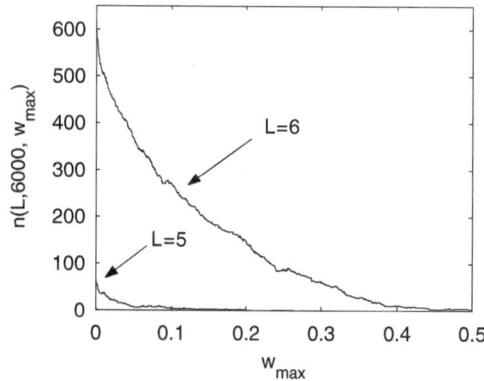

Fig. 9.4 Number of missing ordinal patterns of length L found in a noisy time series of the logistic map with length 6000 vs the uniform noise amplitude w_{max}

Finally, Fig. 9.4 nicely illustrates the resistance of the true forbidden patterns to disappear with increasing noise levels. In this figure, $N = 6000$, $L = 5, 6$, and $0 \leq w_{max} \leq 0.5$.

These numerical simulations suggest the following simple-minded, three-step method to discriminate noisy, deterministic, finite time series from random ones, at least when the noise is white.

(a) Compute the number of missing ordinal L-patterns of adequate length (say $(L + 1)! \leq N$) in sliding windows along the sequence. It is convenient to use segments of variable length N and to draw the corresponding curves, as in Fig. 9.2.
(b) Randomize the sequence, i.e., change the temporal structure of the data in a random way.
(c) Proceed as in step (a) with the randomized sequence.

If the results of (a) and (c) are about the same, the sequence is very likely not deterministic (or the observational noise is so strong as compared to the deterministic signal that the latter has been completely masked). Otherwise, the sequence stems from a deterministic one. Needless to say, the method is more reliable if a statistically significant sample of sequences can be obtained, for instance, by cutting a long sequence into shorter pieces. In the next section we discuss a more quantitative method.

9.3 Detection of Determinism II: Distribution of Visible Ordinal Patterns

Consider once more a univariate or multivariate time series of the form

$$\xi_n = f^n(x_0) + w_n, \tag{9.3}$$

$(0 \le n \le N - 1)$ where w_n is white noise, i.e., outcomes of an independent and identically distributed (i.i.d.) random process. In order to differentiate white noise from a noisy deterministic time series of form (9.3), the perhaps simplest tool consists in counting visible ordinal patterns before and after randomizing the time series under scrutiny; depending on whether the number of visible patterns remains about the same or decreases significantly, we may conclude that the series is random or deterministic, respectively. This is the method discussed in Sect. 9.2.

A more quantitative method calls for performing a chi-square test based on the count of visible ordinal patterns. The *null hypothesis* reads

$$H_0: \quad \text{the } \xi_n \text{ are i.i.d.} \tag{9.4}$$

From a statistical point of view, this method is going to be a test of independence since the alternative to H_0 includes also colored noise.

The method goes as follows. Take sliding windows of size $L \ge 2$, overlapping at a single point (i.e., the last point of a window is the first point of the next one) down the sequence $\xi_0^{N-1} = \xi_0, \ldots, \xi_{N-1}$. For brevity, we call them "non-overlapping" windows. The number of such windows is

$$K = \left\lfloor \frac{N-1}{L-1} \right\rfloor, \tag{9.5}$$

each comprising the entries

$$\mathbf{e}_k = \xi_{kL-k}, \ldots, \xi_{(k+1)L-(k+1)}, \quad 0 \le k \le K - 1.$$

Notice that if the variables $\xi_0, \xi_1, \ldots, \xi_{N-1}$ are independently drawn from the same probability distribution, then the ordinal L-patterns defined by the components of $\mathbf{e}_k \in \mathbb{R}^L$, which we denote by $\pi(\mathbf{e}_k) \in \mathcal{S}_L$, will also be independent and, moreover, uniformly distributed random variables. Therefore, if one or several ordinal patterns are missing in a sample obtained using non-overlapping windows, this might be a statistically significant signal that independence and/or the equality of the distribution are/is not fulfilled.

Given the non-overlapping windows $\{\mathbf{e}_k \in \mathbb{R}^L : k \geq 0\}$ corresponding to an arbitrarily long time series $\{\xi_n : n \geq 0\}$, suppose that some ordinal patterns of length L are missing in the initial segment $\xi_0, \xi_1, \ldots, \xi_{N-1}$. Let v_π be the number of $\mathbf{e}_k$'s such that $\mathbf{e}_k$ is of type $\pi \in \mathcal{S}_L$ (i.e., $\pi(\mathbf{e}_k) = \pi$). Thus, $v_\pi = 0$ means that the L-pattern π has not been observed.

In order to accept or reject the null hypothesis H_0, (9.4), based on our observations, we apply a chi-square goodness-of-fit hypothesis test with statistic [135]

$$
\begin{aligned}
\chi^2(L) &= \sum_{\pi \in \mathcal{S}_L} \frac{(v_\pi - K/L!)^2}{K/L!} \\
&= \frac{L!}{K} \left(\sum_{\pi \in \mathcal{S}_L} v_\pi^2 - 2\frac{K}{L!} \sum_{\pi \in \mathcal{S}_L} v_\pi + \left(\frac{K}{L!}\right)^2 \sum_{\pi \in \mathcal{S}_L} 1 \right) \\
&= \frac{L!}{K} \sum_{\pi \in \mathcal{S}_L} v_\pi^2 - 2K + K \\
&= \frac{L!}{K} \sum_{\pi \in \mathcal{S}_L : \text{visible}} v_\pi^2 - K,
\end{aligned}
\tag{9.6}
$$

since (i) $\sum_{\pi \in \mathcal{S}_L} v_\pi = K$ and (ii) $v_\pi = 0$ if π is missing. Here $K/L!$ is the expected relative frequency of an ordinal L-pattern, if H_0 holds true. In the affirmative case, $\chi^2 = \chi^2(L)$ converges in distribution (as $K \to \infty$) to a chi-square distribution with $L! - 1$ degrees of freedom. Thus, for large K, a test with approximate level α is obtained by rejecting H_0 if $\chi^2 > \chi^2_{L!-1, 1-\alpha}$, where $\chi^2_{L!-1, 1-\alpha}$ is the upper $1 - \alpha$ critical point for the chi-square distribution with $L! - 1$ degrees of freedom [135]. In our case, the hypothetical convergence of χ^2 to the corresponding chi-square distribution may be considered sufficiently good if $v_\pi > 10$ for all visible L-patterns π, and

$$
\frac{K}{L!} > 5.
\tag{9.7}
$$

Notice that since this test is based on distributions, it could happen that a deterministic map has no forbidden L-patterns, thus $v_\pi \neq 0$ for all $\pi \in \mathcal{S}_L$; however, the null hypothesis be rejected because those v_π's are not evenly distributed.

9.4 A Benchmark

A well-known benchmark for independence in time series is the Brock–Dechert–
Scheinkman (BDS) test [38, 193], which is based on the correlation dimension.
Since the numerical simulations below use the algorithm provided in [136], we
follow this reference for the basics of the BDS test.

Let X_t, $t \geq 1$, be i.i.d. random variables, and

$$I_\epsilon(x, y) = \begin{cases} 1 & \text{if } |x - y| < \epsilon, \\ 0 & \text{otherwise.} \end{cases}$$

The probability that two length-m vectors are within ϵ can be estimated by the correlation sum

$$C_{m,n}(\epsilon) = \frac{2}{n(n-1)} \sum_{s=1}^{n} \sum_{t=s+1}^{n} \prod_{j=0}^{m-1} I_\epsilon(X_{s-j}, X_{t-j}).$$

It is shown in [38] that

$$W_{m,n}(\epsilon) = \sqrt{n} \, \frac{C_{m,n}(\epsilon) - C_{1,n}^m(\epsilon)}{\sigma_{m,n}(\epsilon)}$$

converges in distribution to a standard normal distribution. The normalization
$\sigma_{m,n}(\epsilon)$ is given by

$$\sigma_{m,n}^2(\epsilon) = 4 \left[B^m + 2 \sum_{j=1}^{m-1} B^{m-j} C^{2j} + (m-1)^2 C^{2m} - m^2 BC^{2m-2} \right],$$

where C is consistently estimated by $C_{1,n}(\epsilon)$ and B can be estimated by

$$B_n(\epsilon) = \frac{6}{n(n-1)(n-2)} \sum_{t=1}^{n} \sum_{s=t+1}^{n} \sum_{r=s+1}^{n} h_\epsilon(X_t, X_s, X_r),$$

$$h_\epsilon(i, j, k) = \frac{1}{3} \left[I_\epsilon(i, j)I_\epsilon(j, k) + I_\epsilon(i, k)I_\epsilon(k, j) + I_\epsilon(j, i)I_\epsilon(i, k) \right].$$

A statistically significant non-zero value of $W_{m,n}(\epsilon)$ is evidence for determinism
in the univariate time series $\{X_t : t \geq 1\}$.

This method relies on the selection of the parameters m and ϵ. Following the
usual procedure [140], we take $\epsilon = 0.9^j$ with $j = 0, 1, 2, \ldots$. The criterion to say
whether a combination of m and ϵ is "adequate" call for evaluating if a random time
series is accepted as deterministic using this test the number of cases prescribed by
the significance level of the test α.

9.5 Numerical Simulations

As underlying deterministic time series we use projections on the first coordinate of orbits generated by the Lorenz and time-delayed Hénon maps (this amounts in practice to using the standard lexicographical order). The additive noise w_n is modeled as Gaussian white noise,

$$\mathbb{E}(w_m \cdot w_n) = \sigma^2 \delta_{mn}$$

($\mathbb{E}$ stands for expectation value), with different standard deviations σ. Simulations with uniformly distributed noise yield similar results.

Two kinds of results are going to be presented in the two next sections: (i) Plots of the number of missing ordinal patterns as in Sect. 9.2 and (ii) plots of the distribution of the χ^2 statistic. Although the first ones provide only qualitative information, they can eventually complement the information provided by the second ones, as we shall see in the case of the Lorenz map. The specifics of plots (i) and (ii) are as follows.

(i) Let $N_{\max}$ denote the length of the data sequence under scrutiny and let $n(L, N)$ be the number of missing L-patterns in the initial segment $\xi_0, \xi_1, \ldots, \xi_{N-1}$ of variable length $N \leq N_{\max}$. The numbers $n(L, N)$ are determined with *over-lapping* sliding windows of sizes $4 \leq L \leq 7$. In order to make the most of sequences of length $N_{\max} = 8000$, we take this time

$$L! \lesssim N \leq N_{\max}.$$

An average number $\langle n(L, N) \rangle$ is then estimated from 100 sequences.

(ii) *Non-overlapping* windows are used for the chi-square test of independence based on the distributions of ordinal L-patterns, with statistic (9.6)

$$\chi^2 = \chi^2(L) = \frac{L!}{K} \sum_{\pi \in \mathcal{S}_L : \text{visible}} v_\pi^2 - K. \tag{9.8}$$

Here, $K = \left\lfloor \frac{N-1}{L-1} \right\rfloor$ is the number of non-overlapping windows of size L in a data sequence of length N, (9.5). The window sizes in the simulations are $L = 4, 5$. For $L = 4$, the acceptance/rejection thresholds of the null hypothesis (9.5) at levels $\alpha = 0.10, 0.05$ are

$$\chi^2_{23, 0.90} = 32.01, \quad \chi^2_{23, 0.95} = 35.17, \tag{9.9}$$

respectively. For $L \geq 5$, corresponding to degrees of freedom over 100, the following approximation for the thresholds $\chi^2_{L!-1, 1-\alpha}$ is used [135]:

$$\chi^2_{L!-1, 1-\alpha} \approx (L! - 1) \left(1 - \frac{2}{9(L! - 1)} + z_{1-\alpha} \sqrt{\frac{2}{9(L! - 1)}} \right)^3,$$

where $z_{1-\alpha}$ is the upper $1 - \alpha$ critical point for the standard normal distribution, $\mathcal{N}(0, 1)$; in particular, $z_{0.90} = 1.282$ and $z_{0.95} = 1.645$. Thus,

$$\chi^2_{119, 0.90} = 139.15, \quad \chi^2_{119, 0.95} = 145.46. \tag{9.10}$$

Remember from (9.7) that $5L! \lesssim K$ should hold for the chi-square test to be statistically significant. Therefore

$$5L! \lesssim \frac{N}{L-1},$$

i.e., $N \gtrsim 5(L - 1)L!$. In consequence we take sequences of length $N = 1000$ for $L = 4$ and $N = 8000$ for $L = 5$. To plot the χ^2-value distribution, a sample of $10,000$ sequences was used.

The numerical results are summarized in the following two sections.

9.5.1 The Lorenz Map

The *Lorenz map* [193] is defined as

$$x_{n+1} = x_n y_n - z_n, \quad y_{n+1} = x_n, \quad z_{n+1} = y_n. \tag{9.11}$$

It has an attractor with *Kaplan–Yorke dimension* $D_{KY} = 2$ [193]. Assuming the well-tested Kaplan–Yorke conjecture $D_{KY} = D_1$, where D_1 is the *information dimension*, then the *fractal dimension* D_0 satisfies

$$D_0 \geq D_1 = 2.$$

Figure 9.5 shows the return map $\xi_{n+1} = x_{n+1} + w_{n+1}$ vs $\xi_n = x_n + w_n$ for a typical orbit of the Lorenz map on its attractor and additive Gaussian white noise w_n with $\sigma = 0.25$ (SNR$^2 \simeq 10$ dB). The geometry of the attractor has been completely washed out by the noise, but the underlying determinism can still be detected because of the different count of missing ordinal patterns before (Fig. 9.6) and after (Fig. 9.7) switching off the deterministic signal. Not only the count of missing ordinal patterns is different in these two cases, but also their decay rate with N. The different behavior in Fig. 9.6 of the curve $L = 4$, on the one hand, and the curves $L \geq 5$, on the other hand, strongly indicates that the Lorenz map has no forbidden 4-patterns.

Figure 9.8 shows the distribution of the statistic χ^2, (9.8), obtained from $10,000$ projections x_0^{N-1} of orbits of the Lorenz map, contaminated with additive Gaussian noise with $\sigma = 0.25, 0.50$ (SNR $\simeq 10, 4.0$ dB, respectively). Since the rejection

---

[2] SNR is short for "signal-to-noise ratio" and dB is short for "decibel."

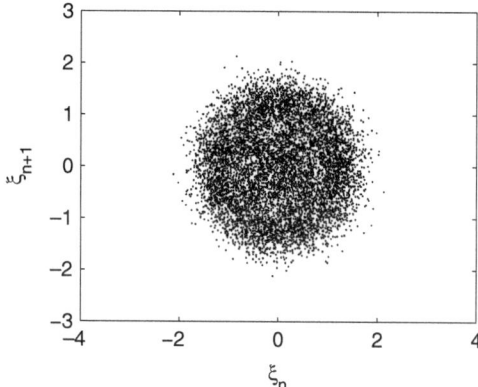

Fig. 9.5 Return map for a time series of the Lorenz map contaminated with Gaussian white noise with $\sigma = 0.25$ (SNR $\simeq$ 10 dB). The structure of the underlying chaotic attractor has been totally blurred. However, the count of missing ordinal patterns is sensibly higher than in the purely random case

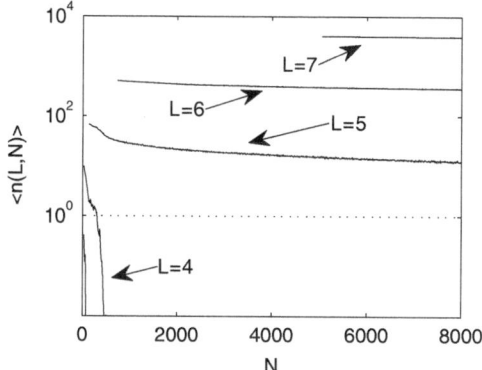

Fig. 9.6 Average number of missing ordinal patterns of length L found in a time series of length N, $\langle n(L, N) \rangle$ (in logarithmic scale), for a noisy series of the Lorenz map with $\sigma = 0.25$ (SNR $\simeq$ 10 dB)

threshold of the null hypothesis H_0 (9.4) at level $\alpha = 0.05$ is $\chi^2_{23, 0.95} = 35.17$ in (a) and $\chi^2_{119, 0.95} = 145.46$ in (b), see (9.9), the χ^2 test clearly detects determinism. It is worth noticing that the rejection of H_0 in case (a) is due to the non-uniform distribution of ν_π since, according to Fig. 9.6, all 4-patterns are visible in noisy time series generated by the Lorenz map with $N \gtrsim 500$ and $\sigma = 0.25$.

Finally, the comparison with the BDS test is shown in Fig. 9.9. There we show the probability P of rejecting the null hypothesis (9.4) for the 27 possible adequate BDS tests on a time series $\xi_0^{N-1} = (x_n + w_n)_{n=0}^{N-1}$ of length $N = 1000$, where now w_n is Gaussian white noise with $0 \leq \sigma \leq 2$. In the same figure it is also

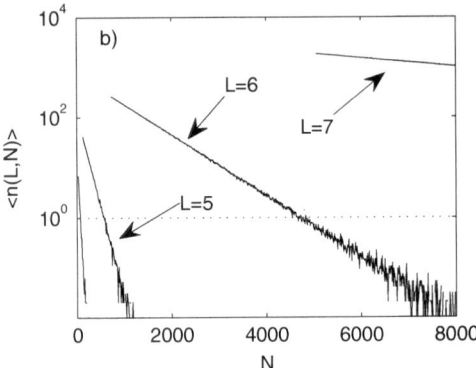

Fig. 9.7 Average number of missing ordinal patterns of length L found in a time series of length N, $\langle n(L,N) \rangle$ (in logarithmic scale), for time series of Gaussian white noise with $\sigma = 0.25$

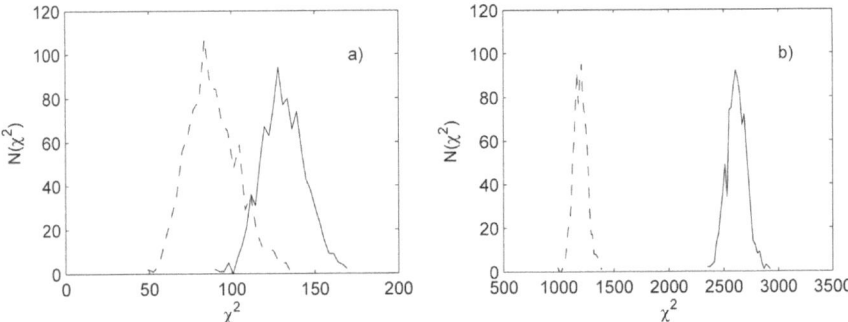

Fig. 9.8 Distribution $N(\chi^2)$ of χ^2 for 10, 000 noisy sequences generated with the Lorenz map, for $L = 4$, $N = 1000$, $\sigma = 0.25$ (*continuous line*) and $\sigma = 0.50$ (*dashed line*) (SNR $\simeq$ 10, 4.0 dB, respectively) (**a**) and for $L = 5$, $N = 8000$, $\sigma = 0.25$ (*continuous line*) and $\sigma = 0.50$ (*dashed line*) (SNR $\simeq$ 10, 4.0 dB, respectively) (**b**)

plotted the probability P of rejecting the null hypothesis using the chi-square test with the same level $\alpha = 0.05$. Notice that the chi-square test correctly rejects the null hypothesis with higher probability than the BDS test in the high-noise regime ($\sigma \geq 1$), and its performance is comparable to the best one of the BDS test in the low-noise regime ($\sigma \leq 1$). Put in a different way, the probability of a false positive is higher with the BDS test. We conclude also from Fig. 9.9 that the BDS test performance strongly depends on the combinations of ϵ and m; for some combinations, this method wrongly accepts the null hypothesis even for small values of σ.

9.5.2 The Delayed Hénon Map

The *time-delayed Hénon map* [194] is defined as

$$x_n = 1 - ax_{n-1}^2 + bx_{n-d}, \qquad (9.12)$$

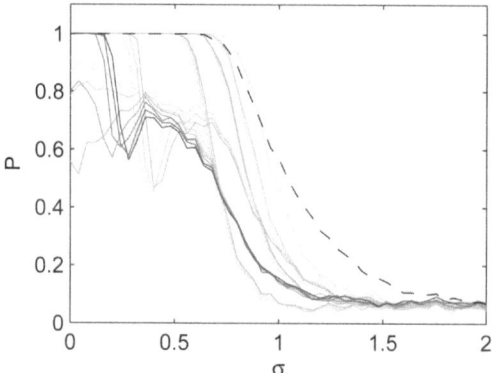

Fig. 9.9 The *continuous lines* indicate the probability of rejecting the null hypothesis H_0 ("the time series is i.i.d.") for a time series projected from the Lorenz map's attractor, contaminated with Gaussian white noise with σ up to $\sigma = 2$, when applying the BDS test with level $\alpha = 0.05$. In total, 27 tests for different combinations of ϵ and m were performed. The lighter the *gray color* is, the bigger is the value of ϵ used (see text for details). The *dashed line* indicates the probability of rejecting H_0 when using the chi-square test based on missing ordinal patterns, with the same level $\alpha = 0.05$. The chi-square test correctly rejects the null hypothesis more often than the BDS test

where a, b are real constants and $d \geq 1$. For $d = 1$, the time-delayed Hénon map is equivalent to the logistic map $x_{n+1} = Ax_{n-1}(1 - x_{n-1})$, with [194]

$$A = \frac{b-1}{2a} \pm \frac{1}{2a}\sqrt{(b-1)^2 + 4a}.$$

For $d = 2$ and $a = 1.4, b = 0.3$, we recover the familiar two-dimensional dissipative Hénon map.

For $a = 1.6$ and $b = 0.1$, Sprott [194] finds the following linear relation between D_{KY} and d over the range $1 \leq d \leq 100$:

$$D_{KY} \cong 0.192d + 0.699.$$

The Kaplan–Yorke conjecture implies now

$$D_0 \geq D_1 = D_{KY} \cong 0.192d + 0.699$$

for the fractal dimension D_0 of the attractor, $1 \leq d \leq 100$. In particular, $D_0 \geq 1.083$ for $d = 2$, $D_0 \geq 10.299$ for $d = 50$, and $D_0 \geq 19.899$ for $d = 100$. Thus, this family of maps provides attractors with a wide range of fractal dimensions.

Figure 9.10 shows the return map ξ_{n+1} vs ξ_n for a typical orbit on the attractor of the time-delayed Hénon map with $d = 50$, both in the absence of noise, $\xi_n = x_n$ (a) and corrupted with Gaussian white noise, $\xi_n = x_n + w_n$, with $\sigma = 0.5$ (SNR $\simeq$ 1.3 dB) (b). Again, the geometry of the attractor has been completely blurred by the

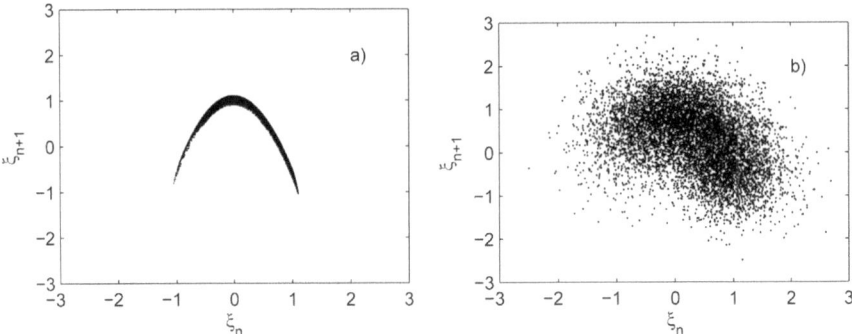

Fig. 9.10 Return map for a time series of the time-delayed Hénon map with $d = 50$ in the absence of noise (**a**) and contaminated with Gaussian white noise with $\sigma = 0.5$ (SNR $\simeq$ 1.3 dB) (**b**). The structure of the underlying chaotic attractor has been totally blurred. Here again the count of missing ordinal patterns is sensibly higher than in the purely random case

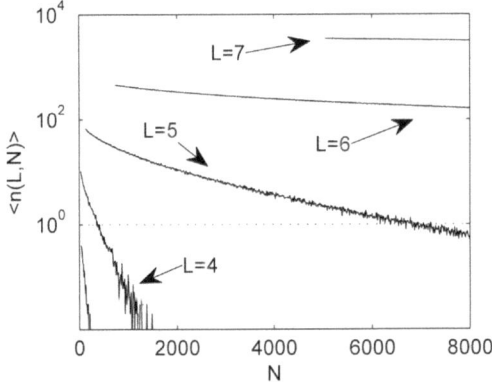

Fig. 9.11 Average number of missing ordinal patterns of length L found in a time series of length N, $\langle n(L,N) \rangle$ (in logarithmic scale), for a noisy series of the time-delayed Hénon map with $\sigma = 0.5$ (SNR $\simeq$ 1.3 dB)

presence of the noise. However, it can be seen in Fig. 9.11 that also in this case, the number of missing ordinal L-patterns found in a time series of length N, $\langle n(L,N) \rangle$, is sensibly larger than in the white noise-only case, Fig. 9.7.

Figure 9.12(a)–(c) depicts the comparison of the chi-square test with the BDS test for $d = 2$, $d = 50$, and $d = 100$, respectively. Again, the probability of a false positive is higher with the BDS test. Since we are interested in the detection of determinism, we may conclude that the chi-square test, based on the distribution of visible ordinal patterns, is more reliable.

In conclusion, the (conditional + dynamical) robustness against additive noise of the forbidden patterns makes them a practical tool to distinguish deterministic, noisy time series from white noise. It is in this sense that we claim that forbidden patterns can be used to detect determinism in noisy time series—determinism as opposite to

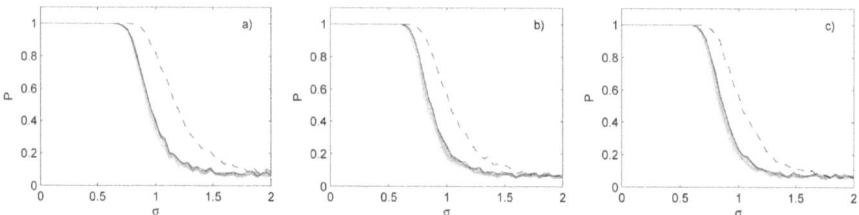

Fig. 9.12 Comparison of the chi-square test and the BDS test applied to projections of the time-delayed Hénon map with $d = 2$ (**a**), $d = 50$ (**b**), $d = 100$ (**c**), and Gaussian white noise with $0 \leq \sigma \leq 2$. The *continuous lines* indicate the probability of rejecting the null hypothesis H_0 ("the time series is i.i.d.") when applying the BDS test with level $\alpha = 0.05$. In total, 27 tests with different combinations of ϵ and m were performed. The lighter the *gray color* is, the bigger is the value of ϵ. The *dashed line* indicates the probability of rejecting H_0 when using the chi-square test with the same level $\alpha = 0.05$. Clearly, the chi-square test rejects the null hypothesis more often than the BDS for all noise values and for the three values of d

statistical independence. In fact, determinism is usually equated to statistical dependence among the observations in applications. On the other hand, the discrimination of deterministic, noisy time series from colored noise seems problematic, although some interesting methods have been proposed; see, e.g., [119] for a method based on nonlinear predictability.

Chapter 10
Space–Time Dynamics

All applications of ordinal analysis hitherto had to do with time series analysis or abstract dynamical systems. A remaining challenge is to expand the applications to physical systems.

In order to tackle the viability of this program, we are going to study the permutation complexity of two simple models of spatially extended physical systems: cellular automata (CA) and coupled map lattices (CMLs). CA were presented in Sect. 1.5. CMLs can be considered as a generalization of the CA; they retain the space coarse graining of the CA, but the state variable take on real values. Despite their apparent simplicity, these are the preferred models when studying the emergence of collective phenomena (such as turbulence, space–time chaos, symmetry breaking, ordering) in systems of many particles interacting nonlinearly. Indeed, their ability to reproduce complex phenomena in, say, fluid dynamics and solid state physics, is impressive. For this reason, they are the ideal choice for our purpose.

10.1 Spatially Extended Systems

Dynamical systems discrete in time as well as in space have been studied to understand physical phenomena while keeping the technical burden at a minimum. The discrete space can be an infinite lattice of dimension 1 (which can be identified with $\mathbb{Z}$) or a finite lattice with periodic or fixed boundary conditions. At each site i of the lattice there is a local variable $x_t(i)$ taking on values in a set S called the *state space*, at every time $t \in \{0, 1, \ldots\} = \mathbb{N}_0$. The change of the state variable $x_t(i)$ from time t to time $t + 1$ depends only on the variables in some fixed vicinity of i at time t.

Unless otherwise explicitly stated, we assume in this chapter the following restrictions for simplicity and computational convenience.

(i) Periodic boundary conditions:

$$x_t(0) = x_t(N) \text{ and } x_t(N + 1) = x_t(1) \tag{10.1}$$

for all $t \geq 0$. These conditions amount to the N sites lying on a ring.

J.M. Amigó, *Permutation Complexity in Dynamical Systems,*
Springer Series in Synergetics, DOI 10.1007/978-3-642-04084-9_10,
© Springer-Verlag Berlin Heidelberg 2010

(ii) Nearest neighbors interaction, i.e.,

$$x_{t+1}(i) = f(x_t(i-1),\, x_t(i),\, x_t(i+1)), \tag{10.2}$$

where $1 \le i \le N$.

Depending on the state space S, there are two well-known instances of such space–time systems: one-dimensional *cellular automata* (CA) if S is finite and one-dimensional *coupled map lattices* (CMLs) if S is an interval of $\mathbb{R}$. Given the formal similarity between both systems (see below for details), it comes as no surprise that they exhibit similar dynamical phenomena, like coherent traveling structures and space–time chaos [58, 110, 46, 47]. Perhaps more surprisingly is the fact that one-dimensional CMLs can be completely described in terms of symbolic dynamical concepts [170, 171]. Along similar lines, we are going to show that CA and CMLs can be handled in a satisfactory way with techniques based on ordinal patterns. In particular, (i) two so-called regularity parameters to be defined below seem to be useful for discriminating Wolfram's complexity classes in the case of CA and (ii) the number of admissible ordinal patterns in the configurations of CMLs separates space–time chaos from regular pattern dynamics.

CA and CMLs are not only related with each other but, in turn, are also related to networks—a subject of much interest in current research [162]. The main difference is the connectivity: while CA and CMLs feature near-neighbor interactions, networks allow also for long-range interactions. For a multidisciplinary introduction to dynamics on complex networks, see [132]. Networks of coupled maps have been studied, e.g., in [145, 104] with reference to synchronization. Whether ordinal analysis is also useful in this more general spatially extended systems is an open question as yet. Nevertheless, in view of the results reported in Sect. 2.4 on synchronization, we conjecture that ordinal analysis will be helpful to characterize the different synchronization regimes.

10.1.1 Cellular Automata

We refer to Sect. 1.1.5 for the generalities on cellular automata (CA). According to restriction (10.2), we consider local maps f with a neighborhood of size 1; furthermore, the state space will be $S = \{0, 1\}$, thus $f : \{0, 1\}^3 \to \{0, 1\}$. Technically, CA correspond to continuous, shift-commuting maps F from a full shift to itself; F is the global transition map induced by f on the configuration space Ω. Thus (Ω, F) is a continuous dynamical system. More generally one can also consider continuous, shift-commuting maps between subshifts of finite type (i.e., shift-invariant subsets of a full shift obtained after excluding a finite set of fixed blocks of symbols) [92, 123].

For brevity, one-dimensional binary CA with a neighborhood of size 1 will be called *elementary*. Elementary CA can be labeled as follows. Given the local rule

$$f(p,\, q,\, r) = \beta,$$

where p, q, r, $\beta \in \{0, 1\}$, order lexicographically the eight different configurations in the neighborhood $\mathcal{U}_1(i) = \{i - 1, i, i + 1\}$, to wit:

$$(0, 0, 0), (0, 0, 1), (0, 1, 0), (0, 1, 1), \ldots, (1, 1, 1).$$

If $\beta_0, \beta_1, \ldots, \beta_7 \in \{0, 1\}$ are the corresponding values of β, then the cellular automaton with the local rule f can be unambiguously identified by the number

$$\text{ID} = \sum_{i=0}^{7} \beta_i 2^i \in \{0, 1, \ldots, 255\}.$$

In other words, there are 256 different elementary CA.

Alternatively, one can argue as follows. To define a local rule, one must specify the update state of the central cell given all possible configurations of its local neighborhood. Since there are eight such configurations and two update states, the number of possible assignments is $2^8 = 256$.

For example, the cellular automaton with local rule

$$\begin{aligned}
f(0, 0, 0) &= 0, & f(1, 0, 0) &= 0, \\
f(0, 0, 1) &= 1, & f(1, 0, 1) &= 1, \\
f(0, 1, 0) &= 1, & f(1, 1, 0) &= 1, \\
f(0, 1, 1) &= 1, & f(1, 1, 1) &= 0
\end{aligned}$$

is coded as the decimal number

$$\begin{aligned}
\text{ID} &= 0 \times 2^0 + 1 \times 2^1 + 1 \times 2^2 + 1 \times 2^3 + 0 \times 2^4 + 1 \times 2^5 + 1 \times 2^6 + 0 \times 2^7 \\
&= 110.
\end{aligned}$$

Conversely, the local rule $f(p, q, r) = \beta$ of an elementary cellular automaton can be obtained from its identification number ID in a recursive form:

$$\beta_0 = \text{ID} \bmod 2,$$
$$\beta_i = \frac{\text{ID} - \beta_0 - \cdots - \beta_{i-1} 2^{i-1}}{2^i} \bmod 2,$$

$1 \le i \le 7$. Let us emphasize that in order to determine the evolution of these CA, all we need are the eight bits β_i—no closed formula for f is necessary. An explicit eight-parameter rule to construct a map $f:\{0, 1\}^3 \to \{0, 1\}$ delivering the right update states β_i for each local configuration can be found in [54, Table 4].

Stephen Wolfram studied exhaustively the asymptotic behavior of all 256 elementary CA. For each local rule and each initial configuration, he calculated the time evolution of the cellular automaton till it exhibited a stable pattern of behavior. Out of all these simulations, Wolfram proposed to classify the elementary cellular automata in four classes [206, 207]. In order of increasing complexity, these classes are the following:

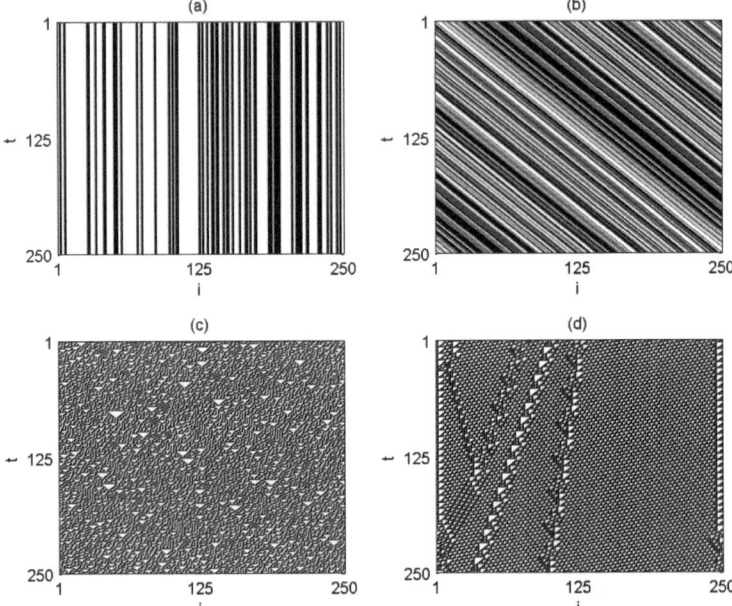

Fig. 10.1 Typical trajectories of elementary CA belonging to the complexity classes W1 (**a**), W2 (**b**), W3 (**c**), and W4 (**d**). The number of cells represented is $N = 250$. Time elapses top to bottom ($T = 250$ iterations represented)

(W1) The configurations converge to a fixed point; Fig. 10.1(a).
(W2) Time evolution yields a sequence of simple stable or periodic structures; Fig. 10.1(b).
(W3) The behavior is "chaotic"; Fig. 10.1(c).
(W4) Time evolution yields localized structures that move around and interact with each other in very complicated ways; Fig. 10.1(d).

A word of caution for the CA practitioners. Real cellular automata are finite deterministic machines, so their configuration space is finite. This means that their evolution is periodic, albeit the period can be very large—so large that this fact may be ignored in simulations.

10.1.2   Coupled Map Lattices

A CML is a discrete-time dynamical system with discrete space and *continuous* states. So one can think of CMLs as generalizations of CA [50], or rather as an intermediate between CA and partial differential equations. CMLs were introduced by Kaneko [106, 107] as a simple test bed for spatiotemporal complexity (turbulence, convection, etc.). For the theoretical aspects of CMLs the reader is referred to the papers of Bunimovich and Sinai, e.g., [42–44].

In dimension 1 the most common choices for the evolution rule (10.2) are

$$x_{t+1}(i) = (1 - \varepsilon)f(x_t(i)) + \varepsilon f(x_t(i - 1))$$

and

$$x_{t+1}(i) = (1 - \varepsilon)f(x_t(i)) + \frac{\varepsilon}{2}\left[f(x_t(i - 1)) + f(x_t(i - 1))\right], \qquad (10.3)$$

which correspond to the so-called *one-way* and *diffusive* CMLs, respectively. Here $0 \leq \varepsilon \leq 1$ so as all coupling coefficients are positive, $i = 1, \ldots, N$ label the sites, and f is a self-map of the state space $I \subset \mathbb{R}$. When the *coupling constant* ε is small, the oscillators will be practically independent of each other, hence the CML will behave similar to an ensemble of uncoupled oscillators. At the other end, strongly coupled oscillators will evolve more or less in a synchronized fashion. Between both cases, we expect to see a variety of behaviors as, so to speak, locally organized dynamics percolates along the lattice. It is the interplay between simple local properties (in our case, the coupling between neighboring oscillators) and the emergence of a complex dynamics on a global scale, what makes the study of CMLs, cellular automata, and the like, so rewarding (Fig. 10.2). For more general evolution rules, see e.g., [192].

The diffusive CML—the only one we consider henceforth— is the discrete analogue of the reaction–diffusion equation with a symmetrical interaction. Additional complexity can be added by allowing the map f to depend on a parameter. Following [108], we shall take the nonlinear ansatz

$$f(x) = 1 - ax^2, \quad x \in [-1, 1] \qquad (10.4)$$

and call $a \in (0, 2]$ the *nonlinearity* of f. Observe that if $x_0(i) \in [-1, 1]$ for $1 \leq i \leq N$, then $x_t(i) \in [-1, 1]$ for all $t \geq 1$ and $1 \leq i \leq N$.

Researchers on this field use to borrow terms from continuum physics like ordered or unordered phase, phase transition, local of global defects. According to [108], the *logistic coupled lattice* (10.3) (10.4) exhibits six major "phases":

(K1) Frozen random patterns; Fig. 10.3(a)
(K2) Pattern selection and suppression of chaos; Fig. 10.3(b)
(K3) Brownian motion of defects; Fig. 10.3(c)
(K4) Defect turbulence; Fig. 10.3(d)
(K5) Pattern competition intermittency; Fig. 10.3(e)
(K6) Fully developed turbulence; Fig. 10.3(f)

These six phases are shown on an a–ε diagram in Fig. 10.2; see [108] for details. Two-dimensional CMLs have been investigated, e.g., in [210, 23, 71].

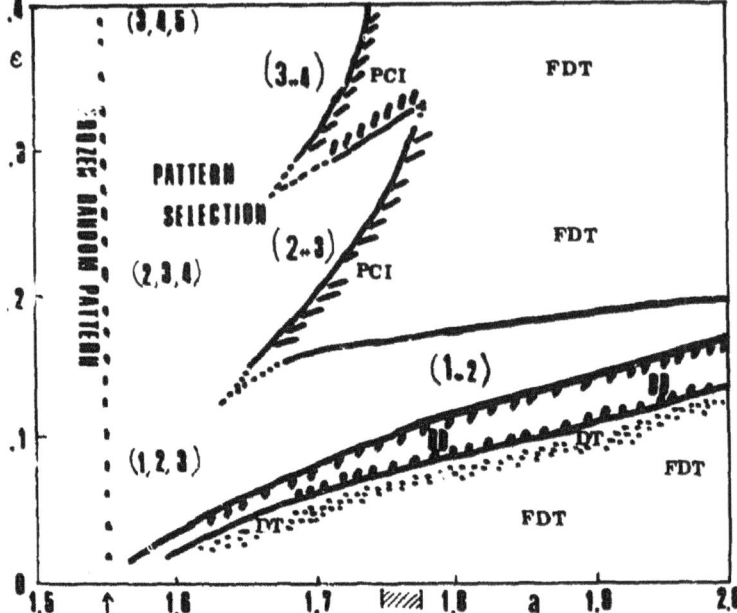

Fig. 10.2 [Reproduced with permission from [108].] Phase diagram of the coupled logistic map (10.4) (*a* varies along the *horizontal axis*, ε along the *vertical*). Here BD, DT, PCI, and FDT are the abbreviations of Brownian motion of defect, defect turbulence, pattern competition intermittency, and fully developed turbulence, respectively. The numbers such as 1,2,3 represent the selected domain sizes

10.2 Applications of Permutation Complexity to Spatiotemporal Dynamics

In this section we are going to show that the ordinal pattern-based approach to time series analysis and abstract dynamical systems works out also with one-dimensional binary cellular automata and one-dimensional coupled logistic lattices. This is a first step to extend ordinal analysis to space–time dynamics.

10.2.1 Topological Entropy of CA

The spatiotemporal complexity of a cellular automaton can be measured by the topological entropy. In Sect. 1.1.5 we mentioned that

$$h_{\text{top}}(F) = \lim_{w \to \infty} \lim_{t \to \infty} \frac{1}{t} \log R(w, t), \tag{10.5}$$

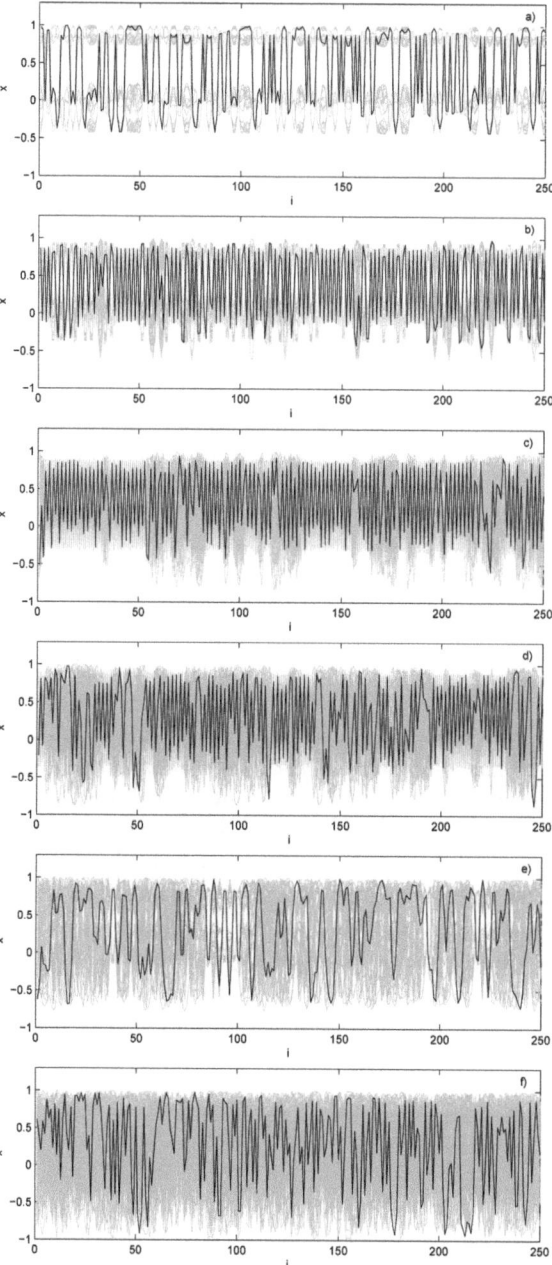

Fig. 10.3 CML space–time plots for (**a**) frozen random patterns ($a = 1.44$, $\varepsilon = 0.1$), (**b**) pattern selection and suppression of chaos ($a = 1.65$, $\varepsilon = 0.1$), (**c**) Brownian motion of defects ($a = 1.86$, $\varepsilon = 0.1$, (**d**) defect turbulence ($a = 1.89$, $\varepsilon = 0.1$), (**e**) pattern competition intermittency ($a = 1.8$, $\varepsilon = 0.3$), and (**f**) fully developed turbulence ($a = 2$, $\varepsilon = 0.3$). Each black line shows the CML state at time $n = 500$, the grey background is the superposition of states at $1 \leq n \leq 499$

where $F:S^{\mathbb{Z}} \to S^{\mathbb{Z}}$ and $R(w, t)$ is the number of distinct rectangles of width w and height (temporal extent) t occurring in a space–time evolution diagram of $(S^{\mathbb{Z}}, F)$; see (1.21) and Fig. 1.4.

Another possibility consists in using the topological permutation entropy $h_{\text{top}}^*(F)$ instead. We shall shortly claim that, under some provisos, the result is going to be the same. But even in a general situation we might wish to link the spatiotemporal complexity of a cellular automaton to the permutation complexity of its time evolution as measured by the topological permutation entropy (in practice, by one or several entropy rates of finite order), or by other quantities based on ordinal patterns. Examples of the latter eventuality are provided by the parameters $\chi_{\text{time}}^2(L)$ and $\chi_{\text{space}}^2(L)$ presented below, absolute and relative frequency distributions of ordinal patterns, and any probability functional whose value is estimated by means of ordinal patterns.

Theorem 15 states that $h_{\text{top}}^*(f) = h_{\text{top}}(f)$ for any *positively expansive* self-map f of an n-dimensional simple domain. We could argue at this point that the proof of Theorem 15 does not rely on any particular property of compact sets in $\mathbb{R}^n$, in order to infer

$$h_{\text{top}}(F) = h_{\text{top}}^*(F) \tag{10.6}$$

for any positively expansive map F on a compact metric space, in particular when F is the global transition map of a one-dimensional cellular automaton. But for our purposes it will suffice to equate $h_{\text{top}}(F)$ with the topological entropy of a topologically conjugate *interval* map. A cellular automaton is said to be *expansive* (correspondingly, *positively expansive*) when its global transition map F is expansive (correspondingly, positively expansive). It is interesting to point out that (i) positively expansive CA only exist in dimension 1 [188] (while expansive interval maps only exist in dimensions greater than 1 [19, Thm. 2.2.31]) and (ii) positively expansive CA are topologically conjugate to one-sided full shifts [62].

So, let us show how to calculate $h_{\text{top}}(F)$ by means of the topological entropy of a two-dimensional interval map. Set $\Omega = S^{\mathbb{Z}}$, where $S = \{0, 1, \ldots, |S| - 1\}$ in the case of a one-dimensional cellular automaton with $|S|$ states, and define similar to (4.20) the map $\phi_{|S|} = (\phi_{|S|}^-, \phi_{|S|}^+): S^{\mathbb{Z}} \to [0, 1]^2$,

$$\phi_{|S|} : \mathbf{x}_t \mapsto (\phi_{|S|}^-(\mathbf{x}_t^-), \phi_{|S|}^+(\mathbf{x}_t^+)), \tag{10.7}$$

where $\mathbf{x}_t = (x_t(i))_{n \in \mathbb{Z}}$, $\mathbf{x}_t^- = (x_t(-i))_{i \in \mathbb{N}}$ is the left sequence of $\mathbf{x}_t$, $\mathbf{x}_t^+ = (x_t(i))_{i \in \mathbb{N}_0}$ is the corresponding right sequence, the component maps $\phi_{|S|}^- : S^{\mathbb{N}} \to [0, 1]$, $\phi_{|S|}^+ : S^{\mathbb{N}_0} \to [0, 1]$ are given by

$$\phi_{|S|}^-(\mathbf{x}_t^-) = \sum_{i=1}^{\infty} \frac{x_t(-i)}{|S|^i}, \quad \phi_{|S|}^+(\mathbf{x}_t^+) = \sum_{i=0}^{\infty} \frac{x_t(i)}{|S|^{i+1}}, \tag{10.8}$$

and the bisequences $\mathbf{x}_t = (\mathbf{x}_t^-, \mathbf{x}_t^+)$ are lexicographically ordered as in (4.19). We already know (Sect. 4.3) that the map $\phi_{|S|}$ is an order isomorphism ($[0, 1]^2$ being lexicographically ordered), up to a measure zero set $\mathcal{N}$ which comprises those bisequences whose left and/or right sequences terminate in $1, 0^\infty$ or $0, (|S| - 1)^\infty$. Furthermore, it is easy to check that $\phi_{|S|}$ is a homeomorphism from $S^{\mathbb{Z}} \backslash \mathcal{N}$ to its range. In other words, the continuous dynamical systems (Ω, F) and $([0, 1]^2, \phi_{|S|} \circ F \circ \phi_{|S|}^{-1})$ are topologically conjugate (modulo 0), hence

$$h_{\text{top}}(F) = h_{\text{top}}(\tilde{F}). \tag{10.9}$$

where $\tilde{F} := \phi_{|S|} \circ F \circ \phi_{|S|}^{-1} : [0, 1]^2 \to [0, 1]^2$ is an interval map.

Suppose, moreover, that F is positively expansive. In this case the same holds for $\tilde{F}$ since positive expansiveness is a topological conjugacy invariant (Sect. B.3.1). Then

$$h_{\text{top}}(\tilde{F}) = h_{\text{top}}^*(\tilde{F}) \tag{10.10}$$

according to Theorem 15. The bottom line from (10.9) and (10.10) is

$$h_{\text{top}}(F) = h_{\text{top}}^*(\tilde{F}) \tag{10.11}$$

for positively expansive (one-dimensional) CA (Ω, F). Finally, to go from (10.11) to (10.6), we only need to invoke that topological permutation entropy is an invariant of order isomorphy (here embodied by the homeomorphism $\phi_{|S|}$); see Theorem 14.

A convenient shortcut in actual calculations is the following. The lexicographical order of bisequences $\mathbf{x} \in S^{\mathbb{Z}}$ and points $(x, y) \in [0, 1]^2$ is determined by the right sequences $\mathbf{x}^+$ and ordinates y, respectively. This means that if the right sequences of a finite orbit $F^t(\mathbf{x}_0)$, $0 \leq t \leq T$, are all different (as usual in numerical simulations), then we may restrict attention to the ordinates of the order-isomorphic orbit $\phi_{|S|} \circ F^t(\mathbf{x}_0) = \phi_{|S|}(\mathbf{x}_t)$. From (10.7) we learn that the ordinate of $\phi_{|S|} \circ F^t(\mathbf{x}_0)$ is

$$\phi_{|S|}^+(F^t(\mathbf{x}_0)^+) = \phi_{|S|}^+(\mathbf{x}_t^+) = \sum_{i=0}^{\infty} \frac{\mathbf{x}_t(i)}{|S|^{i+1}}. \tag{10.12}$$

To check numerically the coincidence of topological permutation entropy and topological entropy for positively expansive CA, we resort to *linear* automata. A one-dimensional CA is said to be linear if its local rule is of the form

$$f(s_t(i - l), s_t(i - l + 1), \ldots, s_t(i + l)) = \sum_{j=-l}^{j=l} \lambda_j s_t(i + j) \bmod |S|. \tag{10.13}$$

For a one-dimensional linear CA, (10.5) yields a closed formula for the topological entropy [62]: if $p_1^{m_1} \cdots p_h^{m_h}$ is the prime factor decomposition of $|S|$, and

$$P_i = \{0\} \cup \{j: \gcd(\lambda_j, p_i) = 1\}, \ L_i = \min P_i, \ R_i = \max P_i,$$

then

$$h_{\text{top}}(F) = \sum_{i=1}^{h} m_i(R_i - L_i) \log p_i. \tag{10.14}$$

Furthermore, it can be proved [141, Theorem 3.2] that a one-dimensional *linear* CA (10.13) is positively expansive if and only if

$$\gcd(|S|, \lambda_{-l}, \ldots, \lambda_{-1}) = \gcd(|S|, \lambda_1, \ldots, \lambda_l) = 1. \tag{10.15}$$

From (10.15) it follows that the local rule $f:\{0, 1\}^3 \rightarrow \{0, 1\}$ with

$$f(s_t(i - 1), s_t(i), s_t(i + 1)) = s_t(i - 1) + s_t(i + 1) \bmod 2$$
$$= s_t(i - 1) \oplus s_t(i + 1) \tag{10.16}$$

$(\lambda_{-1} = \lambda_1 = 1, |S| = 2)$ defines a *positively expansive* CA. According to (10.14),

$$h_{\text{top}}(F) = 2 \log 2 = 2 \text{ bit/symbol.}$$

The topological permutation entropy of the automaton defined by the local rule (10.16) can be now estimated via the ordinal patterns of its global map $F:\{0, 1\}^{\mathbb{Z}} \rightarrow \{0, 1\}^{\mathbb{Z}}$ or alternatively via the ordinal patterns of the interval map $\tilde{F} = \phi_2 \circ F \circ \phi_2^{-1}:[0, 1]^2 \rightarrow [0, 1]^2$. As explained above, it suffices to keep account of the ordinal patterns defined by the ordinates of $\phi_2 \circ F^t(\mathbf{x})$, namely, $\phi_2^+(\mathbf{x}_t^+)$, (10.12).

Figure 10.4 shows different aspects of the cellular automaton (10.16): (a) the time evolution of cells $1 \leq i \leq 250$; (b) the ordinates $\phi_2^+(\mathbf{x}_t^+)$ of the finite orbit $\phi_2(F^t(\mathbf{x}_0)) = \tilde{F}^t(\phi_2(\mathbf{x}_0))$, $0 \leq t \leq 250$, where $x_0(1), \ldots, x_0(250)$ were chosen randomly and extended periodically in both directions; (c) the return map $\phi_2^+(\mathbf{x}_t^+)$ vs $\phi_2^+(\mathbf{x}_{t+1}^+)$ (this graph has seemingly a fractal structure); and (d) the convergence of the topological permutation entropy rates of order L,

$$h_{\text{top}}^*(L, \tilde{F}) = -\frac{1}{L} \log |\{\pi \in \mathcal{S}_L : P_\pi \neq \emptyset\}|,$$

to $h_{\text{top}}(F) = 2$ bit/symbol, with the length of the ordinal patterns. This convergence is fast, also in computation.

10.2.2 Complexity Classes of Elementary CA

Elementary CA with periodic boundary conditions were also extensively studied in a series of papers by Chua and collaborators. According to [55],

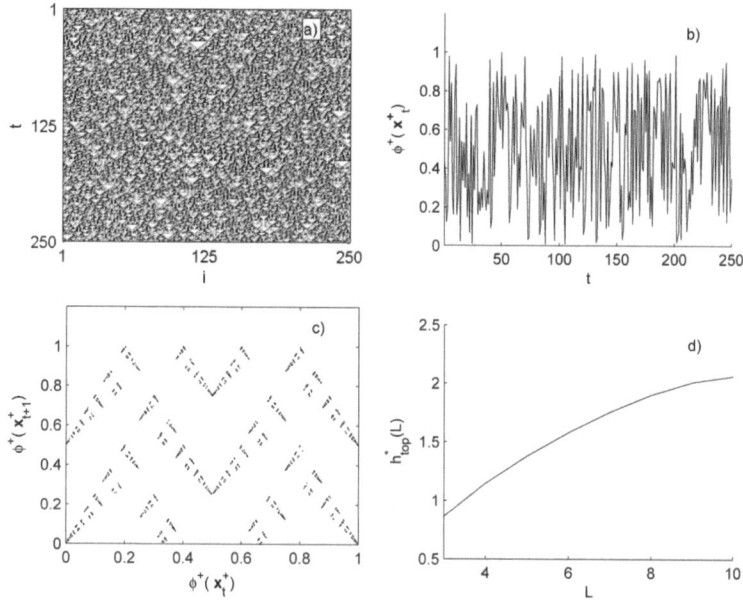

Fig. 10.4 Different aspects of a positively expansive CA (see text). Plot (**d**) shows the convergence of the topological permutation entropy of the automaton to its topological entropy

(1) the cellular automaton with local rule

$$f(p, q, r) = \frac{1}{2}[1 + \text{sign}(2p + 4q + 2r - 5)], \tag{10.17}$$

ID $= 200$, is an instance of *class W1*;

(2) the cellular automaton with local rule

$$f(p, q, r) = p \tag{10.18}$$

(corresponding to the right shift on $\{0, 1\}^{\mathbb{Z}}$), ID $= 240$, belongs to *class W2*;

(3) the cellular automaton with local rule

$$f(p, q, r) = p + q + r + qr \bmod 2, \tag{10.19}$$

ID $= 30$, is *class W3*; and

(4) the cellular automaton with local rule

$$f(p, q, r) = (1 + p)qr + q + r \bmod 2, \tag{10.20}$$

ID $= 110$, belongs to *class W4*. Moreover, this automaton is surely universal in the sense that it can emulate a universal Turing machine [207, p. 1115].

In order to discriminate these four classes, we propose two parameters inspired in the statistic χ^2, used in Sects. 9.3 and 9.5 for detecting determinism in noisy time series. The rationale is as follows. Since the statistic χ^2 is based on ordinal pattern distribution, being small for i.i.d. random processes and large for deterministic processes, we expect that it can also discriminate irregular from regular configurations as time evolves. For this reason, we call them *regularity parameters*.

(a) *Temporal regularity parameter* $\chi^2_{\text{time}}(L)$. In numerical simulations, let

$$\mathbf{x}_t = (x_t(i))_{i=1}^N = x_t(1), x_t(2), \ldots, x_t(N)$$

be the configuration of cells $1 \leq i \leq N$ at time t, $0 \leq t \leq T$. Calculate now $\chi^2(L)$, (9.8), for the multivariate time series

$$\mathbf{x}_0, \mathbf{x}_1, \ldots, \mathbf{x}_T \tag{10.21}$$

using, say, lexicographical order. Alternatively, transform each $\mathbf{x}_t$ into a dyadic rational,

$$\phi(\mathbf{x}_t) = \sum_{i=1}^N \frac{x_t(i)}{2^i} \in [0, 1), \tag{10.22}$$

and calculate $\chi^2(L)$ for the univariate time series

$$\phi(\mathbf{x}_0), \phi(\mathbf{x}_1), \ldots, \phi(\mathbf{x}_T), \tag{10.23}$$

since sequences (10.21) and (10.23) are order isomorphic. Call $\chi^2_{\text{time}}(L)$ the result.

(b) *Spatial regularity parameter* $\chi^2_{\text{space}}(L)$. We want now to calculate the regularity of the univariate time series consisting of the state variables at time t,

$$x_t(1), x_t(2), \ldots, x_t(N),$$

and average the results over all times, $0 \leq t \leq T$. There is a catch though. Statistic (9.8) correspond to i.i.d. random variables taking on real values. In the finite-state case we are considering now, some symbols will necessarily repeat as soon as the length of the sequence exceeds the number of states. For binary variables this implies that not all 2^L *ordinal* patterns of an i.i.d. binary sequence are equiprobable. Indeed, all the $L + 1$ words of length L

$$(0, 0, \ldots, 0, 0), (0, 0, \ldots, 0, 1), (0, 0, \ldots, 1, 1), \ldots, (0, 1, \ldots, 1, 1), (1, 1, \ldots, 1, 1)$$

are of type $\pi_0 = \langle 0, 1, 2, \ldots, L - 1 \rangle$, while each of the remaining $2^L - L - 1$ words defines a distinct ordinal pattern. Therefore, the chi-square statistic χ^2 for windows of size L takes the following form for binary sequences:

$$\chi^2(L) = \frac{\left(\nu_0 - \frac{L+1}{2^L}\right)^2}{(L+1)/2^L} + (2^L - L - 1)\frac{\left(\nu_1 - \frac{1}{2^L}\right)^2}{1/2^L} \qquad (10.24)$$

$$= \frac{\left(2^L\nu_0 - L - 1\right)^2}{2^L(L+1)} + \left(1 - \frac{L+1}{2^L}\right)(2^L\nu_1 - 1)^2,$$

where ν_0 is the number of times the pattern $\pi_0 = \langle 0, 1, 2, \ldots, L-1 \rangle$ has been observed in the sequence and ν_1 is the number of patterns $\pi \in \mathcal{S}_L$, $\pi \neq \pi_0$, observed in the same sequence, when using non-overlapping sliding windows. In sum, in order to obtain the spatial regularity $\chi^2_{\text{space}}(L)$, calculate the parameter $\chi^2(L)$, (10.24), of the univariate time series $(x_t(i))_{i=1}^N$ for each time $0 \leq t \leq T$ and average over them:

$$\chi^2_{\text{space}}(L) = \left\langle \chi^2(L) \right\rangle.$$

In our numerical simulations we chose $N = 250$. To avoid too small samples, we take $L \leq 4$ for $\chi^2_{\text{space}}(L)$. For $\chi^2_{\text{time}}(L)$ we may choose L larger, provided that T is sufficiently long. Furthermore, in order to let transients die out, we forgo the first 5000 iterations.

For the four representatives of the complexity classes W1–W4 given above (ID $= 200, 240, 30$, and 110), we have simulated their time evolution, starting from 100 randomly chosen initial configurations. When the resulting values of $\chi^2_{\text{time}}(5)$ are plotted against $\chi^2_{\text{space}}(4)$ we see, Fig. 10.5, that they cluster in different, non-overlapping regions.

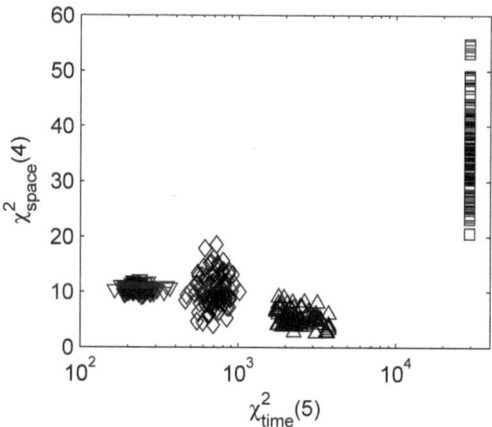

Fig. 10.5 Values of $\chi^2_{\text{time}}(5)$ and $\chi^2_{\text{space}}(4)$ for four CA of different complexity classes and 100 random initial configurations. Symbol assignment: Classes W1 ($\square$), W2 ($\diamondsuit$), W3 (∇), and W4 ($\triangle$)

We have repeated the same exercise with a few more CA and the results are similar, although the clusters of different CA belonging to the same complexity class may lie in different parts of the χ^2_{time}–χ^2_{space} diagram. All this hints that regularity parameters capture the basic features of the different complexity classes of elementary CA.

For the study of the complexity of CA rules by other methods, see, e.g., [103].

10.2.3 Phases of CMLs

The basic difference between CA and CMLs concerns the state space and eventually the appearance of free parameters in the second case (e.g., the nonlinearity a in (10.4)). Therefore, we expect that the same tools used in the last section to study the spatiotemporal complexity of CA will also be useful for CMLs. We shall use the logistic coupled lattice as study case.

So, consider a one-dimensional logistic coupled lattice with N sites (extended periodically in both directions), pick an initial configuration $(x_0(i))_{i=1}^N$, $x_0(i) \in [0, 1]$, and let it evolve during $T_0 = 5000$ time steps according to the diffusive rule (10.3)–(10.4). From T_0 on we assume that the lattice exhibits its asymptotic dynamics.

A first proposal to quantify the complexity of a CML, inspired in the calculation of the topological entropy of positively expansive CA, is the following. At each iteration of the CML, define the symbolic sequence

$$(s_t(i))_{i=1}^N = s_t(1), s_t(2), \ldots, s_t(N) \equiv \mathbf{s}_t, \qquad (10.25)$$

where

$$s_t(i) = \begin{cases} 0 & \text{if } x_t(i) \leq 0, \\ 1 & \text{if } x_t(i) > 0. \end{cases} \qquad (10.26)$$

In this way we get a finite multivariate binary sequence $\mathbf{s}_0, \mathbf{s}_1, \ldots, \mathbf{s}_T$; alternatively, we might prefer to work with the order-isomorphic sequence $\phi(\mathbf{s}_0), \phi(\mathbf{s}_1), \ldots, \phi(\mathbf{s}_T)$, of the dyadic rationals

$$\phi(\mathbf{s}_t) = \sum_{i=1}^N \frac{s_t(i)}{2^i} \in [0, 1). \qquad (10.27)$$

At this point we could count the number of visible ordinal patterns of length L, $N(L)$, of the sequence $\mathbf{s}_0, \mathbf{s}_1, \ldots, \mathbf{s}_T$ or, equivalently, $\phi(\mathbf{s}_0), \phi(\mathbf{s}_1), \ldots, \phi(\mathbf{s}_T)$, and estimate their metric or topological permutation entropy. Other even simpler possibility consists in representing $N(L)$ on the $(a\text{-}\varepsilon)$-plane. This has been done in Fig. 10.6 for $L = 5$ and $N = 250$. As for the nonlinearity a and the coupling constant ε, they are allowed throughout to take 75 values uniformly distributed in similar ranges as in Fig. 10.2, namely, $[1.4, 2]$ and $[0, 0.5]$, respectively. Remarkably, there are two zones of dark/light gray colors in Fig. 10.6 that roughly correspond with the zones

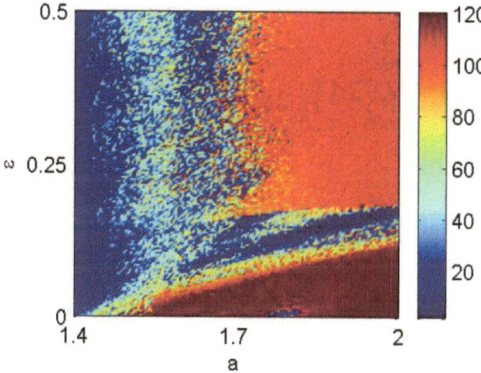

Fig. 10.6 Number of visible ordinal 5-patterns for the logistic coupled lattice as a function of a and ε, obtained from the symbolic sequence $\{\phi(\mathbf{s}_t)\}_{t=1}^{T}$

of space–time chaos and regularity sketched in Fig. 10.2. Note that higher values of $N(L)$ correspond to more complex dynamics.

One further possibility out of many others is to calculate the number of visible ordinal patterns, $N(L)$, in each univariate sequence $\mathbf{x}_t = (x_t(i))_{i=1}^{N}$ and to average the $T+1$ results. In our case, the value of L has to be small because of the condition $L! \ll N = 250$ (so as every ordinal L-pattern has a chance to appear in sliding windows along $\mathbf{x}_t$). The result is shown in Fig. 10.7; note that this figure gives information complementary to that provided by Fig. 10.6. A global increase of regularity (thus a decrease of $N(L)$) is observed as the strength of the coupling ε grows, as expected, but drastic transitions are also observed, corresponding to changes in the dynamics observed previously.

As a benchmark we consider next plots of Lyapunov exponents; these have been used to study various features of CMLs, like synchronization [18]. Figure 10.8

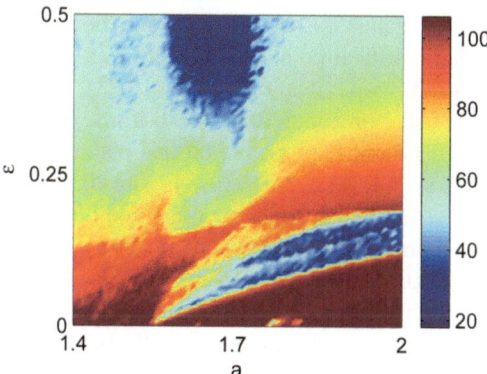

Fig. 10.7 Number of visible ordinal 5-patterns for the logistic coupled lattice as a function of a and ε, obtained from $\mathbf{x}_t$ and averaged over t

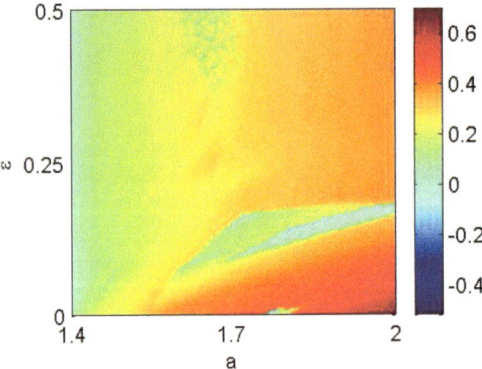

Fig. 10.8 Calculation of the largest Lyapunov exponent of a CML as a function of a and ε

shows a plot of the largest Lyapunov exponent λ calculated for the logistic coupled lattice (10.3)–(10.4) using Wolf's algorithm [204]. It can be observed there
that the boundaries between the different phases of the CML sketched by Kaneko
coincide roughly with abrupt changes in the value of λ. These results are coherent with the results observed in our calculations of $N(L)$ in Figs. 10.6 and 10.7.
Let us point out that the separation between the domains of fully developed turbulence and the rest of phases can be distinguished more clearly in the $N(L)$
plots.

For the sake of completeness we consider also a chain of 60 coupled oscillators,

$$
\begin{aligned}
\dot{u}_i &= 0.5 - 4v_i + \kappa(u_{i+1} + u_{i-1} - 2u_i), \\
\dot{v}_i &= -v_i + 2\max\{u_i - 8\cos t - 16, 0\},
\end{aligned}
\tag{10.28}
$$

with periodic boundary conditions. If we make a stroboscopic map of the variable u_i and plot $u_i(2\pi n)$ against $u_i(2\pi(n+1))$, points lie approximately on a one-
dimensional curve with a critical point at $u_c \approx 6.6$ (Fig. 10.9). Thus, each period
2π we assign to the stroboscopic map of system (10.28), $\{u_i(2\pi n)\}_{i=1}^{60}$, a string of
symbols following the usual procedure ($s_i(n) = 0$ if $u_i(2\pi n) < u_c$, and $s_i(n) = 1$
otherwise), and count the number of visible ordinal patterns of the ensuing binary
multivariate time series $\mathbf{s}_n = (s_1(n), \dots, s_{60}(n))$. Figure 10.10 represents the number
of ordinal 4-patterns, $N(4)$, of such series as a function of the coupling constant κ.
The inlets in this figure are space–time plots of $\{u_i(2\pi n)\}_{i=1}^{60}$ for $n = 1, \dots, 200$ and
three values of κ: $\kappa = 0.008$, $\kappa = 0.1$, and $\kappa = 0.18$ (left to right). Observe that
the decrease of $N(L)$ with κ parallels the diminution of dynamical complexity, in
particular the regularization of the dynamics and/or the reduction of chaotic domains
(i.e., the number of consecutive sites with chaotic dynamics). We conclude that
ordinal analysis might also be suitable to characterize the complexity of oscillator
chains.

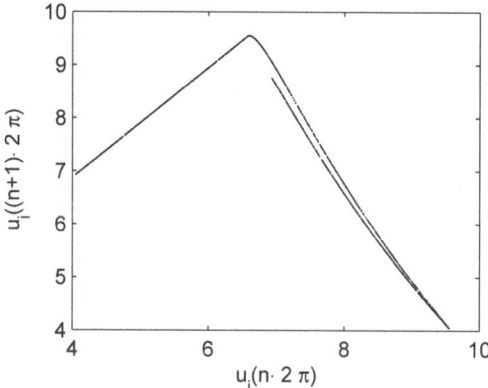

Fig. 10.9 Return map observed by plotting $u_i(n2\pi)$ against $u_i((n+1)2\pi)$ for any of the oscillators of the chain with $\kappa = 0$. Its unimodal appearance allows using symbolic sequences

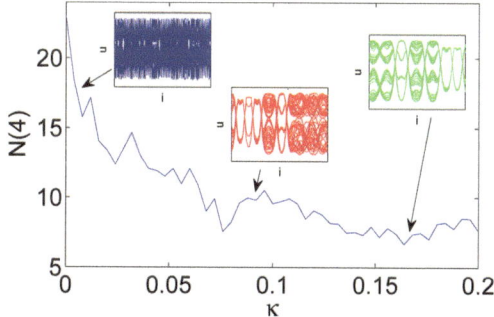

Fig. 10.10 Number of ordinal patterns of length $L = 4$, $N(4)$, found in a time series of length 200. The numbers are actually averages over the results for 20 initial conditions. The decrease of $N(4)$ is consistent with the decrease in complexity of the space–time dynamics shown in the *inlets*, which are space–time plots $\{u_i(\pi 2n)\}_{i=1}^{60}$ for $n = 1, .., 200$ and three values of κ: $\kappa = 0.008$, $\kappa = 0.1$, and $\kappa = 0.18$ (*left* to *right*)

Needless to say, the tools that can be chosen to measure the complexity of a CML are manifold. In the next section we study the use of regularity parameters.

10.2.4 Spatiotemporal Regularity of CMLs

Lastly, we consider the same temporal and spatial regularity parameters proposed for CA. But since the entries of the time series are now real numbers, the parameter $\chi^2(L)$ is given by (9.8) also when calculating $\chi^2_{space}(L)$.

Similarly as in Sect. 10.2.2, we have simulated the evolution of six logistic coupled lattices with $N = 250$ sites, each starting from 100 different random initial configurations. The corresponding parameters a and ε were chosen as in Fig. 10.3, so each lattice was in one of the six phases listed in Sect. 10.1.2. Figure 10.11

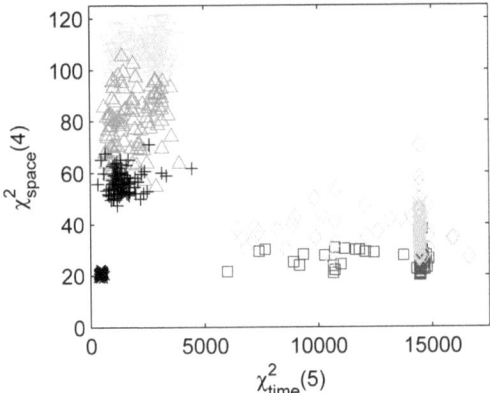

Fig. 10.11 Values of $\chi^2_{\text{time}}(5)$ and $\chi^2_{\text{space}}(4)$ for six logistic coupled lattices in different phases and 100 random initial configurations. Symbol assignment: frozen random patterns ($\square$), pattern selection and suppression of chaos ($\diamond$), Brownian motion of defects (∇), defect turbulence $\triangle$, pattern competition intermittency (+), and fully developed turbulence ($\times$). Different colors in the symbols are used when convenient

summarizes the results again on the plane $\chi^2_{\text{time}}(5)$ vs $\chi^2_{\text{space}}(4)$. The values cluster in different zones, but this time they overlap. The results are coherent with the types of dynamics described by Kaneko [108]. In some cases, overlapping might be due to multistability, i.e., depending on the initial conditions the type of dynamics may greatly vary.

Chapter 11
Conclusion and Outlook

Ordinal (or permutation-based) analysis of dynamical systems originates from the properties of the order relations and order isomorphisms. Thereby it is assumed that the state space of the systems is equipped with a total ordering. The order relations among consecutive elements in the orbits of deterministic or random dynamical systems are then codified in the form of ordinal patterns. The ordinal patterns themselves—whether admissible or forbidden—together with other "higher level" tools based on them, like permutation entropy rates, discrete entropy, frequency or probability distributions, regularity parameters, build the main repertoire of ordinal analysis. Since the sort of properties addressed by ordinal analysis and captured by its tools are not the same as in the usual measure-theoretical and topological approaches, we proposed the term "permutation complexity" to distinguish them.

In the foregoing chapters we have reviewed the theoretical and practical aspects of ordinal analysis. Among the first ones, let us highlight the study of metric (Chap. 6) and topological (Chap. 7) permutation entropies, together with the relation to their standard counterparts. Among the applications, some of them are well established, like the estimation of entropy (Sect. 2.1), complexity analysis of time series (Sect. 2.2), or detection of determinism (Chap. 9). Others like the complexity analysis of spatially extended systems (Chap. 10) are still in an initial stage. An important message to keep regarding all ordinal pattern-based applications is their robustness against observational noise—an asset when analyzing real systems. In particular, deterministic generation is responsible for the persistence of forbidden patterns in very noise data, as shown in Sect. 9.1. Robustness makes ordinal analysis a practical tool.

The reader might be tempted to dismiss ordinal analysis of dynamics as an uninteresting equivalent to well-known symbolic dynamics. In fact, ordinal patterns of dynamical systems do maintain equivalent results with symbolic dynamics, such as the metric and topological entropies we discussed in Chaps. 6 and 7, respectively, but in other ways, there are major distinctions, which are just starting to be explored for permutations. For instance, the canonical tent map and the Bernoulli shift ($f(x) = 2x \mod 1$) are isomorphic under a conventional analysis and in symbolic dynamics are equivalent to an i.i.d. source of white bits. However, under permutation-based analysis, once the state is imbued with total ordering, the class of order isomorphisms is different. Both conventional symbolic dynamics, assuming a generating partition of a map, and ordinal analysis are useful discrete representations of what would otherwise be a dynamical system in continuous space. However, the symbolic dynamics which results from a conventional partitioning is not fundamentally distinguishable from a noisy system; both result in conventional information sources on a discrete alphabet with a positive Shannon entropy. By contrast, the ordinal analysis does show a fundamental distinction between deterministic chaos and noisy systems. With chaos there is a rich structure of forbidden patterns among the ordinal patterns of different length and a hierarchy of consequent derived forbidden patterns (Chap. 3), the nature of which is not shared with conventional symbolic dynamics. More closely impacting the present work, the number of allowed permutations can scale superexponentially, which is fundamentally faster than the exponential scaling which must eventually happen with a noise-free deterministic chaotic system.

As in any research field, work on theory and applications of ordinal analysis is in progress, meaning that the picture is far from complete. In the course of the exposition, we have pointed out different questions which are waiting for answers. I summarized next the most important ones.

One of the basic open problems refers to the relation between a map and the structure of its forbidden patterns. Some natural questions that arise in this context are the following:

- Understand how the allowed or forbidden ordinal patterns (especially the root patterns) depend on the map.
- Given a map, determine the length of its shortest forbidden pattern.
- Describe and/or enumerate (exactly or asymptotically) any of the above classes of ordinal patterns.
- Given a finite or infinite set of, say, root forbidden patterns, find a map with the corresponding ordinal pattern structure.
- More generally, characterize those hierarchies of ordinal patterns for which there exist maps realizing them.

Of course, some of these questions can be answered graphically for simple maps and short pattern lengths. What we seek though are general results, possibly emanating from the structure of periodic points. We reported partial successes along this line for the shifts (Chap. 4) and signed shifts (Chap. 5), but the general case seems exceedingly hard. Even the ordinal structure of a general subshift of finite type (order isomorphic to some piecewise linear maps) seems to be beyond the techniques used in those chapters. A list of more advanced research topics would include the relation of forbidden patterns with the kneading invariants of one-dimensional interval maps or, say, with the directional entropy of cellular automata.

Other interesting (albeit theoretical) problem is the exact relation between the original definition of permutation entropy by Bandt et al. [29], and the definition given in Chaps. 6 and 7. Technically, the difference boils down to the order of two limiting processes (ever longer ordinal patterns and ever finer partitions) in a double limit. In particular, the results of Sects. 6.2 and 6.3 show that both definitions of metric permutation entropy overlap for one-dimensional, piecewise ergodic maps, and numerical simulations advocate a more general coincidence. In any case, the usual computations, with an arithmetic precision fixed by default or by the numerical format chosen, implement our "Kolmogorov-like" approach to permutation entropy.

For practical applications, the numerical tools of the type we discussed in Chap. 9 serve as a way of distinguishing chaos-like dynamics from noise, at least in simulations. This may be useful in the detection of emergent "coherent structures" similar to low-dimensional chaos in what otherwise might be a high degree of freedom system which could be rather noise-like. We comment on the unique property of permutations having a discrete "algebraic" nature permitting some rapid computational methods, without the requirement of estimating a generating partition for each dynamics. We feel that the appropriate tools for analysis of the typically short *observed* time series will require more sophisticated statistical thinking and methods still, just as high-quality estimation of entropies from low-alphabet information sources can be a difficult problem despite the apparent simplicity of the definitions themselves.

In Chap. 9 we also showed that the forbidden pattern-based technique outperforms one of the standard methods for detecting statistical dependence. Similar conclusions were reached in the ordinal analysis of synchronization in [159], see Sect. 2.4. This exercise—comparing a pattern-based technique with the traditional methods—is missing in other applications of ordinal analysis to time series like entropy estimation or complexity study. If the applications refer to natural systems, then the possibilities are virtually unlimited. Real time series appeared only in Sect. 2.2 ("Permutation complexity"), where we considered biomedical data, a recurrent topic in the literature. But, of course, other kinds of real data have also been studied (see Sect. 2.2).

Apart from the future lines of research related to the above-mentioned open problems, other lines of research refer to more recent topics and other follow-up investigations. In Chap. 10 we showed that ordinal analysis provides quantitative tools for and insights into the dynamics of space–time dynamics. This brief account was meant as a corroboration of performances shown in other contexts, as well as a stimulus to further research. Clearly, a survey of permutation complexity in cellular automata and coupled map lattices is a broad field that will require time and ingenuity, especially in the unexplored dimensions 2 and higher. Add to this general networks of coupled map lattices, and you get a long-term research program! But the great challenge is the complexity analysis of physical systems. Simple models, like cellular automata and coupled map lattice, provide a bridge to this more ambitious objective, in that they model non-trivial physical phenomena while being amenable to discrete methods. The situation resembles the study of complex dynamical systems via symbolic dynamics—a quite remarkable technique. The author believes that the interplay between complex dynamical systems and discrete methods is a promising approach also in the case of physical systems. Chapter 10 reported on progress in this direction from the ordinal front. New chapters will follow.

Annex A
Mathematical Framework

This annex is a summary of the mathematical background needed for this book.

A.1 Dynamical Systems

In this book we only consider two kinds of "discrete-time" dynamical systems: continuous and measure-preserving systems. Roughly speaking, the first are the basic objects of topological dynamics and the second ones play a major role in the study of statistical properties.

Definition 7 A continuous (or topological) dynamical system is a pair (M, f), where M is a topological space and $f : M \to M$ a continuous map.

Let Ω be a non-empty set, $\mathcal{B}$ a sigma-algebra of subsets of Ω, and $\mu : \mathcal{B} \to \mathbb{R} \cup \{+\infty\}$ a positive measure on the measurable space $(\Omega, \mathcal{B})$. A typical example of measurable space is a topological space endowed with the *Borel sigma-algebra*, i.e., the sigma-algebra generated by the open sets. The *measure space* $(\Omega, \mathcal{B}, \mu)$ is called a finite-measure space if $\mu(\Omega) < \infty$. A measurable map (function, transformation) $f : \Omega \to \Omega$ is said to preserve the measure μ, or to be μ-preserving, if $\mu(f^{-1}(B)) = \mu(B)$ for all $B \in \mathcal{B}$. Equivalently, the measure μ is said to be f-invariant. Sometimes $(\Omega, \mathcal{B}, \mu)$ is called the *state space* of the dynamic f.

Definition 8 Let $(\Omega, \mathcal{B}, \mu)$ be a finite-measure space and $f : \Omega \to \Omega$ a μ-preserving map. Then $(\Omega, \mathcal{B}, \mu, f)$ is called a measure-preserving dynamical system.

If $(\Omega, \mathcal{B}, \mu, f)$ is a measure-preserving dynamical system, we can assume without loss of generality that $\mu(\Omega) = 1$, i.e., that $(\Omega, \mathcal{B}, \mu)$ is a *probability space*. In this light, Ω is the space of elementary events, $\mathcal{B}$ comprises all outcomes we might be interested in, and $\mu(B)$ is the probability of the outcome $B \in \mathcal{B}$.

Given a measurable map $f : \Omega \to \Omega$, it is very difficult in practice to prove that f preserves the measure μ since, in general, not all elements $B \in \mathcal{B}$ are explicitly known. In general, all we know is a semi-algebra $\mathcal{S}$ generating $\mathcal{B}$. For example, if $\mathcal{B}$ is the Borel sigma-algebra of the interval $[0, 1] \subset \mathbb{R}$ with the standard topology, then $\mathcal{S}$ can be taken to be the collection of all subintervals of $[0, 1]$, or just the collection of subintervals of the forms $[0, b]$ and $(a, b]$, $0 \leq a < b \leq 1$. It can be proved [202] that if (i) $\mathcal{S}$ is a semi-algebra which generates $\mathcal{B}$ and (ii) for every $A \in \mathcal{S}$, $f^{-1}(A) \in \mathcal{B}$ and $\mu(f^{-1}(A)) = \mu(A)$, then f preserves the measure μ.

Exercise 13 Prove that $\mathcal{S} = \{[a, b) : 0 \leq a < b < 1\}$ is a semi-algebra of subsets of the interval $[0, 1)$ that generates the Borel sigma-algebra of $[0, 1)$.

Example 22 Suppose $\Omega = [0, 1)$, $\mathcal{B}$ is the Borel sigma-algebra of $[0, 1)$, and λ is the Lebesgue measure on $[0, 1)$. Furthermore, let $f : \Omega \to \Omega$ be the map given by $f(x) = Nx \bmod 1$, where $N \in \mathbb{Z}$, $|N| \geq 2$. Then f preserves λ. Indeed, for every half-open interval $[a, b) \subset [0, 1)$,

$$f^{-1}([a, b)) = \bigcup_{i=0}^{N-1} \left[\frac{a+i}{N}, \frac{b+i}{N} \right)$$

if $N \geq 2$ and

$$f^{-1}([a, b)) = \bigcup_{i=1}^{|N|} \left(\frac{i-b}{|N|}, \frac{i-a}{|N|} \right]$$

if $N \leq -2$. Hence,

$$\lambda\left(f^{-1}[a, b)\right) = \sum_{i=0}^{N-1} \frac{b-a}{N} = \sum_{i=1}^{|N|} \frac{b-a}{|N|} = b - a = \lambda([a, b)).$$

Example 23 Let the measure space $(\Omega, \mathcal{B}, \mu)$ be as in the previous example and $f : \Omega \to \Omega$ be given now by $f(x) = x + r \bmod 1$, with $r > 0$. This transformation preserves also the Lebesgue measure λ since, for every $[a, b) \subset [0, 1)$,

$$\begin{aligned}
f^{-1}([a, b)) &= [a - r, b - r) && \text{if } a \geq r, \\
f^{-1}([a, b)) &= [a + 1 - r, b + 1 - r) && \text{if } b \leq r, \\
f^{-1}([a, b)) &= [0, b - r) \cup [a + 1 - r, 1) && \text{if } a < r < b.
\end{aligned}$$

In any case,

$$\lambda\left(f^{-1}([a, b))\right) = b - a = \lambda([a, b)).$$

A perhaps more natural way of dealing with this example views f as a rotation on the circle. The f-invariance of λ is then straightforward.

More generally, the Lebesgue measure on $\mathbb{R}^n$ is invariant under translations and rotations in $\mathbb{R}^n$. More sophisticated examples of invariant measures include the Haar measure on a locally compact topological group, the map being the action of the group. In the next section we will meet invariant measures on product spaces.

Exercise 14 Let $f:[0,1) \to [0,1)$ be the Gauss transformation,

$$f(x) = \begin{cases} 0 & \text{if } x = 0, \\ \frac{1}{x} \text{ (mod 1)} & \text{if } x \neq 0. \end{cases}$$

Show that f preserves the measure

$$\mu(B) = \frac{1}{\ln 2} \int_B \frac{dx}{1+x}, \tag{A.1}$$

where B is a Borel set of $[0,1)$. Hint:

$$f^{-1}([a,b)) = \bigcup_{n=1}^{\infty} \left(\frac{1}{b+n}, \frac{1}{a+n} \right].$$

Krylov and Bogolioubov showed that invariant measures exist under quite general conditions.

Theorem 20 [202] *Let Ω be a compact metric space and $f:\Omega \to \Omega$ a continuous map. Then there exists an f-invariant probability measure μ on $(\Omega, \mathcal{B})$, where $\mathcal{B}$ is the Borel sigma-algebra of Ω.*

In general, there can exist more than one f-invariant measure and, besides, some of them can be rather "pathological." For instance, if δ_p is the Dirac measure at p, i.e.,

$$\delta_p(B) = \begin{cases} 1 & \text{if } p \in B \\ 0 & \text{if } p \notin B \end{cases},$$

$B \in \mathcal{B}$, and x is a period-n point for f, then

$$\mu(B) = \frac{1}{n} \sum_{k=0}^{n-1} \delta_{f^k(x)}(B)$$

($f^0(x) := x$ and $f^i(x) = f(f^{i-1}(x))$ for $i \geq 1$) is an atomic measure supported on the points $\{x, f(x), \ldots, f^{n-1}(x)\}$. A set $E \subset \Omega$ is said to be the (unique) *support* of μ if (i) E is closed in Ω, (ii) $\mu(E \cap U) > 0$ if $E \cap U \neq \emptyset$ and U is open in Ω, and (iii) $\mu(E') = 0$, where $E' = \Omega \backslash E$ is the complement of E.

In general, the ordered set $\{f^i(x): i \geq 0\}$ is called the orbit or trajectory of the point (state, initial condition, etc.) $x \in \Omega$ under the "discrete-time" dynamic f and denoted by $\mathcal{O}_f(x)$. In the case of invertible maps, one writes $\mathcal{O}_f^+(x) = \{f^i(x): i \geq 0\}$ for the "forward" orbit, while orbit means $\mathcal{O}_f(x) = \{f^i(x): i \in \mathbb{Z}\}$.

It can happen that for almost all x in a set $U \subset \Omega$ with positive Lebesgue measure, its orbit is bounded and, moreover, the sequences of probability measures

$$\frac{1}{n}\sum_{k=0}^{n-1}\delta_{f^k(x)}$$

converge weakly to a measure μ, i.e., for almost all $x \in U$ and any continuous map $\varphi: \Omega \to \Omega$,

$$\lim_{k \to \infty}\frac{1}{n}\sum_{k=0}^{n-1}\varphi(f^k(x)) = \int_\Omega \varphi d\mu$$

holds. Then μ is an f-invariant measure that is usually called the natural or physical measure for its relevance in physics and computer simulations [72].

An important issue in measure-preserving dynamical systems is the existence of absolutely continuous invariant measures. A measure μ on a topological space Ω is said to be *absolutely continuous* (with respect to the Lebesgue measure dx), if $\mu(dx) = \rho(x)dx =: d\mu$, where the *density function* $\rho: \Omega \to \Omega$ (also called the Radon–Nikodym derivative of μ with respect to the Lebesgue measure, $d\mu/dx$) is continuous. For example, if μ is measure (A.1) on the interval $[0, 1)$ endowed with the Borel sigma-algebra, then

$$\mu(dx) = \frac{1}{\ln 2}\frac{dx}{1+x} \quad \text{or} \quad \frac{d\mu}{dx} = \frac{1}{\ln 2}\frac{1}{1+x}.$$

In general there are few results on the existence of absolutely continuous invariant measures. In the case of self-maps of one-dimensional intervals, there are some general conditions that appear in the usual theorems on existence of such measures.

Recall that a *partition* of a measure space $(\Omega, \mathcal{B}, \mu)$ is a disjoint collection of elements of $\mathcal{B}$ whose union is Ω.

Definition 9 Let $\alpha = \{I_i\}_{i=1}^d$ be a partition of the interval $I = [a, b] \subset \mathbb{R}$ into subintervals I_i. Given the map $f: I \to I$, assume that $f|_{I_i}$ is C^k ($k \geq 1$) for each i.

(a) f is said to be C^k *piecewise expanding* if there exists $\lambda > 1$ such that $|f'(x)| > \lambda$ for all $x \in I_i$ and each i.
(b) f is said to be C^k Markov if $f(\mathring{I}_i) \supset \mathring{I}_j$ whenever $f(\mathring{I}_i) \cap \mathring{I}_j \neq \emptyset$ ("Markov property"), where $\mathring{I}_i$ stands for the interior of I_i, $1 \leq i \leq d$. In this case, α is called a *Markov partition* for f. The matrix $A = (A_{ij})_{1 \leq i, j \leq d}$ with

$$A_{i,j} = \begin{cases} 1 & \textit{if } f(\mathring{I}_i) \supset \mathring{I}_j, \\ 0 & \textit{if } f(\mathring{I}_i) \cap \mathring{I}_j = \emptyset, \end{cases} \tag{A.2}$$

is called the transition matrix for f.

See, for instance, [37, Chap. 5] and [105] for results concerning the existence of absolutely continuous invariant measures for piecewise expanding and/or Markov transformations (complying with additional conditions).

Exercise 15 Prove that the logistic map $g(x) = 4x(1-x), 0 \leq x \leq 1$, has an invariant measure with density function

$$\rho(x) = \frac{1}{\pi\sqrt{x(1-x)}},\tag{A.3}$$

i.e., $\int_0^1 \rho(x)dx = 1$, and

$$\int_{[a,b]} \rho(x)dx = \int_{g^{-1}[a,b]} \rho(x)dx,$$

for all $0 \leq a < b \leq 1$. Figure A.1 shows the plot of the function $\rho(x)$. Is $g(x)$ piecewise expanding? Is $g(x)$ Markovian?

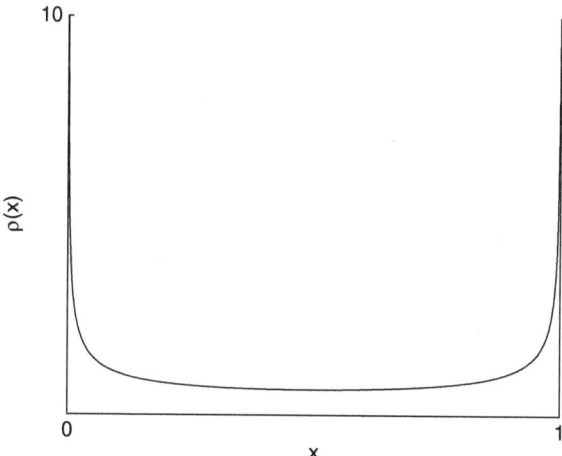

Fig. A.1 The density $\rho(x)$, (A.3)

Once we know that invariant measures are rather abundant objects, suppose that $f: \Omega \to \Omega$ is such that $f^{-1}(B) = B$ for some $B \in \mathcal{B}$. Then $f^{-1}(\Omega \backslash B) = \Omega \backslash B$ and the action of f on Ω can be decomposed into two disjoint pieces: $f|_B$ and $f|_{\Omega \backslash B}$. If f is indecomposable in the previous sense, one says that f is ergodic.

Definition 10 Let $(\Omega, \mathcal{B}, \mu, f)$ be a measure-preserving dynamical system. The map f is said to be ergodic if

$$f^{-1}(B) = B, \quad B \in \mathcal{B} \Rightarrow \mu(B) = 0 \quad \text{or} \quad \mu(B) = 1$$

Alternatively, μ is said to be an ergodic measure for f. Also, the dynamical system $(\Omega, \mathcal{B}, \mu, f)$ is said to be ergodic.

Thus an ergodic measure cannot be decomposed as a (properly weighted or "convex") sum of invariant measures. It might seem that this definition is a far cry from the original Boltzmann's *Ergodenhypothese*, which states that the trajectory of a closed thermodynamic system in the phase space (spanned by the coordinates and conjugate canonical momenta of its constituent particles) covers densely and uniformly the "energy shell," that is, the hypersurface in phase space defined by the restriction that the energy of the system is constant. But it was on the way to lying Boltzmann's proposal on a mathematically sound basis that G. Birkhoff introduced the concept of ergodicity in its modern version. Birkhoff's seminal *ergodic theorem* states the following.

Theorem 21 [202] *If* $(\Omega, \mathcal{B}, \mu, f)$ *is an ergodic dynamical system, then*

$$\lim_{n \to \infty} \frac{1}{n} \sum_{i=0}^{n-1} \varphi(f^i(x)) = \int_{\Omega} \varphi d\mu \quad a.e. \tag{A.4}$$

for all $\varphi \in L^1(\mu)$.

As usual, "a.e." is shorthand for "almost everywhere" with respect to the relevant measure (μ here) and $L^1(\mu)$ is the space of μ-integrable functions. The property assumed by the *Ergodenhypothese* goes by the name of *topological transitivity* in the theory of discrete dynamical systems. A continuous self-map f of a compact metric space Ω is called topologically transitive if there exists some $x \in \Omega$ such that $\mathcal{O}_f(x)$ is dense in Ω (if f is invertible, then $\mathcal{O}_f(x)$ also includes the "backward" iterates $f^{-n}(x)$, $n \in \mathbb{N}$).

Let χ_B denote the *characteristic function* of the set $B \in \mathcal{B}$,

$$\chi_B(x) = \begin{cases} 1 & \text{if } x \in B \\ 0 & \text{if } x \notin B \end{cases}.$$

The substitution $\varphi = \chi_B$ in (A.4) yields then

$$\frac{1}{n} \sum_{i=0}^{n-1} \chi_B(f^i(x)) \to \mu(B) \quad \text{a.e.,}$$

when $n \to \infty$. This means that if $(\Omega, \mathcal{B}, \mu, f)$ is ergodic, then the orbit of almost every initial condition $x \in \Omega$ visits the region B of the state space with asymptotic frequency $\mu(B)$. This resembles the law of large numbers in statistics and, in fact, there are plenty of deep relations between ergodic theory and statistics [31, 67].

Let Ω be a compact metrizable space Ω, and $\mathcal{B}$ the Borel sigma-algebra on Ω. A continuous map $f : \Omega \to \Omega$ is called *uniquely ergodic* if there is only one f-invariant Borel probability measure on Ω. A map f is uniquely ergodic if and only if it has exactly one invariant measure. If f is uniquely ergodic and μ is its invariant measure, then (A.4) holds for all continuous transformations φ and all $x \in \Omega$ [202].

Ergodicity is just but the first step in a series of notions measuring the statistical properties of the orbits generated by the dynamic: ergodicity, mixing, completely positive entropy, etc. Here we will recall only the definition of strong mixing.

Definition 11 The measure-preserving dynamical system $(\Omega, \mathcal{B}, \mu, f)$ is called (strong) mixing if

$$\lim_{n \to \infty} \mu(f^{-n}(A) \cap B) = \mu(A)\mu(B) \tag{A.5}$$

for all $A, B \in \mathcal{B}$.

In contrast to (A.5), f is ergodic if and only if

$$\lim_{n \to \infty} \frac{1}{n} \sum_{i=0}^{n-1} \mu(f^{-i}(A) \cap B) = \mu(A)\mu(B) \tag{A.6}$$

for all $A, B \in \mathcal{B}$. Hence mixing is a stronger condition than ergodicity. In practice it suffices to check (A.6) and (A.5) for $A, B \in \mathcal{S}$, a semi-algebra that generates $\mathcal{B}$.

Sufficient conditions for the existence of *ergodic* absolutely continuous invariant measures can be found, e.g., in [52, Chap. 5] . Mixing piecewise C^2 expanding Markov maps have unique ergodic invariant measures [105].

As in any other area of mathematics, the notion of isomorphism is central. It specifies when two dynamical systems are to be considered equivalent from the point of view of the properties that matter in this theory.

Definition 12 Given the measure-preserving dynamical systems $(\Omega_1, \mathcal{B}_1, \mu_1, f_1)$ and $(\Omega_2, \mathcal{B}_2, \mu_2, f_2)$, we say that f_1 is (metrically) isomorphic to f_2 if there exist $B_1 \in \mathcal{B}_1$, $B_2 \in \mathcal{B}_2$ with $\mu_1(B_1) = \mu_2(B_2) = 1$ such that (i) $f_1(B_1) \subset B_1, f_2(B_2) \subset B_2$ and (ii) there is an invertible, measure-preserving map $\phi: B_1 \to B_2$ with $\phi \circ f_1(x) = f_2 \circ \phi(x)$ for all $x \in B_1$.

The dynamical systems $(\Omega_1, \mathcal{B}_1, \mu_1, f_1)$ and $(\Omega_2, \mathcal{B}_2, \mu_2, f_2)$ are said to be isomorphic. Sometimes ϕ is called an isomorphism "modulo 0" or just "mod 0" (shorthand for modulo measure zero sets), but usually we dispense with measure zero sets without stating it explicitly. In the more general case that ϕ is measure preserving but only surjective, $(\Omega_2, \mathcal{B}_2, \mu_2, f_2)$ is called a *factor* of $(\Omega_1, \mathcal{B}_1, \mu_1, f_1)$ (or $(\Omega_1, \mathcal{B}_1, \mu_1, f_1)$ a *cover* of $(\Omega_2, \mathcal{B}_2, \mu_2, f_2)$) via the factor map ϕ. Two isomorphic maps are obtained from each other by a change of coordinates, so that properties that are independent of such changes of coordinates are invariant. Isomorphism invariants include ergodicity and mixing.

There is a broader (and more technical) concept called *conjugacy* that embraces isomorphism. Both concepts are though equivalent in virtually all probability spaces that one encounters in applications (e.g., compact metric spaces). Indeed, as it turns out, there is essentially only one type of probability space, called a *Lebesgue space*, which is characterized as being measure-theoretically isomorphic to the union of an interval of $\mathbb{R}$ endowed with Lebesgue measure, with at most countably many points of positive measure (called atoms) [177, 202]. In a Lebesgue space, set maps are always induced by point maps. Conjugacy and isomorphy coincide for a Lebesgue space, so both terms can be used interchangeably in that case.

Example 24 The symmetric tent map $\Lambda:[0, 1] \to [0, 1]$,

$$\Lambda(x) = \begin{cases} 2x & 0 \le x \le \frac{1}{2} \\ 2 - 2x & \frac{1}{2} \le x \le 1 \end{cases}, \tag{A.7}$$

preserves the Lebesgue measure $\lambda(dx) = dx$. If, furthermore, $\mu(dx) = \frac{1}{\pi\sqrt{x(1-x)}}dx$ is the natural invariant measure of the logistic map $g:[0, 1] \to [0, 1]$, $g(x) = 4x(1 - x)$ (see (A.3)), then $\phi:([0, 1], \lambda) \to ([0, 1], \mu)$ given by

$$\phi(x) = \sin^2\left(\frac{\pi}{2}x\right) \tag{A.8}$$

is invertible, measure preserving, and it satisfies $g \circ \phi = \phi \circ \Lambda$. Hence, Λ and g are conjugate.

Exercise 16 Show that

$$x_k = \sin^2\left(2^k\xi\right),$$

$\xi \in \mathbb{R}$, is a solution of the logistic recursion (or finite difference equation)

$$x_{k+1} = 4x_k(1 - x_k), \quad k \ge 0,$$

$x_k \in [0, 1]$, with initial condition $x_0 = \sin^2\xi$.

A.2 Shift Systems

Shift systems are dynamical systems which due to their importance as models and prototypes are considered separately in this section. In the simplest and most usual version, the elements of the shift spaces are one-sided or two-sided sequences of N symbols or "letters". Sometimes one has to consider also sequences with elements from an arbitrary (countable or uncountable) "alphabet," and this requires some degree of sophistication. We set out from this more general situation.

First of all, let us recall the definition of a *product measurable space*. For our purposes it is sufficient to consider products of countably many copies of a measurable space $(\Omega, \mathcal{B})$. As index set $\mathbb{K}$ we take without restriction $\mathbb{K} = \mathbb{N}_0 := \{0\} \cup \mathbb{N}$ or $\mathbb{K} = \mathbb{Z}$. Then, $\Pi_{k \in \mathbb{K}}(\Omega, \mathcal{B}) = (\Omega^{\mathbb{K}}, \mathcal{B}_\Pi(\Omega))$, where

$$\Omega^{\mathbb{K}} = \{(\omega_k)_{k \in \mathbb{K}}:\omega_k \in \Omega\}$$

is the set of all one-sided sequences

$$(\omega_k)_{k \in \mathbb{N}_0} = \omega_0, \dots, \omega_k, \dots$$

if $\mathbb{K} = \mathbb{N}_0$, or the set of all two-sided sequences (also called bisequences or doubly infinite sequences)

$$(\omega_k)_{k \in \mathbb{Z}} = \dots, \omega_{-k}, \dots, \omega_0, \dots, \omega_n, \dots$$

if $\mathbb{K} = \mathbb{Z}$, and $\mathcal{B}_{\Pi}(\Omega)$ is the sigma-algebra generated by the semi-algebra $\mathcal{S}$ of *cylinder sets*

$$\prod_{j \in \mathbb{F}} A_j \times \prod_{k \notin \mathbb{F}} \Omega = \{(\omega_k)_{k \in \mathbb{K}} : \omega_j \in A_j \text{ for } j \in \mathbb{F}\}, \tag{A.9}$$

where $\mathbb{F} \subset \mathbb{K}$ is finite and $A_j \in \mathcal{B}$ for $j \in \mathbb{F}$. If $\mathbb{K} = \mathbb{N}_0$ (correspondingly, $\mathbb{K} = \mathbb{Z}$), then we can take $\mathbb{F} = \{0, 1, \dots, n\}$ (correspondingly, $\mathbb{F} = \{-n, \dots, 0, \dots, n\}$), $n \in \mathbb{N}_0$, in (A.9) without restriction.

In most applications we have in mind (for instance, to information theory), $(\Omega, \mathcal{B}) = (S, 2^S)$ with $S = \{0, \dots, N{-}1\}$, $N \geq 2$, and 2^S denoting as usual the family of all subsets of S. In this case, the set of all one-sided sequences of the symbols $0, 1, \dots, N{-}1$,

$$S^{\mathbb{N}_0} = \{(s_n)_{n \in \mathbb{N}_0} : s_n \in S\}, \tag{A.10}$$

is called the (one-sided) *sequence space on N symbols*. Depending on the context, the set of symbols S may receive different names. In the setting of information theory, S is called an *alphabet*, its elements are called *letters*, and sequences $\mathbf{s} = (s_n)_{n \in \mathbb{N}_0}$ are called *messages*. In dynamics, S is sometimes called the state space and its elements, states. Segments (or *words*) of symbols of length L, like $s_k, s_{k+1}, \dots, s_{k+L-1}$, will be shortened as s_k^{k+L-1}.

If S is thought to be a topological space (eventually endowed with the discrete topology), then $S^{\mathbb{N}_0}$ can be promoted to a topological space by means of the product topology, which is generated by the corresponding cylinder sets

$$C_{a_0, \dots, a_n} = \{\mathbf{s} \in S^{\mathbb{N}_0} : s_k = a_k, 0 \leq k \leq n\}, \tag{A.11}$$

where $a_0, \dots, a_n \in S$. (The general definition (A.9) with $A_j = \{a_j\}$ leads to the same topology.) The product topology makes $S^{\mathbb{N}_0}$ compact, perfect (i.e., it is closed and all its points are accumulation points), and totally disconnected. Such topological spaces are sometimes called *Cantor sets* because they are homeomorphic to Cantor's ternary set in the unit interval. By definition, the product sigma-algebra, $\mathcal{B}_{\Pi}(S)$, is generated by the cylinder sets (A.11) and comprises all Borel sets of $S^{\mathbb{N}_0}$.

Moreover, $S^{\mathbb{N}_0}$ is a metrizable space. In fact, there are several (non-equivalent) metrics compatible with the topology of $S^{\mathbb{N}_0}$, the perhaps most popular being

$$d_K(\mathbf{s}, \mathbf{s}') = \sum_{n=0}^{\infty} \frac{\delta(s_n, s'_n)}{K^n}, \tag{A.12}$$

where $\delta(s_n, s'_n) = 1$ if $s_n \neq s'_n$, $\delta(s_n, s_n) = 0$ and $K > 2$. Observe that given $\mathbf{s} \in C_{a_0,\ldots,a_n}$, then $d_K(\mathbf{s}, \mathbf{s}') < \frac{1}{K^n}$ if $\mathbf{s}' \in C_{a_0,\ldots,a_n}$, and $d_K(\mathbf{s}, \mathbf{s}') \geq \frac{1}{K^n}$ if $\mathbf{s}' \notin C_{a_0,\ldots,a_n}$, thus $C_{a_0,\ldots,a_n} = B_{d_K}(\mathbf{s}; \frac{1}{K^n})$, the open ball of radius K^{-n} and center $\mathbf{s}$ in the metric space $(S^{\mathbb{N}_0}, d_K)$. Moreover, every point in $B_{d_K}(\mathbf{s}; \frac{1}{K^n})$ is a center, a property known from non-Archimedean normed spaces (e.g., the rational numbers with p-adic norms [115]).

Exercise 17 1. Prove that the cylinder sets (thus the open balls) are also closed in the product topology. Open and closed sets are sometimes called clopen sets.
2. Prove that the cylinder sets are not connected (i.e., they can be written as a disjoint union of open sets).

Shifting all the symbols of a one-sided sequence to the left one place and dropping the first symbol define a self-map of one-sided sequence spaces which plays an important role in both theory and applications. Formally, the (one-sided) *shift* $\Sigma : S^{\mathbb{N}_0} \to S^{\mathbb{N}_0}$ is defined as

$$\Sigma(s_0, s_1, s_2, \ldots) = (s_1, s_2, s_3, \ldots), \tag{A.13}$$

that is, $\Sigma(\mathbf{s}) = \mathbf{s}'$ with $s'_n = s_{n+1}$. Since $\Sigma^{-1} C_{a_0,\ldots,a_n} = \bigcup_{a \in S} C_{a,a_0,\ldots,a_n}$, Σ is continuous on $(S^{\mathbb{N}_0}, d_K)$, each point $\mathbf{s} \in S^{\mathbb{N}_0}$ having exactly N preimages under Σ. Furthermore, Σ has N fixed points: $\mathbf{s} = a_0^{\infty}, 0 \leq a \leq N - 1$.

In order to make a measure-preserving dynamical system out of $S^{\mathbb{N}_0}$, $\mathcal{B}_{\Pi}(S)$, and Σ, only a Σ-invariant measure is missing. All probability measures on $(S^{\mathbb{N}_0}, \mathcal{B}_{\Pi}(S))$ that make Σ a measure-preserving transformation are obtained in the following way [202]. For any $n \geq 0$ and $a_i \in S, 0 \leq i \leq n$, let a real number $p_n(a_0, \ldots, a_n)$ be given such that (i) $p_n(a_0, \ldots, a_n) \geq 0$, (ii) $\sum_{a_0 \in S} p_0(a_0) = 1$, and (iii) $p_n(a_0, \ldots, a_n) = \sum_{a_{n+1} \in S} p_{n+1}(a_0, \ldots, a_n, a_{n+1})$. If we define now

$$m(C_{a_0,\ldots,a_n}) = p_n(a_0, \ldots, a_n),$$

then m can be extended to a probability measure on $(S^{\mathbb{N}_0}, \mathcal{B}_{\Pi}(S))$. The resulting dynamical system $(S^{\mathbb{N}_0}, \mathcal{B}_{\Pi}(S), m, \Sigma)$ is called the *one-sided shift system*.

If instead of considering (one-sided) sequences $s = (s_n)_{n \in \mathbb{N}_0}$, $s_n \in S = \{0, \ldots, N - 1\}$, we consider two-sided sequences $\mathbf{s} = (s_n)_{n \in \mathbb{Z}}$, we are in the realm of the *two-sided sequence spaces on N symbols*,

$$S^{\mathbb{Z}} = \{(s_n)_{n \in \mathbb{Z}} : s_n \in S\}.$$

The corresponding (invertible) *two-sided shift* on $S^{\mathbb{Z}}$ is defined as $\Sigma : \mathbf{s} \mapsto \mathbf{s}'$ with $s'_n = s_{n+1}$, $n \in \mathbb{Z}$. (Although not strictly correct, we use the same letter Σ for one-sided and two-sided shifts.) The cylinder sets are given now as

$$C_{a_{-n},\ldots,a_0,\ldots,a_n} = \{\mathbf{s} \in S^{\mathbb{Z}} : s_k = a_k, |k| \leq n\}$$

and

$$d_K(\mathbf{s}, \mathbf{s}') = \sum_{n \in \mathbb{Z}} \frac{\delta(s_n, s_n')}{K^{|n|}},$$

$K > 3$, is a metric for $S^{\mathbb{Z}}$. The dynamical system $(S^{\mathbb{Z}}, \mathcal{B}_{\Pi}(S), m, \Sigma)$ is called the *two-sided shift system*.

Exercise 18 Prove that the cylinder set $C_{a_{-n},\ldots,a_0,\ldots,a_n}$ of $S^{\mathbb{Z}}$ coincides with the open ball $B_{d_K}(\mathbf{s}; K^{1-n})$, where $\mathbf{s}$ is any point of $C_{a_{-n},\ldots,a_0,\ldots,a_n}$.

Example 25 (a) Let $\mathbf{p} = (p_0, p_1, \ldots, p_{N-1})$, $N \geq 2$, be a probability vector with non-zero entries (i.e., $p_i > 0$ and $\sum_{i=0}^{N-1} p_i = 1$). Set

$$p_n(a_0, a_1, \ldots, a_n) = p_{a_0} p_{a_1} \cdots p_{a_n}.$$

The resulting measure on $(S^{\mathbb{K}}, \mathcal{B}_{\Pi}(S))$ is called the Bernoulli measure defined by $\mathbf{p}$. The dynamical system $(S^{\mathbb{K}}, \mathcal{B}_{\Pi}(S), m, \Sigma)$, where m is the Bernoulli measure defined by the probability vector $\mathbf{p}$, is called a one-sided (if $\mathbb{K} = \mathbb{N}_0$) or two-sided (if $\mathbb{K} = \mathbb{Z}$) $\mathbf{p}$-*Bernoulli shift*.

(b) Let $\mathbf{p} = (p_0, p_1, \ldots, p_{N-1})$ be a probability vector as in (a) and $P = (p_{ij})_{0 \leq i,j \leq N-1}$ an $N \times N$ stochastic matrix (i.e., $p_{ij} \geq 0$ and $\sum_{j=0}^{N-1} p_{ij} = 1$) such that $\sum_{i=0}^{N-1} p_i p_{ij} = p_j$. Set then

$$p_n(a_0, a_1, \ldots, a_n) = p_{a_0} p_{a_0 a_1} p_{a_1 a_2} \cdots p_{a_{n-1} a_n}.$$

The resulting measure on $(S^{\mathbb{K}}, \mathcal{B}_{\Pi}(S))$ is called the Markov measure defined by $(\mathbf{p}, P)$. The dynamical system $(S^{\mathbb{K}}, \mathcal{B}_{\Pi}(S), m, \Sigma)$, where m is the Markov measure defined by the probability vector $\mathbf{p}$ and the stochastic matrix P, is called a one-sided (if $\mathbb{K} = \mathbb{N}_0$) or two-sided (if $\mathbb{K} = \mathbb{Z}$) $(\mathbf{p}, P)$-*Markov shift*. A $\mathbf{p}$-*Bernoulli shift* can be considered as a $(\mathbf{p}, P)$-Markov shift by taking $p_{ij} = p_j$.

Simple as they might seem, one-sided and two-sided shifts exhibit most of the basic properties of ergodic theory, like ergodicity and strong mixing. In particular, they are easily shown to be *chaotic* in the sense of Devaney [69], i.e., they are sensitive to initial conditions, are strong mixing, and their periodic points are dense. Let us recall at this point the notion of sensitivity to initial conditions.

Definition 13 Given a metric space (M, d), a map $f : M \to M$ is said to be sensitive to initial conditions if there exists $\delta > 0$, called a sensitivity constant, such that for every $x \in \Omega$ and $\varepsilon > 0$ there exists $y \in \Omega$ with $d(x, y) < \varepsilon$ and $d(f^n(x), f^n(y)) \geq \delta$ for some $n \in \mathbb{N}$.

Equivalently, a continuous self-map of a compact metric space is said to be chaotic if it is topologically transitive (that is, it has a dense orbit) and its periodic points are dense [91].

Exercise 19 Prove that the one- and two-sided shifts on N symbols are sensitive to initial conditions, are topological transitive, and their periodic points are dense.

Example 26 Let $\Omega = [0, 1]$, $\mathcal{B}$ the Borel sigma-algebra of $[0, 1]$, λ the corresponding Lebesgue measure, and $E_2: x \mapsto 2x \pmod 1$ the so-called dyadic map. The dynamical system $([0, 1], \mathcal{B}, \lambda, E_2)$ is then isomorphic (up to a measure zero set) to the one-sided $(\frac{1}{2}, \frac{1}{2})$-Bernoulli shift on the symbols $\{0, 1\} = S$. An isomorphism $\phi: S^{\mathbb{N}_0} \to [0, 1]$ is given by

$$(x_0, x_1, \ldots, x_k, ..) \mapsto \sum_{k=0}^{\infty} x_k 2^{-(k+1)}. \tag{A.14}$$

Of course, the map ϕ is not injective in strict sense because the sequences $(x_0, \ldots, x_{n-1}, 0, 1^{\infty})$ and $(x_0, \ldots, x_{n-1}, 1, 0^{\infty})$ are sent to the same point (the upper label "∞" means indefinite repetition); indeed,

$$\sum_{k=0}^{n-1} x_k 2^{-(k+1)} + \sum_{k=n+1}^{\infty} 2^{-(k+1)} = \sum_{k=0}^{n-1} x_k 2^{-(k+1)} + 2^{-(n+1)}.$$

However, since the set of sequences eventually terminating in an infinite string of 0's or 1's is countable, we conclude that $(S^{\mathbb{N}_0}, \mathcal{B}_{\Pi}(S), m, \Sigma)$ and $([0, 1], \mathcal{B}, \lambda, E_2)$ are conjugate modulo 0, i.e., the diagram

$$\begin{array}{ccc} \Sigma: \{0, 1\}^{\mathbb{N}_0} & \to & \{0, 1\}^{\mathbb{N}_0} \\ \phi \downarrow & & \downarrow \phi \\ E_2: [0, 1] & \to & [0, 1] \end{array}$$

is commutative almost everywhere: $E_2 = \phi \circ \Sigma \circ \phi^{-1}$. Observe that there is otherwise a topological obstruction that prevents $S^{\mathbb{N}_0}$ and $[0, 1]$ from being homeomorphic: the first is (homeomorphic to) a Cantor set while, certainly, the second is not.

Exercise 20 Prove that the map $\phi: S^{\mathbb{N}_0} \to [0, 1]$ defined in (A.14) is measure preserving, i.e., $m(\phi^{-1}(I)) = \lambda(I)$ for any interval $I \subset [0, 1]$. It suffices to consider "dyadic" intervals, i.e., intervals of the forms $[0, k_2/2^n]$ and $(k_1/2^n, k_2/2^n]$, $0 \le k_1 < k_2 \le 2^n$, $n \in \mathbb{N}$.

Let us mention in passing the dyadic map $x \mapsto 2x \pmod 1$ is just the first member of the family of *expanding maps* of the circle:

$$E_N: x \mapsto Nx \pmod 1,$$

where N is an integer of absolute value greater than 1. In a way similar to Example 26 one can show that $([0, 1], \mathcal{B}, \lambda, E_N)$ and the $(\frac{1}{N}, \ldots, \frac{1}{N})$-Bernoulli shift are conjugate for $N \ge 2$. In this case, map (A.14) is replaced by $(x_0, x_1, \ldots) \mapsto \sum_{k=0}^{\infty} x_k N^{-(k+1)}$.

Exercise 21 What transformation induces on the sequence space $\{0, 1\}^{\mathbb{N}_0}$ the expanding map E_{-2} via map (A.14)?

A.3 Stochastic Processes and Sequence Spaces

A stochastic (or random) process is a mathematical model for the occurrence of random phenomena as time goes on. This is the case, for example, when a random experiment is repeated over and over again. Put in a formal way, a *stochastic process* is a collection of random variables $\mathbf{X} = \{X_t\}_{t \in \mathcal{T}}$ on a common probability space $(\Omega, \mathcal{B}, \mu)$, called the *sample space*, taking on values in a measurable space $(S, \mathcal{A})$, called the *state space*. Technically this means that $X_t: \Omega \to S$ is a measurable map for all $t \in \mathcal{T}$, i.e., $X_t^{-1}(A) \in \mathcal{B}$ for all $A \in \mathcal{A}$. The index $t \in \mathcal{T}$ is conveniently interpreted as time, the usual choices for $\mathcal{T}$ being (i) $\mathcal{T} = \mathbb{R}$ or $\mathbb{R}_+ = [0, \infty]$, in which case $\mathbf{X}$ is called a continuous-time stochastic process or (ii) $\mathcal{T} = \mathbb{N}_0$ or $\mathbb{Z}$, in which case $\mathbf{X}$ is called a discrete-time stochastic process. The map $t \mapsto X_t(\omega)$ is the realization (sample path, trajectory, etc.) of the process $\mathbf{X}$ associated with the fixed sample point $\omega \in \Omega$. As usual in probability theory and statistics, a realization of a random variable X will be denoted by the same letter in small caps: $X(\omega) = x$.

The stochastic process $\mathbf{X}$ is characterized by its joint (finite-dimensional) probability distributions

$$\mu\{\omega \in \Omega : X_{t_1}(\omega) \in A_1, \ldots, X_{t_r}(\omega) \in A_r\} = \Pr\{X_{t_1} \in A_1, \ldots, X_{t_r} \in A_r\},$$

where $r \geq 1$, $t_1, \ldots, t_r \in \mathcal{T}$ and $A_1, \ldots, A_r \in \mathcal{A}$. If, furthermore, $\mathcal{T}$ is such that $\mathcal{T} + t \in \mathcal{T}$ for any $t \in \mathcal{T}$ (think of $\mathcal{T} = [0, \infty)$ or $\mathcal{T} = \mathbb{N}_0$) and the distribution of the random vector $(X_{t_1+t}, X_{t_2+t}, \ldots, X_{t_r+t})$ does not depend on t for any $r \geq 1$, $t_1, \ldots, t_r \in \mathcal{T}$, then the process $\mathbf{X}$ is called *stationary*. Stationary stochastic processes are also called *information sources* because they are used in information theory to model data sources.

In this book we consider mostly discrete-time, finite-state, one-sided stochastic processes modeling, say, finite-alphabet information sources or arising as symbolic dynamics after dividing the state space of a dynamical system. In this case we use the following notation for the joint probability distributions of the *discrete* random variables $X_0, \ldots, X_n$ with states in (without restriction) $S = \{0, 1, \ldots, N - 1\}$:

$$\mu\{\omega \in \Omega : X_0(\omega) = x_0, \ldots, X_n(\omega) = x_n\} = \Pr\{X_0 = x_0, \ldots, X_n = x_n\}$$
$$= p(x_0, \ldots, x_n), \tag{A.15}$$

and the corresponding notations for the conditional probabilities, etc. Occasionally, these finite-state processes will arise as discretizations or quantizations $\mathbf{X}^\Delta$ of processes $\mathbf{X}$ taking values in a finite interval $I \subset \mathbb{R}^q$ endowed with the Lebesgue measure. Formally this means that there exists a (usually uniform) partition $\delta = \{\Delta_1, \ldots, \Delta_{|\delta|}\}$ of I into a finite number of Lebesgue-measurable subsets (say, subintervals), such that $X_n^\Delta = a_j$ if $X_n^\Delta \in \Delta_j$, where $a_j \in \Delta_j$ is usually set by the precision with which the outputs of $\mathbf{X}$ are measured.

Example 27 A finite-state stochastic process $\mathbf{X} = \{X_n\}_{n \in \mathbb{N}_0}$ is called a Markov process or Markov chain if

$$\Pr\{X_n = x_n | X_{n-1} = x_{n-1}, \ldots, X_0 = x_0\} = \Pr\{X_n = x_n | X_{n-1} = x_{n-1}\},$$

$n \geq 1$, where $x_0, \ldots, x_n \in S = \{0, \ldots, N-1\}$. If, moreover, the conditional probability $\Pr\{X_n = x_n | X_{n-1} = x_{n-1}\}$ does not depend on n, then the Markov process $\mathbf{X}$ is called time homogeneous or time invariant. In this case,

$$P_{i,j} := \Pr\{X_n = j | X_{n-1} = i\},$$

$0 \leq i, j \leq N-1$, is called the transition matrix. We call a probability vector $\mathbf{p} = (p_0, \ldots, p_{N-1})$ an invariant, stationary, or equilibrium probability for $\mathbf{X}$ if $\mathbf{p} = \mathbf{p}P$, that is, if $\mathbf{p}$ is a left eigenvector of P with eigenvalue 1.

Any *stationary* discrete-time stochastic process $\mathbf{X} = \{X_n\}_{n \in \mathbb{K}}$ on a probability space $(\Omega, \mathcal{B}, \mu)$ with state space $(S, \mathcal{A})$ corresponds in a standard way to a shift system $(S^{\mathbb{K}}, \mathcal{B}_\Pi(S), m, \Sigma)$, where $(S^{\mathbb{K}}, \mathcal{B}_\Pi(S))$ is the product measurable space $\Pi_{k \in \mathbb{K}}(S, \mathcal{A})$, via the map $\Phi: \Omega \to S^{\mathbb{K}}$ defined by $(\Phi(\omega))_n = X_n(\omega)$. Here the measure m is the induced or transported probability on the space of possible outputs, $\mathcal{B}_\Pi(S)$, of the random process $\mathbf{X}$:

$$m(B) = \mu(\Phi^{-1}B), \quad B \in \mathcal{B}_\Pi(S), \tag{A.16}$$

that is, $m = \mu \circ \Phi^{-1}$ (note that $\Phi^{-1}B \in \mathcal{B}$ because each X_n is measurable). Moreover, because of the stationarity of $\mathbf{X}$, the probability measure m is shift invariant on cylinder sets and hence on all of $\mathcal{B}_\Pi(S)$.

We will also refer to the shift systems $(S^{\mathbb{K}}, \mathcal{B}_\Pi(S), m, \Sigma)$ as the (sequence space) *model* of the stochastic process or information source $\mathbf{X}$; if S is finite, then we may speak of a *sequence space model*. Models allow to focus on the random process itself as given by the probability distribution of its outputs, dispensing with a perhaps complicated underlying probability space. Depending on the setting or the process being modeled, some particular choices for S and/or $\mathbb{K}$ may be more convenient. For instance, one-sided random processes (i.e., $\mathbb{K} = \mathbb{N}_0$) provide better models than the two-sided processes $\{X_n\}_{n \in \mathbb{Z}}$ for physical information sources that must be turned on at some time. Also, if the source is digital, a finite state space S is the right choice.

Finally, since each information source has associated a dynamical system— its sequence space model—we can eventually assign dynamical properties to the sources. Thus, we say that a source $\mathbf{X}$ is *ergodic*, *mixing*, etc., if its sequence space model $(S^{\mathbb{K}}, \mathcal{B}_\Pi(S), m, \Sigma)$ possesses those properties.

Annex B
Entropy

In this annex we review only the Shannon, Kolmogorov–Sinai, and topological entropies. Standard references include [91, 169, 202].

B.1 Shannon Entropy

One of the most important characterizations one can attach to a random variable and to a stochastic process is its entropy and entropy rate, respectively. We refer to Annex A, Sect. A.3, for the basics of random processes.

B.1.1 The Entropy of a Discrete Random Variable

Let X be a random variable with sample space $(\Omega, \mathcal{B}, \mu)$ and finite state space S. If φ is a real-valued map on S, $\varphi{:}S \to \mathbb{R}$, then $\varphi \circ X = \varphi(X)$ is a random variable with finitely many states $\varphi(S) \subset \mathbb{R}$. The expectation value or average of $\varphi(X)$ will be denoted by $\mathbb{E}\varphi(X)$,

$$\mathbb{E}\varphi(X) = \sum_{x \in S} p(x)\varphi(x),$$

where $p(x)$ is the probability function of X (see (B.21) with $n = 0$).

Definition 14 The (Shannon) entropy of a discrete random variable X on a probability space $(\Omega, \mathcal{B}, \mu)$ is defined by

$$H(X) = -\sum_{x \in S} p(x) \log p(x) = \mathbb{E} \log \frac{1}{p(X)}. \tag{B.1}$$

Whenever convenient, we will write $H_\mu(X)$ to make clear which measure enters into the definition of entropy. Alternatively, one may write $H(p)$ since the entropy depends actually on the probability function $p(x)$ and not on the values taken by X.

(The previous observations hold also for the definitions of different kinds of entropy we will encounter in the sequel.) The logarithm in (B.1) may be taken to any base greater than 1. If the base 2 is used, the entropy comes in units of *bits* (shorthand for "binary digits"). Another usual choice for the logarithm base is Euler's number $e \approx 2.7182818\ldots$, in which case the units of the entropy are called *nats*. Unless otherwise stated, we will henceforth assume the entropy to be in units of bits. Recall that one can change from one logarithmic base a to another base b by means of the formula $\log_b p = \log_b a \log_a p$. By convention, $0 \times \log 0 := \lim_{x \to 0+} x \log x = 0$. Note that $H(X) \geq 0$ because $0 < p(x) \leq 1$ implies $-\log p(x) = \log \frac{1}{p(x)} \geq 0$. On the other hand if $|S|$ denotes the cardinality of the state space S, then $H(X) \leq \log |S|$, as can be easily proved, e.g., using Lagrange multipliers, the highest entropy corresponding to random variables with equiprobable outcomes, that is, $p(x) = 1/|S|$ for all $x \in S$. Observe that Boltzmann's equation (6.1) is nothing else but the entropy for such a flat probability function, $H(X) = \log |S|$, except for the notation (S means entropy in (6.1), while we use S to denote the state space throughout the book) and the physical constant k_B.

Example 28 Suppose that a random variable X takes values $0, 1$ with probabilities $p(0) = p, p(1) = 1 - p(0) = 1 - p$. Then

$$H(X) = -p \log p - (1 - p) \log (1 - p) = H(p). \tag{B.2}$$

The function $H(p)$ is plotted in Fig. B.1. We see that $H(p)$ vanishes when $p = 0$ or $p = 1$, i.e., when the outcome is certain, and it is maximal when $p = 1/2$, i.e., when the uncertainty about the outcome is maximal: $H(1/2) = \log 2 = 1$ bit.

The entropy of a discrete random variable can be given different meanings; see [22] for three interesting interpretations. In information theory one defines

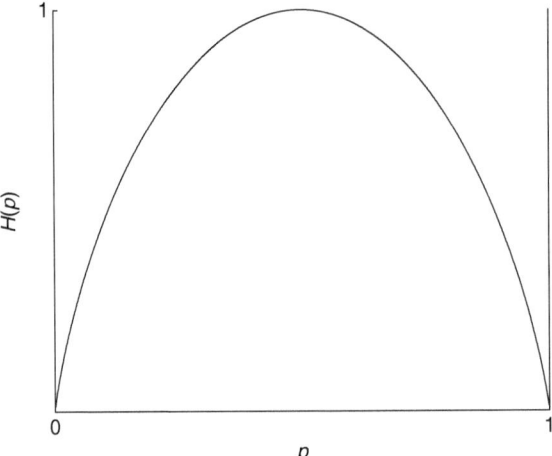

Fig. B.1 The function $H(p)$, (B.2)

$I(X) = -\log p(X)$ to be the *information* of a random variable X with probability function $p(x)$, $-\log p(x)$ being the information conveyed by the outcome $X = x$. Observe that the more rare the event x (that is, the more unlikely the observation of the event x), the more information is gained from its occurrence; one can argue that the most probable events are the less informative ones since their occurrence comes as no surprise. According to Definition 14, $H(X)$ is then the expected value of the information of X: $H(X) = \mathbb{E}I(X)$. Furthermore, if we agree that uncertainty means lack of information, then the entropy can be interpreted as the average uncertainty associated with a random variable or random experiment. In this light, equiprobable events correspond to maximal uncertainty about the outcome.

We turn now to the problem of characterizing the uncertainty associated with more than one random variable.

The *relative entropy* or *Kullback–Leibler distance* between two probability mass functions $p(x)$ and $q(x)$, $x \in S$, is defined as

$$D(p \,\|\, q) = \sum_{x \in S} p(x) \log \frac{p(x)}{q(x)}. \tag{B.3}$$

In this definition, the convention (based on continuity arguments) that $0 \log \frac{0}{q} = 0$ and $p \log \frac{p}{0} = \infty$ is used. From definition (B.3) it follows that $D(p \,\|\, q) \geq 0$ and $D(p \,\|\, q) = 0$ if and only if $p = q$ [59]. On the other hand (and despite of its name), $D(p \,\|\, q)$ is not symmetric in p, q and does not satisfy the triangle inequality. Nonetheless, it is often useful to think of $D(p \,\|\, q)$ as a "distance" between the distributions p and q. The relative entropy $D(p \,\|\, q)$ is a measure of the inefficiency of assuming that the distribution of the random variable X is q when the true distribution is p. For example, if we knew the true distribution p of X, then we could construct a code with average code-word length $H(p)$ (see Sect. 1.1.1, (1.2)). If, instead, we use the code for a distribution q, we would need $H(p) + D(p \,\|\, q)$ bits on the average to describe the random variable X.

Let X and Y be two random variables on a common sample space $(\Omega, \mathcal{B}, \mu)$ but, in general, with different finite state spaces S_1 and S_2, respectively. This corresponds to a situation where two different observations or measurements (with finite precision) are made at the same random experiment. If X and Y have the joint probability function

$$p(x, y) = \mu\{\omega \in \Omega : X(\omega) = x, Y(\omega) = y\} = \Pr(X = x, Y = y)$$

$(x \in S_1, y \in S_2)$, then the *joint entropy* of X and Y is defined as

$$H(X, Y) = -\sum_{x \in S_1} \sum_{y \in S_2} p(x, y) \log p(x, y) = \mathbb{E} \log \frac{1}{p(X, Y)}. \tag{B.4}$$

It is easy to prove that

$$H(X,Y) \le H(X) + H(Y).$$

The generalization of (B.4) to $n \ge 2$ random variables is straightforward and needs no further elaboration.

The joint probability function $p(x,y)$ and the conditional probability function

$$p(y \,|x) = \frac{p(x,y)}{p(x)}$$

allow the definition of two instrumental concepts in information theory: the conditional entropy and the mutual information. The *conditional entropy* of Y given X is

$$H(Y \,|X) = - \sum_{x \in S_1} \sum_{y \in S_2} p(x,y) \log p(y \,|x) = \mathbb{E} \log \frac{1}{p(Y \,|X)}, \qquad \text{(B.5)}$$

and the *mutual information* of X and Y is

$$\begin{aligned} I(X;Y) &= H(X) - H(X \,|Y) = H(Y) - H(Y \,|X) \\ &= H(X) + H(Y) - H(X,Y) \\ &= I(Y;X), \end{aligned} \qquad \text{(B.6)}$$

where we have used the so-called *chain rule* [59]:

$$H(X,Y) = H(X) + H(Y \,|X). \qquad \text{(B.7)}$$

Note that $H(Y \,|X)$ is the average of the uncertainties

$$H(Y \,|X = x) = - \sum_{y \in S_2} p(y \,|x) \log p(y \,|x)$$

weighted with the probabilities $p(x)$, $x \in S_1$. As for the mutual information of two random variables, $I(X;Y)$ is the information about X conveyed by Y (i.e., the information about the realization of X knowing the realization of Y), which is the same as the information about Y conveyed by X, (B.6). Alternatively,

$$I(X;Y) = \mathbb{E} \log \frac{p(X,Y)}{p(X)p(Y)}.$$

Let us mention in passing that the *capacity* of a discrete memoryless channel with input X, output Y, and transition probability $p(Y \,|X)$ is defined as

$$C = \max_{p(x)} I(X;Y),$$

where the maximum is taken over all possible input distributions $p(x)$.

Again, the generalization of these concepts to $n_1 + n_2$ random variables $X_0, \ldots,$ X_{n_1-1} and $Y_0, \ldots, Y_{n_2-1}$ is straightforward. In particular, the (joint) entropy of the random vector $X_0^{n-1} = X_0, \ldots, X_{n-1}$, where, say, all components can take the same states $x_i \in S$, is given by

$$H(X_0, \ldots, X_{n-1}) = - \sum_{x_0, \ldots, x_{n-1} \in S} p(x_0, \ldots, x_{n-1}) \log p(x_0, \ldots, x_{n-1})$$

$$= \mathbb{E} \log \frac{1}{p(X_0, \ldots, X_{n-1})},$$

where $p(x_0, \ldots, x_{n-1})$ is the joint probability function of $X_0, \ldots, X_{n-1}$.

Exercise 22 By iteration of the two-variable rules $p(X, Y) = p(X)p(Y|X)$ and (B.7) prove the general chain rule for the joint entropy: given the random variables $X_0, \ldots, X_{n-1}$ with a joint probability function $p(x_0, \ldots, x_{n-1})$, then

$$p(X_0, \ldots, X_{n-1}) = \prod_{i=0}^{n-1} p(X_i | X_{i-1}, \ldots, X_0) \tag{B.8}$$

and

$$H(X_0, \ldots, X_{n-1}) = \sum_{i=0}^{n-1} H(X_i | X_{i-1}, \ldots, X_0), \tag{B.9}$$

with the conventions $p(X_0 | X_{-1}) := p(X_0)$ and $H(X_0 | X_{-1}) := H(X_0)$.

B.1.2 The Entropy Rate of a Discrete-Time Finite-State Stochastic Process

Definition 15 The entropy rate of a finite-state random process $\mathbf{X} = \{X_n\}_{n \in \mathbb{N}_0}$ on a probability space $(\Omega, \mathcal{B}, \mu)$ is defined by

$$h(\mathbf{X}) = \lim_{n \to \infty} \frac{1}{n} H(X_0, \ldots, X_{n-1}), \tag{B.10}$$

provided the limit exists.

Sometimes the terms

$$h(X_0, \ldots, X_{n-1}) = \frac{1}{n} H(X_0, \ldots, X_{n-1})$$

($n \geq 2$) are called the *entropy rates of order n* of $\mathbf{X}$. Hence, $h(X_0, \ldots, X_{n-1})$ or, more compactly written, $h(X_0^{n-1})$ is the average uncertainty per symbol (time unit, channel use, etc. depending on the interpretation of n) about n consecutive outcomes of the random experiment modeled by $\mathbf{X}$. If we repeat the experiment an arbitrarily long number of times, these average uncertainty rates eventually converge to a limit— Shannon's entropy rate $h(\mathbf{X})$.

Although $h(X_0^{n-1})$ and, consequently, $h(\mathbf{X})$ are actually entropy *rates*, the term "rate" is generally omitted—also in other types of entropy. We follow sometimes this common usage, since this does not lead to misunderstandings.

Lemma 12 *For a stationary stochastic process* $\mathbf{X} = \{X_n\}_{n \in \mathbb{N}_0}$, *the sequence of conditional entropies* $H(X_n | X_{n-1}, \dots, X_0)$ *is decreasing.*

Proof Indeed,

$$
\begin{aligned}
H(X_{n+1} | X_n, \dots, X_1, X_0) &\leq H(X_{n+1} | X_n, \dots, X_1) \\
&= H(X_n | X_{n-1}, \dots, X_0),
\end{aligned}
$$

where the inequality follows from the fact that conditioning reduces uncertainty, and the equality follows from the stationarity of $\mathbf{X}$. □

Theorem 22 *For a stationary stochastic process* $\mathbf{X} = \{X_n\}_{n \in \mathbb{N}_0}$,

$$
h(\mathbf{X}) = \lim_{n \to \infty} H(X_n | X_{n-1}, \dots, X_0). \tag{B.11}
$$

Proof First of all, limit (B.11) converges because, according to Lemma 12, the positive sequence $H(X_n | X_{n-1}, \dots, X_0)$ is decreasing. Furthermore, by the chain rule (B.9),

$$
h(X_0, \dots, X_n) = \frac{1}{n+1} \sum_{i=0}^{n} H(X_i | X_{i-1}, \dots, X_0).
$$

By Cesáro's mean theorem ("If $a_n \to a$ and $b_n = \frac{1}{n+1} \sum_{i=0}^{n} a_i$, then $b_n \to a$"),

$$
h(\mathbf{X}) = \lim_{n \to \infty} h(X_0, \dots, X_n) = \lim_{n \to \infty} H(X_n | X_{n-1}, \dots, X_0).
$$

□

From Lemma 12 it follows that the convergence of the entropy rates of order n, $h(X_0, \dots, X_{n-1})$, to $h(\mathbf{X})$ is monotonically decreasing:

$$
h(X_0) \geq h(X_0, X_1) \geq \cdots \geq h(X_0, \dots, X_{n-1}) \geq \cdots. \tag{B.12}
$$

Thus, when estimating the entropy rate of a stationary random process by its entropy rate of order n, the estimation always exceeds the true value. Intuitively speaking, with increasing n we see more and more correlations among the variables $X_0, \dots, X_{n-1}$ and this reduces our uncertainty about the next observation X_n. We turn back to this point in Example 31.

In an information-theoretical setting and in applications (Sect. A.3), one can think of a stationary stochastic process $\mathbf{X} = \{X_n\}_{n \in \mathbb{N}_0}$ as a data source. Its realizations are then the messages output by the source. This is illustrated in Fig. B.2. Here x_0 can be considered the current and last letter of the message, the other letters having been output in the past, the greater the index, the earlier in time.

$$\ldots x_n \ldots x_1 x_0 \quad \longleftarrow \quad \boxed{ \mathbf{X} }$$

Fig. B.2 A data source $\mathbf{X}$ outputs a message x_0^∞

B.2 Kolmogorov–Sinai Entropy

B.2.1 Deterministic Systems

A partition of a probability space $(\Omega, \mathcal{B}, \mu)$ is a collection $\alpha = (A_i)_{i \in J}$ of disjoint sets $A_i \in \mathcal{B}$, with a countable index set J, such that $\bigcup_{i \in J} \mu(A_i) = 1$. If J is finite, α is called a finite partition. If α is a finite partition of $(\Omega, \mathcal{B}, \mu)$, then the collection of all elements of $\mathcal{B}$ which are unions of elements of α is a finite sub-sigma-algebra of $\mathcal{B}$ which we denote by $\mathcal{B}(\alpha)$. We write $\alpha \leq \beta$, where α, β are two finite partitions of $(\Omega, \mathcal{B}, \mu)$, to mean that each element of α is a union of elements of β. In this case, β is called a *refinement* of α. We have $\alpha \leq \beta$ iff $\mathcal{B}(\alpha) \subset \mathcal{B}(\beta)$.

Definition 16 Let $\alpha = \{A_1, \ldots, A_{|\alpha|}\}$ be a finite partition of $(\Omega, \mathcal{B}, \mu)$. The entropy of the partition α is the number

$$H_\mu(\alpha) = -\sum_{i=1}^{|\alpha|} \mu(A_i) \log \mu(A_i).$$

The same considerations concerning the base of the logarithm we made after the definition of Shannon's entropy, Definition 14, apply here as well. By the same token, $H(\alpha)$ is a measure of the information gained (or the uncertainty removed) by performing a random experiment whose outcomes have probabilities $\mu(A_1), \ldots, \mu(A_{|\alpha|})$.

Sometimes it is convenient to quantify the "coarseness" of a partition. Roughly speaking, if we assign a "size" to each $A \in \alpha$, then we can take the maximum of those sizes as the coarseness of α. The resulting parameter is called the *norm of the partition* α and denoted by $\|\alpha\|$. In metric spaces (X, d), one can take $\|\alpha\| = \max_{A \in \alpha} \mathrm{diam}(A)$, where $\mathrm{diam}(A) = \sup\{d(x, y) : x, y \in A\}$ is called the "diameter" of A.

If $f : \Omega \to \Omega$ is a measure-preserving function on the probability space $(\Omega, \mathcal{B}, \mu)$, we denote by $f^{-n}\alpha$ the partition $\{f^{-n}A_1, \ldots, f^{-n}A_{|\alpha|}\}$. Furthermore, given two finite partitions $\alpha = \{A_1, \ldots, A_{|\alpha|}\}$ and $\beta = \{B_1, \ldots, B_{|\beta|}\}$ of $(\Omega, \mathcal{B}, \mu)$, we denote by $\alpha \vee \beta$ their *least common refinement*,

$$\alpha \vee \beta = \{A \cap B : A \in \alpha, B \in \beta, \mu(A \cap B) > 0\}.$$

More general refinements, like

$$\alpha \vee f^{-1}\alpha \vee \cdots \vee f^{-(n-1)}\alpha = \bigvee_{i=0}^{n-1} f^{-i}\alpha,$$

are defined recursively.

Definition 17 Let $(\Omega, \mathcal{B}, \mu, f)$ be a measure-preserving dynamical system. If α is a finite partition of $(\Omega, \mathcal{B}, \mu)$, then

$$h_\mu(f, \alpha) = \lim_{n \to \infty} \frac{1}{n} H_\mu \left(\bigvee_{i=0}^{n-1} f^{-i}\alpha \right) \tag{B.13}$$

is called the metric entropy of f with respect to α.

In this setting, consider now a finite-state random process $\mathbf{X}^\alpha = \{X_n^\alpha\}_{n \in \mathbb{N}_0}$, with $X_n^\alpha : \Omega \to S = \{0, \ldots, |\alpha| - 1\}$, defined as follows:

$$X_n^\alpha(\omega) = i \quad \text{iff} \quad f^n(\omega) \in A_i \in \alpha. \tag{B.14}$$

Note that $X_{n+1} = X_n \circ f$, thus $X_n = X_n \circ f^n$. Then

$$\Pr\{X_0^\alpha - i_0, \ldots, X_n^\alpha - i_n\} = \mu\{\omega \subset \Omega : \omega \subset A_{i_0}, f(\omega) \in A_{i_1}, \ldots, f^n(\omega) \in A_{i_n}\}$$
$$= \mu\{A_{i_0} \cap \cdots \cap f^{-n}A_{i_n}\}, \tag{B.15}$$

$n \geq 0$, and similarly,

$$\Pr\{X_k^\alpha = i_0, \ldots, X_{n+k}^\alpha = i_n\} = \mu\{f^{-k}(A_{i_0} \cap \cdots \cap f^{-n}A_{i_n})\}$$
$$= \Pr\{X_0^\alpha = i_0, \ldots, X_n^\alpha = i_n\}$$

because of the f-invariance of μ. We conclude that $\mathbf{X}^\alpha$ is a stationary process, which is called the *symbolic dynamics* of $(\Omega, \mathcal{B}, \mu, f)$ with respect to the partition ("coarse graining" or "quantization") α. Depending on the context, $\mathbf{X}^\alpha$ is also called a *coding map* (dynamical systems) or a collection of *simple observations* with respect to f with precision $\|\alpha\|$ (information theory). Moreover, it follows from (B.15) that

$$h_\mu(f, \alpha) = h_\mu(\mathbf{X}^\alpha). \tag{B.16}$$

This not only proves that limit (B.13) does exist but also that the *entropy rates of order n* of f with respect to α,

$$h_\mu^{(n)}(f,\alpha) = \frac{1}{n}H_\mu\left(\bigvee_{i=0}^{n-1}f^{-i}\alpha\right),$$

decrease to $h_\mu(f,\alpha)$ when $n \to \infty$ (remember (B.12)).

Definition 18 Let $(\Omega, \mathcal{B}, \mu, f)$ be a measure-preserving dynamical system and α a finite partition of $(\Omega, \mathcal{B}, \mu)$. Then,

$$h_\mu(f) = \sup_\alpha h_\mu(f,\alpha) \tag{B.17}$$

is called the metric entropy (or just, the entropy) of the map f with respect to μ.

Sometimes $h_\mu(f)$ is called the Kolmogorov–Sinai entropy or the measure-theoretic entropy too. To streamline the notation, the subscript μ may be dropped from $H_\mu(\alpha)$, $h_\mu(f,\alpha)$, and $h_\mu(f)$, as we generally do, if the probability measure is clear from the context.

The isomorphic invariance is one of the fundamental properties of entropy.

Theorem 23(a) *If the dynamical systems* $(\Omega_1, \mathcal{B}_1, \mu_1, f_1)$ *and* $(\Omega_2, \mathcal{B}_2, \mu_2, f_2)$ *are isomorphic, then* $h(f_1) = h(f_2)$.
(b) *If* $(\Omega_2, \mathcal{B}_2, \mu_2, f_2)$ *is a factor of* $(\Omega_1, \mathcal{B}_1, \mu_1, f_1)$, *then* $h(f_2) \leq h(f_1)$.

It should be obvious from definitions (B.13) and (B.17) that the exact calculation of $h(f)$ from scratch is, in general, unfeasible. There are though a few results that, depending on the specifics of the dynamical system in question, can come to the rescue. We mention a few next.

A finite partition α of $(\Omega, \mathcal{B}, \mu)$ is called a *generating partition* or a *generator for* a μ-*preserving transformation* $f:\Omega \to \Omega$ if (i)

$$\bigvee_{n=-\infty}^{\infty} f^{-n}\mathcal{B}(\alpha) = \mathcal{B} \text{ (modulo } \mu\text{-zero sets)} \tag{B.18}$$

when f is invertible (i.e., f is an automorphism) or (ii)

$$\bigvee_{n=0}^{\infty} f^{-n}\mathcal{B}(\alpha) = \mathcal{B} \text{ (modulo } \mu\text{-zero sets)} \tag{B.19}$$

when f is non-invertible (i.e., f is an endomorphism). This means that for any $B \in \mathcal{B}$, there is a $B' \in \bigvee_{n=-\infty}^{\infty}f^{-n}\mathcal{B}(\alpha)$ or $B' \in \bigvee_{n=0}^{\infty}f^{-n}\mathcal{B}(\alpha)$, respectively, such that $\mu(B \triangle B') = 0$. If f is invertible and the stronger condition (B.19) holds, then α is called a strong or one-sided generator for f. Equivalent definitions of generators and one-sided generators by means of partition refinements converging to the point partition $\epsilon = \{\{x\}:x \in \Omega\}$ were given in Sect. 1.3.

Example 29 Since the sigma-algebra $\mathcal{B}_\Pi(S)$ of the one-sided and two-sided shift spaces are generated by the cylinder sets

$$C_{a_0,\dots,a_k} = \{\mathbf{s} = (s_n)_{n \in \mathbb{N}_0} : s_0 = a_0, \dots, s_k = a_k\} = \bigcap_{i=0}^{k} \Sigma^{-i} C_{a_i}$$

and

$$C_{a_{-k},\dots,a_0,\dots,a_k} = \{\mathbf{s} = (s_n)_{n \in \mathbb{Z}} : s_{-k} = a_{-k}, \dots, s_k = a_k\} = \bigcap_{i=-k}^{k} \Sigma^{-i} C_{a_i},$$

respectively, it follows that the partition

$$\gamma = \{C_a : a \in S\}$$

is a generator of both the one-sided and two-sided shifts.

Generating partitions can be found numerically; see, e.g., [40] for a general method based on relaxation algorithms. For higher dimensional maps, numerical techniques have been proposed for the dissipative Hénon map [87], the standard map [53], two-dimensional hyperbolic maps [26], etc. A method based on unstable period orbits was proposed in [63]. The construction of one-dimensional maps possessing generating partitions was studied in [99].

Theorem 24 *(Kolmogorov–Sinai Theorem) Let* $(\Omega, \mathcal{B}, \mu, f)$ *be a dynamical system.*

(a) If f *is an automorphism and* α *is a generator or a one-sided generator for* f, *then*
   $h(f) = H(f, \alpha)$.
(b) If f *is an endomorphism and* α *is a generator for* f, *then* $h(f) = H(f, \alpha)$.

The case of automorphisms with one-sided generators is uninteresting since then one can show that $h(f) = 0$ [202]. More interestingly, *Krieger's theorem* states that if f is an ergodic automorphism with $h(f) < \infty$, then f has a generator [67, 130, 169]. Although Krieger's proof is non-constructive, Smorodinsky [191] and Denker [65] provided methods to construct a two-sided generator for ergodic and aperiodic automorphisms. Denker's construction could even be extended by Grillenberger [66] to all aperiodic automorphisms. The existence of generators for endomorphisms was proved by Kowalski under different assumptions [128, 129]. At variance with the previous case, the construction of one-sided generators for endomorphisms remains an open problem till this very day; see [182] for some progress in this issue.

Example 30 Using the fact that the cylinder sets C_a are generators for the one-sided and two-sided $(\mathbf{p}, P)$-Markov shifts Σ on N symbols, one can prove

$$h_\mu(\Sigma) = -\sum_{i,j=1}^{N} p_i P_{ij} \log P_{ij}, \tag{B.20}$$

where μ is the Markov measure defined by $(\mathbf{p}, P)$ (see Example 25 (b)). Upon substituting $P_{ij} = p_j$ in (B.20), we get for $\mathbf{p}$-Bernoulli shifts

$$h_\mu(\Sigma) = -\sum_j^N p_j \log p_j,$$

where μ is the Bernoulli measure defined by $\mathbf{p}$ (see Example 25 (a)).

A second practical way of calculating (or, at least, estimating) the entropy is provided by the following theorem.

Theorem 25 [169, Ch. 5, Prop. 3.6] *Let* $(\Omega, \mathcal{B}, \mu, f)$ *be a measure-preserving dynamical system. If* $\alpha_0 \le \alpha_1 \le \cdots$ *is an increasing sequence of finite partitions of* $(\Omega, \mathcal{B}, \mu)$ *and* $\vee_{n=0}^\infty \mathcal{A}(\alpha_n) = \mathcal{B}$ *up to sets of measure 0, then*

$$\lim_{n\to\infty} h_\mu(f, \alpha_n) = h_\mu(f).$$

A third practical method calls for Pesin's theorem and Lyapunov exponents. Since this topic would take us too far away, we refer the interested reader to the specialized literature [142, 52, 72]. Due to the important role that the Lyapunov exponent(s) play in nonlinear dynamics, several numerical schemes have been developed to calculate them [193]. On the other hand, Pesin's theorem and its generalizations require the invariant measure to possess some properties—but invariant measures are in many interesting cases unknown. This fact limits the application of this method. For the calculation of the metric entropy in some one-dimensional systems, see [105].

B.2.2 Random Systems

Let $\mathbf{X} = \{X_n\}_{n\in\mathbb{N}_0}$ be a stationary stochastic process on a probability space $(\Omega, \mathcal{B}, \mu)$, taking on values in $S = \{0, \ldots, N-1\}$. In Sect. A.3 it is shown that $\mathbf{X}$ can be associated in a canonical way with a shift system $(S^{\mathbb{N}_0}, \mathcal{B}_\Pi(S), m, \Sigma)$, called its sequence space model, via $\Phi: \Omega \to S^{\mathbb{N}_0}$, $(\Phi(\omega))_n = X_n(\omega)$. The joint probability function $p(x_0, \ldots, x_{n-1})$ of the random process $\mathbf{X}$ is related to the measure of the cylinder sets $C_{x_0,\ldots,x_{n-1}}$, $x_0, \ldots, x_{n-1} \in S$, of the sequence space model in the following way:

$$\begin{aligned}
p(x_0, \ldots, x_{n-1}) &= \mu\left\{\omega \in \Omega : X_0(\omega) = x_0, \ldots, X_{n-1}(\omega) = x_{n-1}\right\} \\
&= \mu\left\{\Phi^{-1}\left\{\mathbf{s} \in S^{\mathbb{N}_0} : s_0 = x_0, \ldots, s_{n-1} = x_{n-1}\right\}\right\} \\
&= m\left\{C_{x_0,\ldots,x_{n-1}}\right\} \\
&= m\{C_{x_0} \cap \cdots \cap \Sigma^{-(n-1)} C_{x_{n-1}}\}.
\end{aligned}$$

Since the partition $\gamma = \{C_{x_0} : x_0 \in S\}$ is a generator of Σ (Example 29), we have

$$h_\mu(\mathbf{X}) = -\lim_{n\to\infty} \frac{1}{n} \sum_{x_0,\ldots,x_{n-1}\in S} p(x_0,\ldots,x_{n-1}) \log p(x_0,\ldots,x_{n-1})$$

$$= -\lim_{n\to\infty} \frac{1}{n} H_m \left(\bigvee_{i=0}^{n-1} \Sigma^{-i}\gamma \right)$$

$$= h_m(\Sigma, \gamma)$$

$$= h_m(\Sigma)$$

by Theorem 24 (b). In words, the Shannon entropy rate of a stochastic process $\mathbf{X} = \{X_n\}_{n\in\mathbb{N}_0}$ coincides with the Kolmogorov–Sinai entropy rate of its sequence space model.

An important property of ergodic processes is the so-called *asymptotic equipartition property* or *Shannon–McMillan–Breiman theorem*.

Theorem 26 (*Shannon–McMillan–Breiman*) *If* $\mathbf{X} = \{X_n\}_{n\in\mathbb{N}_0}$ *is a finite-valued stationary ergodic process, then* $-\frac{1}{n}\log p(X_0,\ldots,X_{n-1})$ *converges in probability to the entropy rate* $h(\mathbf{X})$.

Example 31 The sequence space model of a finite-state, time-homogeneous Markov chain $\mathbf{X} = \{X_n\}_{n\in\mathbb{N}_0}$ (Example 27) with transition matrix $P_{i,j}$, $0 \le i,j \le N-1$, and stationary probability vector $\mathbf{p}$ is the one-sided $(\mathbf{p}, P)$-Markov shift $\Sigma_{\mathbf{p},P}$. Therefore,

$$h(\mathbf{X}) = h(\Sigma_{\mathbf{p},P}) = -\sum_{i,j=0}^{N-1} p_i P_{ij} \log P_{ij}.$$

For the specific case

$$P = \begin{pmatrix} 1 - p_{01} & p_{01} \\ p_{10} & 1 - p_{10} \end{pmatrix} = \begin{pmatrix} 0.9 & 0.1 \\ 0.1 & 0.9 \end{pmatrix},$$

the stationary probability is

$$\mathbf{p} = \left(\frac{p_{10}}{p_{01} + p_{10}}, \frac{p_{01}}{p_{01} + p_{10}} \right) = \left(\frac{1}{2}, \frac{1}{2} \right).$$

The upper curve in Fig. B.3 shows the entropy rates of order n, $h(X_0,\ldots,X_{n-1})$, closing in on the true value $h(\mathbf{X}) = 0.469$ bits/symbol (horizontal line). The lower curve shows what happens in practice when $h(\mathbf{X})$ is estimated numerically in a naive way. Here the probabilities $p(x_0,\ldots,x_{n-1})$ were estimated by the frequencies of the word $x_0,\ldots,x_{n-1}$ in a sequence of 10,000 draws. In the left part of the experimental curve, we see the entropy rates of successive order $n = 1, 2, \ldots$ converging from above to the true value. For $n \approx 20$, the numerical values provide accurate estimates of the entropy. For greater lengths, the estimates tend toward zero along the parabola

$$h(n) = \frac{\log(N - n + 1)}{n}$$

due to undersampling.

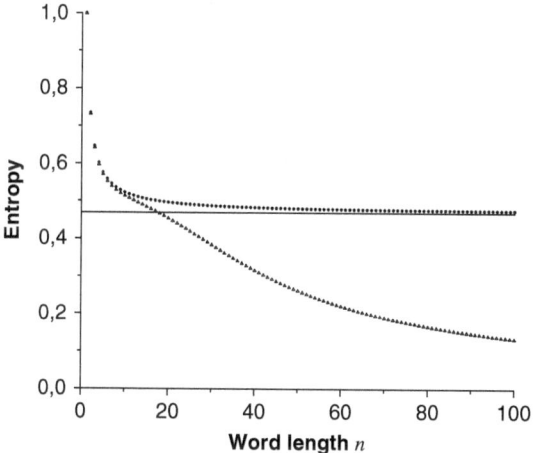

Fig. B.3 The *upper dotted line* shows the convergence of the entropy rate of order n to the true value, 0.469 bits/symbol (*horizontal line*), for an arbitrarily long sequence generated by a two-state Markov chain with transition probabilities $p_{01} = p_{10} = 0.1$. The *lower dotted line* shows what happens in practice due to undersampling

A particular case is of interest. Consider now not a general stationary stochastic process but the symbolic dynamics $\mathbf{X}^\alpha = \{X_n^\alpha\}_{n \in \mathbb{N}_0}$ of the system $(\Omega, \mathcal{B}, \mu, f)$ with respect to a partition $\alpha = \{A_1, \ldots, A_{|\alpha|}\}$ (see (B.14)), and let $(S^{\mathbb{N}_0}, \mathcal{B}_\Pi(S), m, \Sigma)$ be the sequence space model of $\mathbf{X}^\alpha$; hence $S = \{1, \ldots, |\alpha|\}$ and

$$m(C_{a_0, a_1, \ldots, a_n}) = \mu(A_{a_0} \cap f^{-1}A_{a_1} \cap \cdots \cap f^{-n}A_{a_n})$$

for any cylinder set $C_{a_0, \ldots, a_n} = \{\mathbf{s} \in S^{\mathbb{N}_0} : s_0 = a_0, \ldots, s_n = a_n\}$, with $a_0, \ldots, a_n \in S$. In this setting, the following question arises. When are the dynamical systems $(\Omega, \mathcal{B}, \mu, f)$ and $(S^{\mathbb{N}_0}, \mathcal{B}_\Pi(S), m, \Sigma)$ isomorphic (via $\Phi^\alpha : \Omega \to S^{\mathbb{N}_0}$, $(\Phi^\alpha(\omega))_n = X_n^\alpha(\omega)$)? Since $\{C_a : a \in S\}$ is a generator for Σ and $(\Phi^\alpha)^{-1}C_a = A_a$ for every $a \in S$, we need clearly that

$$\{(\Phi^\alpha)^{-1}C_a : 1 \le a \le |\alpha|\} = \{A_a : 1 \le a \le |\alpha|\} = \alpha$$

is also a generator for f. In other words, a generator for f gives a natural isomorphism between $(\Omega, \mathcal{B}, \mu, f)$ and the sequence space model associated with its symbolic dynamics. By Krieger's theorem we conclude that any ergodic, invertible dynamical system with finite entropy can be represented as a two-sided shift system. This result is useful in that it provides prototypes of ergodic, finite-entropy systems.

B.3 Topological Entropy

Topological entropy for continuous self-maps of compact topological spaces was introduced by Adler, Koheim, and McAndrews by means of open covers [3]. Later Dinaburg [70] and Bowen [36] found alternative approaches via separating and spanning sets in (not necessarily compact) metric spaces.

B.3.1 Generalities

Recall that a continuous or topological dynamical system is a pair (M,f), where M is a topological space and $f:M \rightarrow M$ is a continuous map. As compared to measure-theoretical dynamical systems, there is here no measurable structure involved (although M can be thought to be endowed with the Borel sigma-algebra); instead, continuity enters the scenario. Sometimes, continuity is weakened to piecewise continuity, especially in conjunction with other properties like piecewise monotonicity.

Furthermore, in this section (M,d) denotes a metric space and $f:M \rightarrow M$ a uniformly continuous map. If, moreover, M is compact, then f needs only to be continuous (since every continuous self-map of a compact space is uniformly continuous).

Definition 19 Let K be a compact topological space, α an open cover of K, and $N(\alpha)$ the number of sets in a finite subcover of α with smallest cardinality. The entropy of the cover α is then defined as $H(\alpha) = \log N(\alpha)$.

If α is an open cover of K and $f:K \rightarrow K$ is continuous, then $f^{-1}\alpha$ is the open cover consisting of all sets $f^{-1}A$, $A \in \alpha$.

Definition 20 If α is an open cover of the compact space K and $f:K \rightarrow K$ is continuous, then the entropy of f relative to α is given by

$$h(f,\alpha) = \lim_{n\to\infty} \frac{1}{n} H\left(\bigvee_{i=0}^{n-1} f^{-i}\alpha \right) \tag{B.21}$$

and the topological entropy of f is given by

$$h(f) = \sup_{\alpha} h(f,\alpha). \tag{B.22}$$

It can be proved that the limit in (B.21) exists and the supremum in (B.22) can be taken over *finite* open covers of K.

In a metric space (M,d), the alternative definitions of topological entropy via spanning and separating sets may be more useful.

Definition 21 Let $n \in \mathbb{N}$, $\varepsilon > 0$, and $K \subset M$ compact. A subset $A \subset M$ is said to (n,ε)-span K with respect to $f:M \rightarrow M$ if for each $x \in K$ there exists $y \in A$ such that

$$\max_{0 \le i \le n-1} d(f^i(x), f^i(y)) \le \varepsilon.$$

Furthermore, let $r_n(\varepsilon, K)$ denote the smallest cardinality of any (n, ε)-spanning set for K with respect to f.

Definition 22 The topological entropy of $f : M \to M$ is

$$h_d(f) = \sup_K \lim_{\varepsilon \to 0} \limsup_{n \to \infty} \frac{1}{n} \log r_n(\varepsilon, K), \tag{B.23}$$

where the supremum is taken over all compact subsets of M.

The definition of topological entropy by means of separating sets is as follows.

Definition 23 Let $n \in \mathbb{N}$, $\varepsilon > 0$, and $K \subset M$ compact. A subset $A \subset K$ is said to be (n, ε)-separated with respect to $f : M \to M$ if $x, y \in A$, $x \ne y$, implies

$$\max_{0 \le i \le n-1} d(f^i(x), f^i(y)) > \varepsilon.$$

Furthermore, let $s_n(\varepsilon, K)$ denote the largest cardinality of any (n, ε)-separated subset of K with respect to f.

Thus, an (n, ε)-separated subset of Ω is a kind of microscope that allows us to distinguish orbits of length n up to a precision ε.

Definition 24 The topological entropy of $f : M \to M$ is

$$h_d(f) = \sup_K \lim_{\varepsilon \to 0} \limsup_{n \to \infty} \frac{1}{n} \log s_n(\varepsilon, K), \tag{B.24}$$

where the supremum is taken over all compact subsets of M.

If M is compact, then $h_d(f)$ can be shown [202] not to depend on the metric d (thus, it will be denoted by $h_{\text{top}}(f)$) and, moreover, definitions (B.23) and (B.24) can be simplified to

$$h_{\text{top}}(f) = \lim_{\varepsilon \to 0} \limsup_{n \to \infty} \frac{1}{n} \log r_n(\varepsilon, M) = \lim_{\varepsilon \to 0} \limsup_{n \to \infty} \frac{1}{n} \log s_n(\varepsilon, M). \tag{B.25}$$

Both $r_n(\varepsilon, M)$ and $s_n(\varepsilon, M)$ can be interpreted as the number of orbits of length n up to an error ε. For $\varepsilon \ll 1$,

$$e^{nh(f)} \sim r_n(\varepsilon, M) \quad \text{and} \quad e^{nh(f)} \sim s_n(\varepsilon, M),$$

where $\sim$ stands for "asymptotically as $n \to \infty$" (assuming the convergence of $\frac{1}{n} \log r_n(\varepsilon, M)$ and $\frac{1}{n} \log s_n(\varepsilon, M)$ in this limit), so the topological entropy measures the asymptotic exponential growth rate with n of the number of orbits of length n, up to error ε.

Definition 25 Let $f_1:M_1 \rightarrow M_1$ and $f_2:M_2 \rightarrow M_2$ be continuous maps of metric spaces and suppose that there exists a continuous surjective map $\phi:M_1 \rightarrow M_2$ such that $\phi \circ f_1 = f_2 \circ \phi$. Then we say that f_1 is topologically semiconjugate to f_2 or that f_2 is a factor of f_1 via the topological semi-conjugacy or factor map ϕ. In the case that ϕ is a homeomorphism, then f_1 and f_2 are said to be topologically conjugate and ϕ is said to be a topological conjugacy.

In particular, if two maps are metrically conjugate via a (measure-preserving) homeomorphism, then they are also topologically conjugate. Such is the case of the logistic and symmetric tent maps via the homeomorphism A.8 (Example 24). The qualifiers "topological" and "topologically" may be dropped if it is clear that they refer to a topological system.

Thus, conjugate maps are obtained from each other by a continuous change of coordinates. Therefore, properties that are independent of such changes of coordinates will be invariant under topological conjugacy, e.g., sensitivity to initial conditions, topological transitivity, number of periodic orbits of a given period.

Just as metric entropy is an invariant of metric conjugacy, so is topological entropy an invariant of topological conjugacy.

Theorem 27 *Let f_1 and f_2 be continuous self-maps of compact spaces. If f_1 and f_2 are topologically conjugate, then $h(f_1) = h(f_2)$. More generally, if f_2 is a factor of f_1, then $h(f_2) \leq h(f_1)$.*

Exercise 23 Show that the quadratic transformations $f_1(x) = vx(1 - x)$ on $[0, 1]$, $0 < v \leq 4$, and

$$f_2(y) = \tfrac{1}{2}(y^2 - v^2 + 2v)$$

on $[-v, v]$ are topologically conjugate via the homeomorphism

$$\phi(x) = v(1 - 2x) = f_1'(x).$$

In spite of not involving a measure-theoretical structure, topological entropy is tightly related to metric entropy through the following *variational principle*.

Theorem 28 *Let M be a compact metric space endowed with the Borel sigma-algebra $\mathcal{B}$, and $f:M \rightarrow M$ a continuous map. Then*

$$h_{top}(f) = \sup h_\mu(f), \qquad (B.26)$$

where the supremum is taken over all f-invariant measures μ on the measurable space $(M, \mathcal{B})$.

Note that the set of f-invariant measures invoked in the variational principle (B.26) is non-empty by Theorem 20. Moreover, the supremum in (B.26) can be restricted to ergodic measures [202],

$$h_{\text{top}}(f) = \sup_{\mu \in E(M,f)} h_\mu(f), \qquad (B.27)$$

where $E(M,f)$ is the set of f-invariant, ergodic measures on $(M, \mathcal{B})$. Measures μ such that $h_{\text{top}}(f) = h_\mu(f)$ are called *measures with maximal entropy* for obvious reasons.

In Sect. A.1 we defined the concept of generator of a measure-preserving transformation. In topological dynamics, there is also a concept of generator that plays a similar role with respect to the topological entropy. Given a compact metric space M and a map $f{:}M \to M$, a finite open cover $\alpha = \{A_1, \ldots, A_{|\alpha|}\}$ of M is said to be a *generator* for f if

(a) in case f is invertible, for any bisequence $(a_i)_{i \in \mathbb{Z}}$, $1 \le a_i \le |\alpha|$, the intersection

$$\bigcap_{i=-\infty}^{\infty} f^{-i}A_{a_i}$$

   contains at most one point or

(b) in case f is non-invertible, for any sequence $(a_i)_{i \in \mathbb{N}_0}$, $1 \le a_i \le |\alpha|$, the intersection

$$\bigcap_{i=0}^{\infty} f^{-i}A_{a_i}$$

   contains at most one point.

The topological dynamical systems that admit a generator have a simple characterization.

Definition 26 Let M be a compact metric space. A homeomorphism (correspondingly, a continuous map) $f{:}M \to M$ is said to be *expansive* if there exists $\delta > 0$, called an *expansivity constant* for f, such that

$$d(f^n(x), f^n(y)) \le \delta$$

for all $n \in \mathbb{Z}$ (correspondingly, $n \in \mathbb{N}_0$) implies $x = y$. Expansive non-invertible maps and homeomorphisms for which the expansiveness condition holds already for non-negative iterates are collectively called *positively expansive* maps.

Alternatively, if $x \ne y$ and δ is an expansivity constant for f, then there exists $n \in \mathbb{Z}$ (correspondingly, $n \in \mathbb{N}_0$) with $d(f^n(x), f^n(y)) > \delta$. Notice that expansiveness differs from sensitive dependence in that *all* nearby points eventually separate by at least δ (for sensitive dependence it suffices this to occur for a single point in each neighborhood of the other). Intuitively, the orbits of an expansive map f can be resolved to any desired precision by taking n sufficiently large. Expansive maps *f* have some nice properties like having a countable number of periodic points, and at least one invariant measure with maximal entropy [202]. Examples of expansive maps include the shift transformations and the hyperbolic toral automorphisms. On the other hand, there are no expansive maps of closed one-dimensional intervals [19, Thm. 2.2.31] nor expansive homeomorphisms of the circle [202]. Expansiveness and positively expansiveness are topological conjugacy invariants.

Theorem 29 *Let* $f{:}M \to M$ *be a map of the compact metric space* (M,d). *Then* f *is expansive if and only if* f *has a generator.*

Observe that the cylinder sets C_a are generators both in the measure-theoretical and in the topological senses because, among other considerations, they build a partition and an open cover at the same time. Therefore, shifts on sequence spaces are expansive transformations. Expansiveness is an invariant of topological conjugacy.

Theorem 30 *If* $f{:}M \to M$ *be an expansive map of the compact metric space* (M,d) *and* α *is a generator for* f, *then* $h_{top}(f) = h(f,\alpha)$.

Example 32 Let $S = \{0,\dots,k-1\}$ and Σ be the shift on the bisequence space $S^{\mathbb{Z}} = \{(s_n)_{n\in\mathbb{Z}}\}$. Then Σ has topological entropy $\log N$. Indeed, apply Theorem 30 with α comprising the cylinder sets $C_j = \{(x_n)_{n\in\mathbb{Z}}{:}x_0 = j\}$ to obtain

$$h_{top}(\Sigma) = \lim_{n\to\infty} \frac{1}{n} \log N \left(\bigvee_{i=0}^{n-1} \Sigma^{-i}\alpha \right) = \lim_{n\to\infty} \frac{1}{n} \log k^n = \log k.$$

Thus, if μ_0 is the Bernoulli measure on $(S^{\mathbb{Z}}, \mathcal{B}_{\Pi}(S))$ defined by the probability vector $\mathbf{p}_0 - (\frac{1}{k},\dots,\frac{1}{k})$, we have

$$h_{\mu_0}(\Sigma) = \log k = h_{top}(\Sigma).$$

This illustrates the existence of (in this case, unique) measures of maximal entropy. The result in the one-sided case is the same.

Example 33 Let $S = \{0,\dots,k-1\}$, $A = (a_{ij})_{i,j=0}^{k-1}$ be a $k \times k$ matrix whose entries a_{ij} are either 0's or 1's, and

$$\Omega_A = \{\omega \in S^{\mathbb{Z}}{:}a_{\omega_n\omega_{n+1}} = 1 \text{ for } \forall n \in \mathbb{Z}\}.$$

The space Ω_A is closed and shift invariant. The restriction

$$\Sigma_A := \Sigma|_{\Omega_A}$$

is called the two-sided *topological Markov chain* determined by the matrix A, a *Markov subshift*, or a *subshift of finite type* (see Sect. 1.1.2). One-sided topological Markov chains are defined analogously over $S^{\mathbb{N}_0}$. The matrix A is said to be irreducible if for any pair i,j there is $n > 0$ such that $a_{ij}^{(n)} > 0$, where $a_{ij}^{(n)}$ are the entries of A^n. If A is irreducible and Σ_A is a one-sided or two-sided topological Markov chain, then [202]

$$h_{top}(\Sigma_A) = \log \lambda, \tag{B.28}$$

where λ is the largest positive eigenvalue of A. A topological Markov chain Σ_A has a unique measure (called its Parry measure) of maximal topological entropy.

It can be proved [31, Sect. 4.3] that a C^2 piecewise expanding Markov map f is topologically conjugate (modulo 0) to the one-sided topological Markov chain Σ_A, where A is the transition matrix for f. Therefore, piecewise expanding Markov maps admit a symbolic description.

Example 34 Consider the *rooftop map f* defined by

$$f(x) = \begin{cases} ax + c & \text{if } 0 \leq x \leq c, \\ (1-b)x & \text{if } c \leq x \leq 1, \end{cases}$$

$a > 1$, $b > 1$, and $c = \frac{1}{1+a}$; see Fig. B.4. Set $I_1 = [0, c)$ and $I_2 = [c, 1]$. Then f is C^∞ on I_1 and I_2 (lateral derivatives at the endpoints),

$$|f'(x)| = \begin{cases} a & \text{if } x \in I_1, \\ b & \text{if } x \in I_2, \end{cases}$$

and

$$f(\mathring{I}_1) = \mathring{I}_2, \quad f(\mathring{I}_2) \supset \mathring{I}_1 \cup \mathring{I}_2.$$

It follows that f is a smooth piecewise expanding Markov map with transition matrix

$$A = \begin{pmatrix} 0 & 1 \\ 1 & 1 \end{pmatrix},$$

see (B.2). Finally, from (B.28) we get

$$h_{top}(f) = h_{top}(\Sigma_A) = \log \frac{1+\sqrt{5}}{2}.$$

B.3.2 Topological Entropy of One-Dimensional Maps

Topological entropy, as metric entropy, is in general difficult to calculate and even to estimate. An exception worth mentioning because of its importance in applications is the case of one-dimensional interval maps.

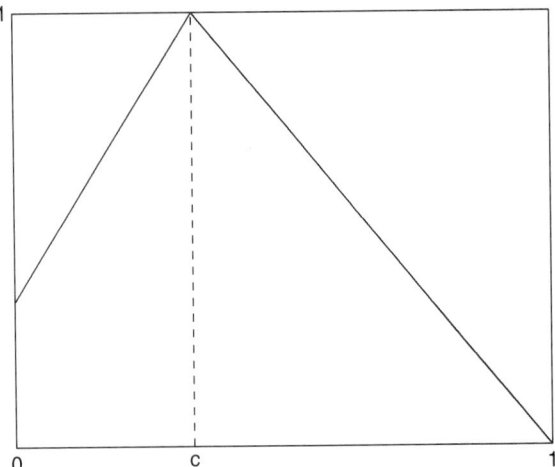

Fig. B.4 Rooftop map

Definition 27 Given an interval $I \subset \mathbb{R}$, a map $f{:}I \rightarrow I$ is said to be piecewise monotone if there is a finite partition of I into subintervals, such that f is continuous and monotone on each of those subintervals.

If $f{:}I \rightarrow I$ is piecewise monotone, there are different expressions for its topological entropy $h_{\text{top}}(f)$ that allow calculating it analytically in many cases. For instance [4, 155],

$$h_{\text{top}}(f) = \lim_{n \rightarrow \infty} \frac{1}{n} \log \text{lap}(f^n) \tag{B.29}$$

and

$$h_{\text{top}}(f) = \lim_{n \rightarrow \infty} \frac{1}{n} \log \left| \{ x \in I{:}f^n(x) = x \} \right|, \tag{B.30}$$

where $\text{lap}(f^n)$ is the number of pieces of monotonicity of f^n (called laps of f^n) and $|\cdot|$ stands for the cardinality.

Other expressions of $h(f)$ are related to the notion of variation [4, 155]:

$$h_{\text{top}}(f) = \lim_{n \rightarrow \infty} \frac{1}{n} \log^+ \text{var}(f^n), \tag{B.31}$$

where, as usual, $\log^+ x = \max\{0, \log x\}$. Let us recall that the variation of a function $\varphi{:}I \rightarrow \mathbb{R}$ is given as

$$\text{var}(\varphi) = \sup \left\{ \sum_{i=1}^{s} |\varphi(x_i) - \varphi(x_{i-1})| \right\},$$

where the supremum is taken over all finite sequences $x_0 < x_1 < \cdots < x_s$ of elements of I. If φ is piecewise monotone, then (i) $\mathrm{var}(\varphi) < \infty$, (ii) φ has finite derivative φ' almost everywhere on I, and (iii) φ' is integrable on I [95]. In this case,

$$\mathrm{var}(\varphi) = \int_I |\varphi'(x)|\, dx. \tag{B.32}$$

Note that, for a piecewise monotone map φ, $\mathrm{var}(\varphi)$ is closely related to the length of the graph of φ,

$$\mathrm{len}(\varphi) = \int_I \sqrt{1 + |\varphi'(x)|^2}\, dx.$$

Indeed, since

$$|\varphi'(x)| < \sqrt{1 + |\varphi'(x)|^2} \leq |\varphi'(x)| + 1 \tag{B.33}$$

for all $x \in I$, we have

$$\mathrm{var}(\varphi) < \mathrm{len}(\varphi) \leq \mathrm{var}(\varphi) + \mathrm{len}(I), \tag{B.34}$$

upon integration of (B.33) over the interval I ($\mathrm{len}(I)$ denotes the length of I). It follows

$$\lim_{n \to \infty} \frac{1}{n} \log^+ \mathrm{len}(f^n) = \lim_{n \to \infty} \frac{1}{n} \log^+ \mathrm{var}(f^n) = h(f), \tag{B.35}$$

since $\lim_{n \to \infty} \frac{1}{n} \log^+ \mathrm{len}(I) = 0$.

Corollary 10 *If f is a continuous, piecewise monotone interval map of constant slopes $\pm s$, then*

$$h_{top}(f) = \log^+ s.$$

This result is very interesting for the following reason. If f is a continuous, piecewise monotone interval map and $h_{top}(f) = \log \beta > 0$, then f is semiconjugate to some continuous, piecewise monotone interval map of constant slopes $\pm\beta$ (via a non-decreasing map) [4]. If, moreover, f is topologically transitive, then "semiconjugate" can be replaced by "conjugate" in the previous statement (and the condition $h_{top}(f) > 0$ can be dropped because it is automatically satisfied).

Finally, let us mention that there are efficient algorithms for the numerical estimation of the topological entropy of piecewise monotone interval maps; see, for example, [27] for an algorithm that converges rapidly and provides both upper and lower bounds.

References

1. H.D.I. Abarbanel, *Analysis of Observed Chaotic Data*. Springer, New York, 1996.
2. M. Abraimowitz and I.A. Stegun (Eds.), *Handbook of Mathematical Functions*. Dover, New York, 1972.
3. R.L. Adler, A.G. Koheim, and M.H. McAndrews, Topological entropy, *Transactions of the American Mathematical Society* **114** (1965) 309–319.
4. L. Alsedà, J. Llibre, and M. Misiurewicz, *Combinatorial Dynamics and Entropy in Dimension One*. World Scientific, Singapore, 2000.
5. G. Alvarez, M. Romera, G. Pastor, and F. Montoya, Gray codes and 1D quadratic maps, *Electronic Letters* **34** (1998) 1304–1306.
6. J.M. Amigó, J. Szczepanski, E. Wajnryb, and M.V. Sanchez-Vives, Estimating the entropy of spike trains via Lempel-Ziv complexity, *Neural Computation* **16** (2004) 717–736.
7. J.M. Amigó, M.B. Kennel, and L. Kocarev, The permutation entropy rate equals the metric entropy rate for ergodic information sources and ergodic dynamical systems, *Physica D. Nonlinear Phenomena* **210** (2005) 77–95.
8. J.M. Amigó, L. Kocarev, and J. Szczepanski, Order patterns and chaos, *Physics Letters A* **355** (2006) 27–31.
9. J.M. Amigó and M.B. Kennel, Variance estimators for the Lempel-Ziv entropy rate estimators, *Chaos* **16** (2006) 043102.
10. J.M. Amigó, L. Kocarev, and I. Tomovski, Discrete entropy, *Physica D. Nonlinear Phenomena* **228** (2007) 77–85.
11. J.M. Amigó and M.B. Kennel, Topological permutation entropy, *Physica D. Nonlinear Phenomena* **231** (2007) 137–142.
12. J.M. Amigó, S. Zambrano, and M.A.F. Sanjuán, True and false forbidden patterns in deterministic and random dynamics, *Europhysics Letters* **79** (2007) 50001.
13. J.M. Amigó, L. Kocarev, and J. Szczepanski, Discrete Lyapunov exponent and resistance to differential cryptanalysis, *IEEE Transactions on Circuits and Systems II* **54** (2007) 882–886.
14. J.M. Amigó, S. Elizalde, and M.B. Kennel, Forbidden patterns and shift systems, *Journal of Combinatorial Theory, Series A* **115** (2008) 485–504.
15. J.M. Amigó, S. Zambrano, and M.A.F. Sanjuán, Combinatorial detection of determinism in noisy time series, *Europhysics Letters* **83** (2008) 60005.
16. J.M. Amigó and M.B. Kennel, Forbidden ordinal patterns in higher dimensional dynamics, *Physica D. Nonlinear Phenomena* **237** (2008) 2893–2899.
17. J.M. Amigó, The ordinal structure of the signed shift transformations, *International Journal of Bifurcation and Chaos* **19** (2009) 3311–3327.
18. C. Anteneodo, A.M. Batista, and R.L. Viana, Synchronization threshold in coupled logistic map lattices, *Physica D. Nonlinear Phenomena* **223** (2006) 270–275.
19. N. Aoki and K. Hiraide, *Topological theory of dynamical systems*. North Holland, Amsterdam, 1994.

20. D.K. Arrowsmith and C.M. Place, *Dynamical Systems*. Chapman and Hall, Boca Raton, 1996.

21. D. Arroyo, G. Alvarez, and J.M. Amigó, Estimation of the control parameter from symbolic sequences: Unimodal maps with variable critical point, *Chaos* **19** (2009) 023125.

22. R.B. Ash, *Information Theory*. Dover Publications, New York, 1990.

23. H. Atmanspacher and H. Scheingraber, Inherent global stabilization of unstable local behavior in coupled map lattices, *International Journal of Bifurcation and Chaos* **15** (2005) 1665–1676.

24. N. Ay and J.P. Crutchfield, Reductions of hidden information sources, *Journal of Statistical Physics* **120** (2005) 659–684.

25. E. Babson and E. Steingrímsson, Generalized permutation patterns and a classification of the Mahonian statistics, Séminaire Lotharingien de Combinatoire **44** (2000), Article B44b, 18.

26. A. Bäcker and N. Chernov, Generating partitions for two-dimensional hyperbolic maps, *Nonlinearity* **11** (1998) 79–87.

27. N.J. Balmforth, E.A. Spiegel, and C. Tresser, Topological entropy of one-dimensional maps: Approximations and bounds, *Physical Review Letters* **72** (1994) 80–83.

28. C. Bandt and B. Pompe, Permutation entropy: A natural complexity measure for time series, *Physical Review Letters* **88** (2002) 174102.

29. C. Bandt, G. Keller, and B. Pompe, Entropy of interval maps via permutations. *Nonlinearity* **15** (2002) 1595–1602.

30. C. Bandt and F. Shiha, Order patterns in time series, *Journal of Time Series Analysis* **28** (2007) 646–665.

31. A. Berger, *Chaos and Chance*, Walter de Gruyter, Berlin, 2001.

32. G.D. Birkhoff, Proof of a recurrence theorem for strongly transitive systems, *Proceedings of the National Academy of Science* **17** (1931) 650.

33. F. Blanchard, P. Kurka, and A. Maass, Topological and measure-theoretical properties of one-dimensional cellular automata, *Physica D. Nonlinear Phenomena* **103** (1997) 86–99.

34. S. Boccaletti and D.L. Valladares, Characterization of intermittent lag synchronization, *Physical Review E* **62** (2000) 7497–7500.

35. L. Boltzmann, Über die mechanischen Analogien des zweiten Hauptsatzes der Thermodynamik, *Journal für reine und angewandte Mathematik* **100** (1887) 201.

36. R.E. Bowen, Entropy for group endomorphisms and homogeneous spaces, *Transactions of the American Mathematical Society* **153** (1971) 401–414.

37. A. Boyarsky and P. Gora, *Laws of Chaos*. Birkhäuser, Boston, 1997.

38. W.A. Brock, W.D. Dechert, J.A. Scheinkman, and B. LeBaron, A test for independence based on the correlation dimension, *Econometrics Reviews* **15** (1996) 197–235.

39. A.A. Brudno, Entropy and complexity of trajectories of a dynamical system. *Transactions of the Moscow Mathematical Society* **44** (1983) 127–151.

40. M. Buhl and M.B. Kennel, Statistically relaxing to generating partitions for observed time-series data, *Physical Review E* **71** (2005) 046213: 1–14.

41. J. Bunge and M. Fitzpatrick, Estimating the Number of Species: A Review, *Journal of the American Statistical Association* **88** (1993) 364–373.

42. L.A. Bunimovich and Y.G. Sinai, Space-time chaos in coupled map lattices, *Nonlinearity* **1** (1988) 491–518.

43. L.A. Bunimovich and Y.G. Sinai, Statistical mechanics of coupled map lattices, In: K.Kaneko (Ed.), *Theory and Applications of Coupled Map Lattices*. Wiley, New York, 1993.

44. L.A. Bunimovich, Coupled map lattices: Some topological and ergodic properties, *Physica D. Nonlinear Phenomena* **103** (1997) 1–17.

45. Y. Cao, W. Tung, J.B. Gao, V.A. Protopopescu, and L.M. Hively, Detecting dynamical changes in time series using the permutation entropy, *Physical Review E* **70** (2004) 046217.

46. R. Carretero-González, Low dimensional travelling interfaces in coupled map lattices, *International Journal of Bifurcations and Chaos* **7** (1997) 2745–2754.

47. R. Carretero-González, D.K. Arrowsmith, and F. Vivaldi, One-dimensional dynamics for traveling fronts in coupled map lattices, *Physical Review E* **61** (2000) 1329–1336.

48. K. Cattell and J.C. Muzio, Synthesis of one-dimensional linear hybrid cellular automata, *IEEE Transactions on Computer-Aided Design of Integrated circuits and Systems* **15** (1996) 325–335.

49. A. Chao, Nonparametric estimation of the number of classes in a population, *Scandinavian Journal of Statistics, Theory and Applications* **9** (1984) 265–270.

50. H. Chaté and P. Manneville, Coupled map lattices as cellular automata, *Journal of Statistical Physics* **56** (1989) 357–370.

51. B.V. Chirikov and F. Vivaldi, An algorithmic view of pseudochaos, *Physica D. Nonlinear Phenomena* **129** (1999) 223–235.

52. G.H. Choe, *Computational Ergodic Theory*. Springer Verlag, Berlin, 2005.

53. F. Christiansen and A. Politi, Generating partition for the standard map, *Physical Review E* **51** (1995) R3811.

54. L.O. Chua, V.I. Sbitnev, and S. Yoon, A nonlinear dynamics perspective of Wolfram's New Kind of Science –Part II: Universal neuron, *International Journal of Bifurcation and Chaos* **13** (2003) 2377–2491.

55. L.O. Chua, V.I. Sbitnev, and S. Yoon, A nonlinear dynamics perspective of Wolfram's new kind of science –Part IV: From Bernoulli shift to $1/f$ spectrum, *International Journal of Bifurcation and Chaos* **15** (2005) 1045–1183.

56. R.W. Clarke, M.P. Freeman, and N.W. Watkins, Application of computational mechanics to the analysis of natural data: An example in geomagnetism, *Physical Review E* **67** (2003) 016203.

57. P. Collet and J.P. Eckmann, *Iterated Maps on the Interval as Dynamical Systems*, 5th printing. Birkhäuser, Boston, 1997.

58. M. Courbage, D. Mercier, and S. Yasmineh, Traveling waves and chaotic properties in cellular automata, *Chaos* **9** (1999) 893–901.

59. T.M. Cover and J.A. Thomas, *Elements of Information Theory*, 2nd edition. New York, John Wiley & Sons, 2006.

60. J.P. Crutchfield and K. Young, Inferring statistical complexity, *Physical Review Letters* **63** (1989) 105–108.

61. R. Dahlhaus, J. Kurths, P. Maass, and J. Timmer, *Mathematical Methods in Time Series Analysis and Digital Image Processing*. Springer Verlag, Berlin, 2008.

62. M. D'amico, G. Manzini, and L. Margara, On computing the entropy of cellular automata, *Theoretical Computer Science* **290** (2003) 1629–1646.

63. R. Davidchack, Y.C. Lai, E.M. Bollt, and M. Dhamala, Estimating generating partitions by unstable periodic orbits, *Physical Review E* **61** (2000) 1353–1356.

64. K. Denbigh, How subjective is entropy? In: H.S. Leff and A.F. Rex (Ed.), *Maxwell's Demon, Entropy, Information, Computing*, pp. 109–115. Princeton University Press, Princeton,1990.

65. M. Denker, Finite generators for ergodic, measure-preserving transformations, *Zeitschrift für Wahrscheinlichkeitstheorie und verwandte Gebiete* **29** (1974) 45–55.

66. M. Denker, C. Grillenberger, and K. Sigmund, *Ergodic Theory on Compact Spaces*. Springer Lecture Notes in Math. **527**, Springer Verlag, Berlin, 1976.

67. M. Denker and W.A. Woyczynski, *Introductory Statistics and Random Phenomena*. Birkhäuser, Boston, 1998.

68. M. Denker, *Einführung in die Analysis Dynamischer Systeme*. Springer Verlag, Berlin, 2005.

69. R.L. Devaney, *Chaotic Dynamical Systems* (2nd edition). Westview Press, Boulder, 2003.

70. E.I. Dinaburg, The relation between topological entropy and metric entropy, *Soviet Mathematics* **11** (1970) 13–16.

71. Y. Dobyns and H. Atmanspacher, Characterizing spontaneous irregular behavior in coupled map lattices, *Chaos, Solitons & Fractals* **24** (2005) 313–327.

72. J.P. Eckmann and D. Ruelle, Ergodic theory of chaos and strange attractors, *Review of Modern Physics* **57** (1985) 617–656.

73. J.P. Eckmann, S.O. Kamphorst, and D. Ruelle, Recurrence plots of dynamical systems, *Europhysics Letters* **4** (1987) 973–977.

74. S. Elizalde and M. Noy, Consecutive patterns in permutations, *Advances in Applied Mathematics* **30** (2003) 110–125.

75. S. Elizalde, Asymptotic enumeration of permutations avoiding generalized patterns, *Advances in Applied Mathematics* **36** (2006) 138–155.

76. S. Elizalde, The number of permutations realized by a shift, *SIAM Journal of Discrete Mathematics* **23** (2009) 765–786.

77. R. Érdi, *Complexity Explained.* Springer Verlag, Berlin, 2007.

78. A. Fernández, J. Quintero, R. Hornero, P. Zuluaga, M. Navas, C. Gómez, J. Escudero, N. García-Campos, J. Biederman, and T. Ortiz, Complexity analysis of spontaneous brain activity in attention-deficit/hyperactivity disorder: Diagnosis implications, *Biological Psychiatry* **65** (2009) 571–577.

79. J. Ford, G. Mantica, and G.H. Ristow, The Arnold's cat: Failure of the correspondence principle, *Physica D. Nonlinear Phenomena* **50** (1991) 493–520.

80. A.M. Fraser and H.L. Swinney, Independent coordinates for strange attractors from mutual information, *Physical Review A* **33** (1986) 1134–1140.

81. J.B. Gao and H.Q. Cai, On the structures and quantification of recurrence plots, *Physics Letters A* **270** (2000) 75–87.

82. Y. Gao, I. Kontoyiannis, and E. Bienenstock, Estimating the entropy of binary time series: Methodology, some theory and a simulation study, *Entropy* **10** (2008) 71–99.

83. J. García-Ojalvo, J.M. Sancho, and L. Ramírez-Piscina, Generation of spatiotemporal colored noise, *Physical Review A* **46** (1992) 4670–4675.

84. M. Gardner, The fantastic combinations of John Conway's new solitaire game "life", *Scientific American* **223** (1970) 120–123.

85. A. Golestani, M.R. Jahed Motlagh, K. Ahmadian, A.H. Omidvarnia, and N. Mozayani, A new criterion to distinguish stochastic and deterministic time series with the Poincaré section and fractal dimension, *Chaos* **19** (2009) 013137.

86. S.W. Golomb, *Bulletin of the American Mathematical Society* **70** (1964) 747 (research problem 11).

87. P. Grassberger and H. Kantz, Generating partitions for the dissipative Hénon map, *Physics Letters A* **113** (1985) 235–238.

88. P. Grassberger, Finite sample corrections to entropy and dimension estimates, *Physics Letters A* **128** (1988) 369–373.

89. R.M. Gray, *Entropy and Information Theory.* Springer Verlag, New York, 1990.

90. F. Gu, X. Meng, E. Shen, and Z. Cai, Can we measure consciousness with EEG complexities?, *International Journal of Bifurcations and Chaos* **13** (2003) 733–742.

91. B. Hasselblatt and A. Katok, *A First Course in Dynamics.* Cambridge University Press, Cambridge, 2003.

92. G.A. Hedlund, Endomorphisms and automorphisms of the shift dynamical system, *Mathematical Systems Theory* **3** (1969) 320–375.

93. H. Herzel, Complexity of symbol sequences, *Systems, Analysis, Modelling, Simulations* **5** (1988) 435–444.

94. H. Herzel, A.O. Schmitt, and W. Ebeling, Finite sample effects in sequence analysis, *Chaos, Solitons & Fractals* **4** (1994) 97–113.

95. E. Hewitt and K. Stromberg, *Real and Abstract Analysis.* Springer Verlag, New York 1965.

96. F.C. Hoppensteadt, *Analysis and Simulation of Chaotic Systems* (2nd edition). Springer Verlag, New York, 2000.

97. K. Hiraide, Nonexistence of positively expansive maps on compact connected manifolds with boundary, *Proceedings of the American Mathematical Society* **110** (1990) 565–568.

98. M.W. Hirsch, S. Smale, and R.L. Devaney, *Differential Equations, Dynamical Systems, and an Introduction to Chaos.* Academic Press, San Diego, 2003.

99. C.S. Hsu and M.C. Kim, Construction of maps with generating partitions for entropy evaluation, *Physical Review A* **31** (1985) 3253–3265.

100. J. Hughes, J. Hellman, T.H. Rickets, and B.J.M. Bohannan, Counting the uncountable: Statistical approaches to estimating microbial diversity, *Applied and Environ. Microbiology* **67** (2001) 4399–4406.

101. L.P. Hurd, J. Kari, and K. Culik, The topological entropy of cellular automata is uncomputable, *Ergodic Theory and Dynamical Systems* **12** (1992) 255–265.

102. Y. Ishii and D. Sands, Monotonicity of the Lozi family near the tent-maps, *Communications in Mathematical Physics* **198** (1998) 397–406.

103. N. Israeli and N. Goldenfeld, Coarse-graining of cellular automata, emergence, and the predictability of complex systems, *Physical Review E* **73** (2006) 1–17.

104. S. Jalan, J. Jost, and F.M. Atay, Symbolic synchronization and the detection of global properties of coupled dynamics from local information, *Chaos* **16** (2006) 033124.

105. O. Jenkinson and M. Pollicott, Entropy, exponents and invariant densities for hyperbolic systems: Dependence and computation. In: M. Brin, B. Hasselblatt, and Y. Pesin (Eds.), *Modern Dynamical Systems and Applications*. pp. 365–384 Cambridge University Press, Cambridge, 2004.

106. K. Kaneko, Transition from torus to chaos accompanied by frequency lockings with symmetry breaking, *Progress in Theoretical Physics* **69** (1983) 1427–1442.

107. K. Kaneko, Period-doubling of kink-antikink patterns, quasiperiodicity in anti-ferro-like structures and spatial intermittency in coupled logistic lattice, *Progress in Theoretical Physics* **72** (1984) 480–486.

108. K. Kaneko, Pattern dynamics in spatiotemporal chaos, *Physica D. Nonlinear Phenomena* **34** (1989) 1–41.

109. K. Kaneko, Spatiotemporal chaos in one- and two-dimensional coupled map lattices, *Physica D. Nonlinear Phenomena* **37** (189) 60–82.

110. K. Kaneko, Chaotic traveling waves in a coupled map lattice, *Physica D. Nonlinear Phenomena* **68** (1993) 299–317.

111. H. Kantz, Quantifying the closeness of fractal measures, *Physical Review E* **49** (1994) 5091–5097.

112. H. Kantz and T. Schreiber, *Nonlinear Time Series Analysis*. Cambridge University Press, Cambridge, 1997.

113. N.J. Kasdin, Discrete simulation of colored noise and stochastic processes and $1/f^\alpha$ power law noise generation, *Proceedings of the IEEE* **83** (1995) 802–827.

114. A. Katok and B. Hasselbaltt, *Introduction to the Theory of Dynamical Systems*. Cambridge University Press, Cambridge, 1998.

115. S. Katok, *p-adic Analysis compared with real*. American Mathematical Society, Providence, 2007.

116. K. Keller and K. Wittfeld, Distances of time series components by means of symbolic dynamics, *International Journal of Bifurcation and Chaos* **14** (2004) 693–703.

117. K. Keller and M. Sinn, Ordinal analysis of time series, *Physica A* **356** (2005) 114–120.

118. K. Keller, H. Lauffer, and M. Sinn, Ordinal analysis of EEG time series, *Chaos and Complexity Letters* **2** (2007) 247–258.

119. M.B. Kennel and S. Isabelle, Method to distinguish possible chaos from colored noise and to determine embedding parameters, *Physical Review A* **46** (1992) 3111–3118.

120. M.B. Kennel, Statistical test for dynamical nonstationarity in observed time-series data, *Physical Review E* **56** (1997) 316–321.

121. M.B. Kennel and A.I. Mees, Context-tree modeling of observed symbolic dynamics, *Physical Review E* **66** (2002) 056209.

122. M.B. Kennel, J. Shlens, H.D.I. Abarbanel, and E.J. Chichilnisky, Estimating entropy rates with Bayesian confidence intervals, *Neural Computation* **17** (2005) 1531–1576.

123. B.P. Kitchens, *Symbolic Dynamics*. Springer Verlag, Berlin, 1998.

124. L. Kocarev and J. Szczepanski, Finite-space Lyapunov exponents and pseudo-chaos, *Physical Review Letters* **93** (2004) 234101.

125. L. Kocarev, J. Szczepanski, J.M. Amigó, and I. Tomovski, Discrete Chaos – Part I: Theory, *IEEE Transactions on Circuits and Systems I* **53** (2006) 1300–1309.

126. A.N. Kolmogorov, Entropy per unit time as a metric invariant of automorphism, *Doklady of Russian Academy of Sciences* **124** (1959) 754–755.

127. I. Kontoyiannis, P.H. Algoet, Y.M. Suhov, and A.J. Wyner, Nonparametric entropy estimation for stationary processes and random fields, with applications to English text. *IEEE Transactions on Information Theory* **44** (1998) 1319–1327.

128. Z.S. Kowalski, Finite generators of ergodic endomorphisms, *Colloquium Mathematicum* **49** (1984) 87–89.

129. Z.S. Kowalski, Minimal generators for aperiodic endomorphisms, *Commentationes Mathematicae Universitatis Carolinae* **36** (1995) 721–725.

130. W. Krieger, On entropy and generators of measure-preserving transformations, *Transactions of the American Mathematical Society* **149** (1970) 453–464.

131. A.P. Kurian and S. Puttusserypady, Self-synchronizing chaotic stream ciphers, *Signal Processing* **88** (2008) 2442–2452.

132. J. Kurths, D. Maraun, C.S. Zhou, G. Zamora-López, and Y. Zou, Dynamics in complex systems, *European Review* **17** (2009), 357–370.

133. J.C. Lagarias, Pseudorandom numbers, *Statistical Science* **8** (1993) 31–39.

134. A. Lasota and J.A. Yorke, On the existence of invariant measures for piecewise monotonic transformations, *Transactions of the American Mathematical Society* **186** (1973), 481–488.

135. A.M. Law and W.D. Kelton, *Simulation, Modeling, and Analysis*, 3rd edition. McGraw-Hill, Boston, 2000.

136. B. LeBaron, A fast algorithm for the BDS statistics, *Studies in Nonlinear Dynamics & Econometrics* **2** (1997) 53–59.

137. A. Lempel and J. Ziv, On the complexity of an individual sequence, *IEEE Transactions on Information Theory* **IT-22** (1976) 75–78.

138. M. Li and P. Vitányi, *An Introduction to Kolmogorov Complexity and Its Applications*. Springer Verlag, New York, 1997.

139. D. Lind and B. Marcus, *Symbolic Dynamics and Coding*. Cambridge University Press, Cambridge, 2003.

140. T. Liu, C.W.J. Granger, and W.P. Heller, Using the correlation exponent to decide whether an economic series is chaotic. *Journal of Applied Econometrics,* Supplement: Special Issue on Nonlinear Dynamics and Econometrics (Dec., 1992) S25–S39.

141. G. Manzini and L. Margara, A complete and efficiently computable topological classification of linear cellular automata over $\mathbb{Z}_m$, *Theoretical Computer Science* **221** (1999) 157–177.

142. R. Mañé, *Ergodic Theory and Differentiable Dynamics*. Springer Verlag, Berlin, 1987.

143. M.T. Martin, A. Plastino, and O.A. Rosso, Generalized statistical complexity measures: Geometrical and analytical properties, *Physica A* **369** (2006) 439–462.

144. N. Marwan, M.C. Romano, M. Thiel, and J. Kurths, Recurrence plots for the analysis of complex systems, *Physics Reports* **438** (2007) 237–329.

145. C. Masoller and A.C. Martí, Random delays and the synchronization of chaotic maps, *Physical Review Letters* **94** (2005) 134102.

146. M. Matilla-García, A non-parametric test for independence based on symbolic dynamics, *Journal of Economic Dynamic & Control* **31** (2007) 3889–3903.

147. M. Matilla-García and M. Ruiz Marín, A non-parametric independence test using permutation entropy, *Journal of Econometrics* **144** (2008) 139–155.

148. M. Matsumoto and T. Nishimura, Mersenne Twister: A 623-dimensionally equidistributed uniform pseudo-random number generator, *ACM Trans. on Modeling and Computer Simulation* **8** (1998) 3–30.

149. W. Meier and O. Staffelbach, The self-shrinking generator. In: Proc. of Eurocrypt'94, Lecture Notes in Computer Science. vol. 950, pp. 205–214. Springer Verlag, Berlin, 1994.

150. W. de Melo and S. van Strien, *One-Dimensional Dynamics*. Springer Verlag, Berlin, 1993.
151. A.J. Menezes, P.C. van Oorschoot, and S.A. Vanstone, *Handbook of Applied Cryptography*. CRC Press, Boca Raton, 1997.
152. M.E. Mera and M. Morán, Geometric noise reduction for multivariate time series, *Chaos* **16** (2006) 013116.
153. N. Metropolis, M. Stein, and P. Stein, On finite limit sets for transformations on the unit interval, *Journal of Combinatorial Theory, Series A* **15**, 25–44 (1973).
154. J. Milnor, Non-expansive Hénon Maps, *Advances in Mathematics* **69** (1988) 109–114.
155. M. Misiurewicz and W. Szlenk, Entropy of piecewise monotone mappings, *Studia Mathematica* **67** (1980) 45–63.
156. M. Misiurewicz, Strange attractors for the Lozi mappings. In: R.G. Helleman (Ed.), *Nonlinear Dynamics*, Vol. **357**, pp. 348–358 The New York Academy of Science, New York 1980.
157. M. Misiurewicz, Permutations and topological entropy for interval maps, *Nonlinearity* **16** (2003) 971–976.
158. M. Mitchell, *Complexity —A Guided Tour*. Oxford University Press, New York, 2009.
159. R. Monetti, W. Bunk, T. Aschenbrenner, and F. Jamitzky, Characterizing synchronization in time series using information measures extracted from symbolic representations, *Physical Review E* **79** (2009) 046207.
160. M. Morse and G.A. Hedlund, Symbolic Dynamics, *American Journal of Mathematics* **60** (1938) 815–866.
161. J. von Neumann, The general and logical theory of automata. In: L.A. Jeffress (Ed.), *Cerebral Mechanisms in Behavior*. Wiley, New York, 1951.
162. M. Newman, A.L. Barabási, and D.J. Watts, *The Structure and Dynamics of Networks*. Princeton University Press, Princeton, 2006.
163. E. Olbrich, N. Bertschinger, N. Ay, and J. Jost, How should complexity scale with system size?, *The European Physical Journal B* **63** (2008) 407–415.
164. G.J. Ortega and E. Louis, Smoothness implies determinism in time series: A measure based approach, *Physical Review Letters* **81** (1998) 4345–4348.
165. E. Ott, *Chaos in Dynamical Systems*. Cambridge University Press, Cambridge, 2002.
166. N.H. Packard, J.P. Crutchfield, J.D. Farmer, and R.S. Shaw, Geometry from a time series, *Physical Review Letters* **45** (1980) 712–716.
167. L. Paninski, Estimation of entropy and mutual information, *Neural Computation* **15** (2003) 1191–1253.
168. H.O. Peitgen, H. Jürgens, and D. Saupe, *Chaos and Fractals*. Springer Verlag, New York, 2004.
169. K. Petersen, *Ergodic Theory*. Cambridge University Press, Cambridge, 1983.
170. S.D. Pethel, N.J. Corron, and E. Bollt, Symbolic dynamics of coupled map lattices, *Physical Review Letters* **96** (2006) 034105.
171. S.D. Pethel, N.J. Corron, and E. Bollt, Deconstructing spatiotemporal chaos using local symbolic dynamics, *Physical Review Letters* **99** (2007) 214101.
172. J. Piepzryk, T. Hardjorno, and J. Seberry, *Fundamentals of Computer Security*. Springer Verlag, Berlin, 2003.
173. W.H. Press, S.A. Teukolsky, W.T. Vetterling, and B.P. Flannery, *Numerical Recipes: The Art of Scientific Computing*. Cambridge University Press, Cambridge, 2007.
174. R.C. Robinson, *An Introduction to Dynamical Systems*. Pearson Prentice Hall, Upper Saddle River NJ, 2004.
175. M.G. Rosenblum, A.S. Pikovsky, and J. Kurths, Phase synchronization of chaotic oscillators, *Physical Review Letters* **76** (1997) 1804–1807.
176. O.A. Rosso, H.A. Larrondo, M.T. Martin, A. Platino, and M.A. Fuentes, Distinguishing noise from chaos, *Physical Review Letters* **99** (2007) 154102.
177. D.J. Rudolph, *Fundamentals of Measurable Dynamics*. Oxford University Press, Oxford, 1990.
178. http://topo.math.u-psud.fr/ sands/Programs/Lozi/index.html.

179. A.N. Sarkovskii, Coexistence of cycles of a continuous map of a line into itself, *Ukrainian Mathematical Journal* **16** (1964) 61–71.

180. P.R. Scalassara, M.E. Dajer, C. Dias Maciel, C. Capobianco Guido, and J.C. Pereira, Relative entropy measures applied to healthy and pathological voice characterization, *Applied Mathematics and Computation* **207** (2009) 95–108.

181. A.O. Schmitt, H. Herzel, and W. Ebeling, A new method to calculate higher-order entropies from finite samples, *Europhysics Letters* **23** (1993) 303–309.

182. O. Schmitt, *Remarks on the Generator-Problem* (Thesis). University of Göttingen, 2001.

183. T. Schreiber, Detecting and analyzing nonstationarity in a time series using nonlinear cross predictions, *Physical Review Letters* **78** (1997) 843–846.

184. R. Sexl and J. Blackmore (Eds.), *Ludwig Boltzmann - Ausgewahlte Abhandlungen* (Ludwig Boltzmann Gesamtausgabe, Band 8). Vieweg, Braunschweig, 1982.

185. C.R. Shalizi and J.P. Crutchfield, Computational mechanics: Pattern and prediction, structure and simplicity, *Journal of Statistical Physics* **104** (2001) 817–879.

186. C.E. Shannon, A mathematical theory of communication, *Bell System Technical Journal* **27** (1948) 379–423, 623–653.

187. L.A. Shepp and S.P. Lloyd, Ordered cycle length in a random permutation, *Transactions of the American Mathematical Society* **121** (1966) 340–357.

188. M.A. Shereshevsky, Expansiveness, entropy and polynomial growth for groups acting on subshifts by automorphisms. *Indagationes Mathematicae* **4** (1993) 203–210.

189. Y.G. Sinai, On the Notion of Entropy of a Dynamical System, *Doklady of Russian Academy of Sciences* **124** (1959) 768–771.

190. M. Sinn and K. Keller, Estimation of ordinal pattern probabilities in fractional Brownian motion, arXiv:0801.1598.

191. M. Smorodinsky, *Ergodic Theory, Entropy* (Lectures Notes in Mathematics) Vol. **214**. Springer Verlag, Berlin, 1971.

192. D. Sotelo Herrera and J. San Martín, Analytical solutions of weakly coupled map lattices using recurrence relations, *Physics Letters A* **373** (2009) 2704–2709.

193. J.C. Sprott, *Chaos and Time-Series Analysis*. Oxford University Press, Oxford, 2003.

194. J.C. Sprott, High-dimensional dynamics in the delayed Hénon map. *Electronic Journal of Theoretical Physics* **3** (2006) 19–35.

195. S.P. Strong, R. Koberle, R.R. de Ruyter van Steveninck, and W. Bialek, Entropy and information in neural spike trains. *Physical Review Letters* **80** (1998) 197–200.

196. J. Szczepanski, J.M. Amigó, E. Wajnryb, and M.V. Sanchez-Vives. Application of Lempel-Ziv complexity to the analysis of neural discharges, *Network: Computation in Neural Systems* **14** (2003) 335–350.

197. F. Takens, Detecting strange attractors in turbulence, In: D. Rand and L.S. Young (Eds.), *Dynamical Systems and Turbulence*, Lecture Notes in Mathematics, vol. 898. Springer, Berlin, 1981, pp. 366–381.

198. T. Toffoli and N. Margolus, *Cellular Automata Machines*. The MIT Press, Cambridge MA, 1987.

199. S. Ulam, Random process and transformations, Proceedings of the International Congress of Mathematicians 2 (1952), 264–275.

200. D.B. Vasconcelos, S.R. Lopes, R.L. Viana, and J. Kurths, Spatial recurrence plots, *Physical Review E* **73** (2006) 056207.

201. S.B. Volchan, What is a Random Sequence, *The American Mathematical Monthly* **109** (2002) 46–63.

202. P. Walters, *An Introduction to Ergodic Theory*. Springer Verlag, New York, 2000.

203. L. Wang and N.D. Kazarinoff, On the universal sequence generated by a class of unimodal functions, *Journal of Combinatorial Theory, Series A* **46** (1987) 39–49.

204. A. Wolf, J.B. Swift, H.L. Swinney, and J.A. Vastano, Determining Lyapunov exponents from a time series, *Physica D. Nonlinear Phenomena* **16** (1985) 285–317.

205. S. Wolfram, Computation theory of cellular automata, *Communications in Mathematical Physics* **96** (1984) 15–57.
206. S. Wolfram, Universality and complexity in cellular automata, *Physica* **10D** (1984) 1–35.
207. S. Wolfram, *A New Kind of Science*. Wolfram Media, Champaign, 2002.
208. X-S. Zhang, R.J. Roy, and E.W. Jensen, EEG complexity as a measure of depth anesthesia for patients, *IEEE Transactions on Biomedical Engineering* **48** (2001) 1424–1433.
209. J. Zhang and M. Small, Complex networks from pseudoperiodic time series: Topology versus dynamics. *Physical Review Letters* **96** (2006) 238701.
210. G.C. Zhuang, J. Wang, Y. Shi, and W. Wang, Phase synchronization and its cluster feature in two-dimensional coupled map lattices, *Physical Review E* **66** (2002) 046201.
211. J. Ziv and A. Lempel, Compression of individual sequences via variable-rate coding *IEEE Transactions on Information Theory* **IT-24** (1978) 530–536.
212. L. Zunino, D.G. Pérez, M.T. Martín, M. Garavaglia, A. Plastino, and O.A. Rosso, Permutation entropy of fractional Brownian motion and fractional Gaussian noise, *Physics Letters A* **372** (2008) 4768–4774.
213. L. Zunino, D.G. Pérez, M.T. Martín, M. Garavaglia, A. Plastino, and O.A. Rosso, Fractional Brownian motion, fractional Gaussian noise, and Tsallis permutation entropy, *Physica A* **387** (2008) 6057–6068.

Index